D1090466

INORGANIC SYNTHESES

Volume 34

●●●●●●●

Board of Directors

Future Volumes

35 ALFRED P. SATTELBERGER *Los Alamos National Lab*
36 THOMAS B. RAUCHFUSS *University of Illinois, Urbana-Champaign*
37 TOBIN J. MARKS *Northwestern University*

International Associates

Editor-in-Chief

JOHN R. SHAPLEY

University of Illinois, Urbana-Champaign

●●●

INORGANIC SYNTHESES

Volume 34

JOHN WILEY & SONS, INC. PUBLICATION

Published by John Wiley & Sons, Inc., Hoboken, New Jersey.
Published simultaneously in Canada.

For general information on our other products and services please contact our Customer Care Department within the U.S. at 877-762-2974, outside the U.S. at 317-572-3993 or fax 317-572-4002.

Wiley also publishes its books in a variety of electronic formats. Some content that appears in print, however, may not be available in electronic format.

Library of Congress Catalog Number: 39:23015

ISBN 0-471-64750-0

Printed in the United States of America

10 9 8 7 6 5 4 3 2 1

PREFACE

The present volume of *Inorganic Syntheses* continues the pattern of the last three volumes in the series, namely, specific thematic chapters along with other contributions that together reflect the diversity of inorganic synthetic activities in modern research.

The five chapters in this volume are arranged in a rough order of increasing complexity for the compounds described. Chapter 1 is a collection of syntheses for main group compounds, some interesting in their own right and others primarily for their use in metal complexes. Chapter 2 details procedures for largely mononuclear organometallic and coordination complexes, with central elements ranging across the periodic table and with a wide variety of ligand types. In contrast, Chapter 3 has a specific focus on transition metal compounds containing carbonyl ligands. Chapter 4 illustrates an explosively developing research theme in which the cyanide ligand is used as a linking agent in the assembly of polynuclear metal complexes with the purpose, for example, of achieving unique magnetic properties. Finally, procedures for other types of polynuclear and cluster compounds are displayed in Chapter 5. The articles in this volume will provide tested syntheses of compounds targeted for ongoing research. However, I trust, as in my own experience with *Inorganic Syntheses*, that they also will stimulate new research ideas, the results of which will serve to nurture future volumes in the series.

A volume of this sort does not happen without the contributions of many people, first and foremost, of course, the submitters and the checkers of the individual articles. I appreciate their patience when progress appeared to be slow and their quick responses when urgency was requested. Several individuals deserve explicit acknowledgment for their critical support of this project: Heinrich Vahrenkamp for his insight and effort in soliciting the cyanide-related syntheses in Chapter 4; Herb Kaesz for his interest in involving me with *Inorganic Syntheses*, initially with Volume 26, and for his sharp editorial eye regarding many articles that appear here; Marcetta Darensbourg for her experienced advice and counsel throughout the process of planning and assembling this volume; Stan Ching for his prompt and efficient handling of the manuscripts as they were submitted; and Julie Sides and Maureen Buxton for their truly invaluable secretarial assistance. I thank also the members of the Editorial Board for their many useful comments and suggestions regarding the submitted manuscripts.

Finally, I dedicate this volume to the memory of John A. Osborn (1940–2000), who first showed me the fun to be had in exploring inorganic syntheses.

John R. Shapley
Urbana, Illinois

NOTICE TO CONTRIBUTORS AND CHECKERS

The *Inorganic Syntheses* series is published to provide all users of inorganic substances with detailed and reliable procedures for the preparation of important and timely compounds. Thus the series is the concern of the entire scientific community. The Editorial Board hopes that all chemists will share in the responsibility of producing *Inorganic Syntheses* by offering their advice and assistance in both the formulation and the laboratory evaluation of outstanding syntheses. Help of this kind will be invaluable in achieving excellence and pertinence to current scientific interests.

There is no rigid definition of what constitutes a suitable synthesis. The major criterion by which syntheses are judged is the potential value to the scientific community. An ideal synthesis is one that presents a new or revised experimental procedure applicable to a variety of related compounds, at least one of which is critically important in current research. However, syntheses of individual compounds that are of interest or importance are also acceptable. Syntheses of compounds that are readily available commercially at reasonable prices are not acceptable. Corrections and improvements of syntheses already appearing in *Inorganic Syntheses* are suitable for inclusion.

The Editorial Board lists the following criteria of content for submitted manuscripts. Style should conform with that of previous volumes of *Inorganic Syntheses*. The introductory section should include a concise and critical summary of the available procedures for synthesis of the product in question. It should also include an estimate of the time required for the synthesis, an indication of the importance and utility of the product, and an admonition if any potential hazards are associated with the procedure. The Procedure section should present detailed and unambiguous laboratory directions and be written so that it anticipates possible mistakes and misunderstandings on the part of the person who attempts to duplicate the procedure. Any unusual equipment or procedure should be clearly described. Line drawings should be included when they can be helpful. All safety measures should be stated clearly. Sources of unusual starting materials must be given, and, if possible, minimal standards of purity of reagents and solvents should be stated. The scale should be reasonable for normal laboratory operation, and any problems involved in scaling the procedure either up or down should be discussed. The criteria for judging the purity of the final product should be delineated clearly. The Properties section should supply and discuss those physical and chemical characteristics that are relevant to judging the purity of the product and to permitting its handling and use in an intelligent manner. Under References, all pertinent literature citations should be listed in order. A style sheet is available from the Secretary of the Editorial Board.

The Editorial Board determines whether submitted syntheses meet the general specifications outlined above. Every procedure will be checked in an independent laboratory, and publication is contingent on satisfactory duplication of the syntheses. For online access to information and requirements, see: *www.inorgsynth.com.*

Each manuscript should be submitted in duplicate to the Secretary of the Editorial Board, Professor Stanton Ching, Department of Chemistry, Connecticut College, New London, CT 06320. The manuscript should be typewritten in English. Nomenclature should be consistent and should follow the recommendations presented in *Nomenclature of Inorganic Chemistry*, 2nd ed., Butterworths & Co, London, 1970 and in *Pure and Applied Chemistry*, Volume 28, No. 1 (1971). Abbreviations should conform to those used in publications of the American Chemical Society, particularly *Inorganic Chemistry.*

Chemists willing to check syntheses should contact the editor of a future volume or make this information known to Professor Ching.

TOXIC SUBSTANCES AND LABORATORY HAZARDS

Chemicals and chemistry are by their very nature hazardous. Chemical reactivity implies that reagents have the ability to combine. This process can be sufficiently vigorous as to cause flame, an explosion, or, often less immediately obvious, a toxic reaction.

The obvious hazards in the syntheses reported in this volume are delineated, where appropriate, in the experimental procedure. It is impossible, however, to foresee every eventuality, such as a new biological effect of a common laboratory reagent. As a consequence, *all* chemicals used and *all* reactions described in this volume should be viewed as potentially hazardous. Care should be taken to avoid inhalation or other physical contact with all reagents and solvents used in this volume. In addition, particular attention should be paid to avoiding sparks, open flames, or other potential sources that could set fire to combustible vapors or gases.

A list of 400 toxic substances may be found in the *Federal Register*, Volume 40, No. 23072, May 28, 1975. An abbreviated list may be obtained from *Inorganic Syntheses*, Vol. 18, p. xv, 1978. A current assessment of the hazards associated with a particular chemical is available in the most recent edition of *Threshold Limit Values for Chemical Substances* and *Physical Agents in the Workroom Environment* published by the American Conference of Governmental Industrial Hygienists.

The drying of impure ethers can produce a violent explosion. Further information about this hazard may be found in *Inorganic Syntheses*, Volume 12, p. 317.

CONTENTS

INORGANIC SYNTHESES

Volume 34

Chapter One

MAIN GROUP COMPOUNDS

1. DIBORANE(4) COMPOUNDS

Submitted by M. J. GERALD LESLEY,* NICHOLAS C. NORMAN,* and CRAIG R. RICE*
Checked by DAVID W. NORMAN† and R. TOM BAKER†

Diborane(4) compounds are key reagents in transition-metal-catalyzed diboration[1] and Suzuki–Miyaura coupling reactions.[2] Two of the most widely used compounds are the pinacolate derivative $B_2(pin)_2$ (pin = $OCMe_2CMe_2O$)[3] and the catecholate species $B_2(cat)_2$ (cat = 1,2-$O_2C_6H_4$),[4] both of which are prepared from tetrakis(dimethylamino) diborane(4), $B_2(NMe_2)_4$, described initially by Brotherton.[5] Detailed preparations for $B_2(pin)_2$ and $B_2(NMe_2)_4$ have been described in Ref. 3b. Here we present a slightly different preparation for $B_2(NMe_2)_4$ together with details of the synthesis of 1,2-$B_2Cl_2(NMe_2)_2$[6] and $B_2(cat)_2$. Simple modifications of the $B_2(cat)_2$ synthesis given here allow for the preparation of a host of diol-derived diborane(4) compounds.

A. TETRAKIS(DIMETHYLAMINO)DIBORANE(4), $B_2(NMe_2)_4$

$$BCl_3 + 2\,B(NMe_2)_3 \longrightarrow 3\,BCl(NMe_2)_2$$

$$2\,BCl(NMe_2)_2 + 2\,Na \longrightarrow B_2(NMe_2)_4$$

*School of Chemistry, University of Bristol, Cantock's Close, Bristol, BS8 ITS, UK.
†Chemical Science and Technology Division, Los Alamos National Laboratory, Los Alamos, NM 87545.

Inorganic Syntheses, Volume 34, edited by John R. Shapley
ISBN 0-471-64750-0

General Comments

All solvents were freshly distilled under dinitrogen from an appropriate drying agent immediately prior to use. Glassware was either kept in an oven at 150°C overnight or flame-dried under vacuum. All manipulations were carried out under dinitrogen using standard Schlenk techniques.

Procedure

■ **Caution.** *Sodium can spontaneously ignite on exposure to water or air.* BCl_3 *fumes vigorously in air, producing HCl. These reagents should only be handled in a fume hood.*

To a two-necked 250-mL round-bottomed flask equipped with a sidearm and a magnetic stirring bar and mounted on a magnetic stirrer, $B(NMe_2)_3$ (Aldrich, 33 mL, 0.2 mol) is added under a constant stream of nitrogen and stirring is started. The flask is then cooled to about −78°C by means of an external dry-ice ethanol bath, and a solution of BCl_3 in heptane (Aldrich, 100 mL of a 1.0 M solution) is added. (*Note*: It is important that heptane rather than hexane be used, since yields are much lower when hexane is employed.) The reaction mixture is then stirred at low temperature for 1 h, after which time the cooling bath is removed, and the reaction flask and contents are allowed to warm to room temperature. Stirring is continued for a further 3 h. After this time the reaction mixture forms a pale yellow solution with small quantities of a white solid. [*Note*: Checkers state that at this stage the reaction can be assayed by ^{11}B NMR, although this is rarely necessary if freshly distilled or purchased BCl_3 and $B(NMe_2)_3$ are used. The solid can be removed by filtration, although its presence in subsequent steps seems not to affect the yield.] Clean metallic sodium (6.9 g, 0.3 mol) is then slowly added in small pieces, and afterward a Liebig condenser is attached to the reaction flask. The magnetic stirrer is replaced with a stirrer–heating mantle, and the reaction mixture is then refluxed for 16 h, resulting in a brown solution and a purple precipitate. After cooling to room temperature, the condenser is removed, and the reaction mixture is then filtered through a glass frit into a separate flask. The filter cake is washed with hexanes (2 × 10 mL). All solvent is removed from the filtered reaction solution by vacuum pumping (standard vacuum pump) at room temperature, affording a brown oil. Fractional vacuum distillation (0.2 mmHg) of the brown oil affords small quantities of unreacted $B(NMe_2)_3$ (1.5 g, 10%) as the first fraction (20–25°C) followed by a second fraction (38–42°C) forming $B_2(NMe_2)_4$ (16.8 g, 85%) (checkers' yields 79 and 87%) as a colorless liquid.

Properties

$B_2(NMe_2)_4$ is a thermally stable but moisture-sensitive colorless liquid (bp 70–73°C at 3 mmHg) that is soluble in most common solvents. Samples can be stored in a sealed tube at 4°C for indefinite periods. ^1H and ^{11}B NMR spectra ($CDCl_3$) at room temperature show singlets at δ 2.60 and 34.9 (broad) ppm, respectively (the latter referenced to $BF_3 \cdot Et_2O$). Traces of $B(NMe_2)_3$, if present, can be identified by its ^{11}B NMR signal at 25.4 ppm.

B. 1,2-DICHLORO-1,2-BIS(DIMETHYLAMINO)DIBORANE(4), $B_2Cl_2(NMe_2)_2$

$$B_2(NMe_2)_4 + 4\,HCl \longrightarrow 1,2\text{-}B_2Cl_2(NMe_2)_2 + 2\,[NH_2Me_2]Cl$$

Procedure

■ **Caution.** *HCl in Et_2O is corrosive and lachrymatory. All manipulations should be carried out in a fume hood and gloves worn.*

This procedure is similar to that described in Ref. 6. To a rapidly stirred solution of $B_2(NMe_2)_4$ (3.31 g, 16.7 mmol) in Et_2O (25 mL) in a 100-mL Schlenk flask, a solution of HCl in Et_2O (66.8 mL of a 1.0 M solution) is added over a period of 10 min, during which time a large amount of a white precipitate forms. The reaction mixture is allowed to stir for 1 h, after which time it is filtered. The solid filtercake is washed with Et_2O (2 × 10 mL), and all solvent is then carefully removed from the colorless filtrate by vacuum pumping (standard vacuum pump) at room temperature affording a colorless oil. Vacuum distillation of this oil (48°C, 0.2 mmHg) affords pure 1,2-$B_2Cl_2(NMe_2)_2$ as a colorless liquid (2.08 g, 69%).*

Properties

1,2-$B_2Cl_2(NMe_2)_2$ is a thermally stable but moisture-sensitive colorless liquid (bp 48°C at 0.2 mmHg) that is soluble in most common solvents. Samples can be stored in a sealed tube at 4°C for indefinite periods. The ^1H NMR spectrum ($CDCl_3$) at room temperature shows two singlets at δ 2.90 (s, 6H) and 2.87

*Checkers indicate that product can be easily lost if care is not exercised in removing the solvent. Best results (75 and 80% yields for two preparations) were obtained by removing Et_2O at 0°C.

(s, 6H). The ^{11}B NMR spectrum shows a broad singlet at 35.7 ppm. Additional data on the reactions of $B_2(NMe_2)_4$ with HCl can be found in Refs. 6 and 7, including the preparation of the synthetically useful compound $[B_2Cl_4(NHMe_2)_2]$.[4]

C. 2,2′-BIS(1,3,2-BENZODIOXABOROLE), $B_2(1,2\text{-}O_2C_6H_4)_2$

$$B_2(NMe_2)_4 + 2\,1,2\text{-}(HO)_2C_6H_4 \longrightarrow [B_2(1,2\text{-}O_2C_6H_4)_2(NHMe_2)_2] + 2\,NHMe_2$$

$$[B_2(1,2\text{-}O_2C_6H_4)_2(NHMe_2)_2] + 2\,NHMe_2 + 4\,HCl \longrightarrow B_2(1,2\text{-}O_2C_6H_4)_2 + 4\,[NH_2Me_2]Cl$$

Procedure

A magnetically stirred solution of catechol (1.22 g, 11.1 mmol) in Et_2O (20 mL) is prepared in a 100-mL Schlenk flask and cooled to 0°C. A solution of $B_2(NMe_2)_4$ (1.0 g, 5.0 mmol) in Et_2O (5 mL) is then added, and stirring is continued for 20 min. After this time the solution is allowed to warm to room temperature, and stirring is continued overnight. At this stage the reaction mixture contains a large amount of a white precipitate of the amine adduct $[B_2(1,2\text{-}O_2C_6H_4)_2(NHMe_2)_2]$. The reaction mixture is then recooled to 0°C, and a solution of HCl in Et_2O (20 mL of a 1.0 M solution) is added, with stirring continued for 20 min. This amount of HCl is a slight excess over the amount required to remove all the amine present (free or coordinated), specifically, 4 equiv relative to the $B_2(NMe_2)_4$; the use of a larger excess can lead to reductions in the yield of $B_2(cat)_2$. The reaction mixture is then allowed to warm to room temperature and stir overnight, after which time a white precipitate of $[NH_2Me_2]Cl$ is present. All volatiles are then removed by vacuum pumping, affording a white solid. Addition of hot toluene, filtration, and subsequent concentration of the filtrate then affords white crystals of crude $B_2(cat)_2$. Crystallization is completed by cooling the mixture to -30°C. The mother liquor is then removed by syringe, and the white solid is dried by vacuum pumping (alternatively, if a drybox is being used, the solid may be isolated by filtration). Washing with MeCN (2×1 mL) and then with hexanes (2×1 mL), and finally drying by vacuum affords analytically pure $B_2(cat)_2$ (1.01 g, 85%) (checkers' yields 88 and 93%.) as a white solid.

Properties

$B_2(cat)_2$ is a white solid that is stable in air for brief periods but can be stored indefinitely in sealed flasks under a nitrogen atmosphere. It is soluble in hot toluene, $CHCl_3$, CH_2Cl_2, and tetrahydrofuran (THF), although solutions are susceptible to hydrolysis. The ^1H NMR spectrum ($CDCl_3$) at room temperature shows multiplets at δ 7.42 (m, 4H) and 7.21 (m, 4H) and the ^{11}B NMR spectrum ($CDCl_3$) exhibits a broad singlet at 28.8 ppm. Commonly encountered minor decomposition products in the crude solid (which are removed by washing with MeCN) are $B_2(cat)_3$ (δ ^{11}B 16.6),[8] (cat)BOB(cat) (δ ^{11}B 18.1)[9] and $[NH_2Me_2][B(cat)_2]$ [δ ^{11}B (sharp) 14.0]. Traces of $[B_2Cl_4(NHMe_2)_2]$ [δ ^{11}B (sharp) 9.5][4] may also be present.

References

1. T. B. Marder and N. C. Norman, *Top. Catal.* **5**, 63 (1998).
2. N. Miyaura and A. Suzuki, *Chem. Rev.* **95**, 2457 (1995).
3. (a) H. Nöth, *Z. Naturforsch., Teil B* **39**, 1463 (1984); (b) T. Ishiyama, M. Murata, T. Akiko, and N. Miyaura, *Org. Synth.* **77**, 176 (2000).
4. F. J. Lawlor, N. C. Norman, N. L. Pickett, E. G. Robins, P. Nguyen, G. Lesley, T. B. Marder, J. A. Ashmore, and J. C. Green, *Inorg. Chem.* **37**, 5282 (1998); W. Clegg, M. R. J. Elsegood, F. J. Lawlor, N. C. Norman, N. L. Pickett, E. G. Robins, A. J. Scott, P. Nguyen, N. J. Taylor, and T. B. Marder, *Inorg. Chem.* **37**, 5289 (1998).
5. A. L. McCloskey, R. J. Brotherton, J. L. Boone, and H. M. Manasevit, *J. Am. Chem. Soc.* **82**, 6245 (1960). See also R. J. Brotherton, in *Progress in Boron Chemistry*, H. Steinberg and A. L. McCloskey, eds., Macmillan, New York, 1964, Chapter 1.
6. H. Nöth and W. Meister, *Z. Naturforsch., Teil B* **17**, 714 (1962).
7. S. C. Malhotra, *Inorg. Chem.* **3**, 862 (1964).
8. S. A. Westcott, H. P. Blom, T. B. Marder, R. T. Baker, and J. C. Calabrese, *Inorg. Chem.* **32**, 2175 (1993).
9. W. Gerrard, M. F. Lappert, and B. A. Mountfield, *J. Chem. Soc.* 1529 (1959).

2. SODIUM TETRAKIS(3,5-BIS(TRIFLUOROMETHYL) PHENYL)BORATE, $Na[B(3,5-(CF_3)_2C_6H_3)_4]$

Submitted by DANIEL L. REGER,* CHRISTINE A. LITTLE,* JAYDEEP J. S. LAMBA,* and KENNETH J. BROWN*
Checked by JENNIFER R. KRUMPER,† ROBERT G. BERGMAN,† MICHAEL IRWIN,‡ and JOHN P. FACKLER, Jr.‡

The anion $[B\{3,5-(CF_3)_2C_6H_3\}_4]^-$ is a highly soluble, noncoordinating counter-ion that has gained considerable attention.[1] The anion is typically prepared as the sodium[1a] or potassium[2] salt, which are extremely versatile reagents that can be converted to the acid form, $[H(OEt_2)_2]$ $[B\{3,5-(CF_3)_2C_6H_3\}_4]$,[1a] the silver salt $Ag[B\{3,5-(CF_3)_2C_6H_3\}_4]$,[2,3] or the thallium salt $Tl[B\{3,5-(CF_3)_2C_6H_3\}_4]$[4] for

*Department of Chemistry and Biochemistry, University of South Carolina, Columbia, SC 29208.
†Department of Chemistry, University of California Berkeley, Berkeley, CA 94720.
‡Department of Chemistry, Texas A & M University, College Station, TX 77843.

further reactions. A significant disadvantage in previous preparations of the sodium and potassium salts is that the Grignard reagent prepared in situ is potentially explosive.[5] We have dramatically improved the synthesis of $Na[B\{3,5\text{-}(CF_3)_2C_6H_3\}_4]$.[6] The critical modification is to add $NaBF_4$ to the reaction mixture before forming the Grignard reagent from magnesium and 1-Br-3,5-$(CF_3)_2C_6H_3$. Presumably, most of the potentially explosive Grignard reagent formed first in the reaction is consumed immediately by the $NaBF_4$ present and does not build up in solution, thereby reducing the danger of explosion. Also, 1,2-dibromoethane is added at the beginning of the reaction to ensure that the Grignard reaction initiates immediately on the dropwise addition of 1-Br-3,5-$(CF_3)_2C_6H_3$. This preparation of $Na[B\{3,5\text{-}(CF_3)_2C_6H_3\}_4]$ is safer and leads to pure product in very high yield.

■ **Caution.** *The $(CF_3)_2C_6H_3MgBr$ generated in situ in this preparation is potentially explosive.[5] The reaction solution should not be allowed to go to dryness before addition of the solution of Na_2CO_3 in water. Care should be taken to ensure that the reaction vessels are tightly sealed before initiating the Grignard reaction by using a heat gun because of the flammability of the ether solvent. Benzene is a human carcinogen.*

General Considerations

The Grignard reaction is carried out under a nitrogen atmosphere by using standard Schlenk techniques. All solvents are dried and distilled prior to use. Magnesium metal turnings from EM Science and Na_2SO_4 (anhydrous) from Fisher Scientific are used as received. The compounds $NaBF_4$, 1,2-dibromoethane, and 3,5-bis(trifluoromethyl)bromobenzene are purchased from Aldrich and used as received. The unopened bottle of $NaBF_4$ is opened and subsequently stored in a Vacuum Atmospheres HE-493 drybox. Decolorizing carbon, NORIT 211, is purchased from Acros and used as received.

Procedure

A 500-mL three-necked flask containing a stirring bar is fitted with a reflux condenser, a pressure-equalizing addition funnel, and a glass stopper. This apparatus is flame-dried under vacuum, cooled under nitrogen, and then charged with Mg turnings (1.01 g, 41.7 mmol) against a stream of nitrogen. The Mg turnings are stirred under nitrogen in the empty flask for 2 h.* $NaBF_4$ (0.70 g, 6.4 mmol) is

*One of the checkers activated the Mg turnings by submerging them in 1.0 M HNO_3 until a metallic shine was visible. The solution was decanted, and the turnings were washed 3 times with distilled water, with acetone, and then 2 times with diethyl ether.

weighed in the glovebox and then is transferred quickly to the reaction flask against a stream of nitrogen. Also, 150 mL Et_2O is added via syringe. Then 1,2-dibromoethane (0.49 mL, 5.7 mmol) is added through the addition funnel, and the flask is heated for several minutes with a heat gun to initiate reaction.* The heating is stopped and the solution stirred until the refluxing stops. Tiny bubbles coming off the Mg turnings indicates that the reaction has initiated, and the solution slowly becomes cloudy. Then 3,5-bis(trifluoromethyl)bromobenzene (6.2 mL, 36 mmol), diluted with 50 mL Et_2O, is added dropwise over ~30 min. The addition causes the solution to gently reflux, and the solution slowly darkens to light brown by the end of the addition. The reaction mixture is heated with a heating mantle to continue the reflux for an additional 30 min. The heat is then removed, and the reaction mixture is stirred overnight at room temperature under nitrogen.

The reaction mixture is added to Na_2CO_3 (16 g) in water (200 mL), stirred for 30 min, and then filtered. (*Note*: Typical benchtop procedures can be used for the workup at this point, but an inert atmosphere must be used for the Dean–Stark trap azeotropic distillation step and manipulations of the solid after that point.) The ether layer is separated, and the aqueous layer is extracted with ether (3× 50 mL). The combined organic solution is dried over sodium sulfate (10 g) and then treated with decolorizing charcoal (2 g). The mixture is filtered and the ether is removed with a rotary evaporator. The remaining oily solid is suspended in 200 mL of benzene, and the water is removed with a Dean–Stark trap by azeotropic distillation (under nitrogen) for 2 h. The solvent volume is reduced to ~50 mL via distillation of the benzene and removal through the stopcock on the Dean–Stark trap. The remaining solution is then cooled to room temperature, and the solvent is removed via cannula filtration.† The residual light tan solid‡ is dried under vacuum overnight (5.04 g, 5.6 mmol, 89%). ^1H NMR (acetone d_6), δ 7.79 (8H, br, O–H), 7.67 (4H s, p-H).

Anal. (for a sample protected from air). Calcd. for $C_{32}H_{12}F_{24}BNa$: C, 43.37; H, 1.36. Found: C, 43.48; H, 1.61.

Properties

The compound $Na[B\{3,5\text{-}(CF_3)_2C_6H_3\}_4]$ is a light brown solid, mp 300–306°C. Elemental analysis shows that pure $Na\{B[3,5\text{-}(CF_3)_2C_6H_3]_4\}$ obtained in this procedure picks up two equivalents of water when handled in air.

*One of the checkers added both the dibromoethane and subsequently the ether solution of bis(trifluoromethyl)bromobenzene via syringe through a rubber septum stopper in place of the addition funnel.

†Filtration of the benzene solution removes unreacted starting material and other soluble impurities.

‡The tan solid obtained after drying by the benzene azeotrope procedure is much less soluble in CH_2Cl_2 than the oily solid present before.

Acknowledgment

The submitters thank the National Science Foundation (CHE-0110493) for support.

References

1. (a) M. Brookhart, B. Grant, and J. A. F. Vople, *Organometallics* **11**, 3920 (1992); (b) J. Ledford, C. S. Shultz, D. P. Gates, P. S. White, J. M. DeSimone, and M. Brookhart, *Organometallics* **20**, 5266 (2001); (c) A. Llamazares, H. W. Schmalle, and H. Berke, *Organometallics* **20**, 5277 (2001); (d) A. E. Enriquez, P. S. White, and J. L. Templeton, *J. Am. Chem. Soc.* **123**, 4992 (2001); (e) S. Reinartz, P. S. White, M. Brookhart, and J. L. Templeton,. *J. Am. Chem. Soc.* **123**, 12724 (2001); (f) D. M. Tellers and R. G. Bergman, *Organometallics* **20**, 4819 (2001); (g) X. Fang, J. G. Watkin, B. L. Scott, and G. L. Kubas, *Organometallics* **20**, 3351 (2001); (h) W. B. Sharp, P. Legzdins, and B. O. Patrick, *J. Am. Chem. Soc.* **123**, 8143 (2001).
2. W. E. Buschmann and J. S. Miller, *Inorg. Chem.* **33**, 83 (2002).
3. Y. Hayashi, J. J. Rohde, and E. J. Corey, *J. Am. Chem. Soc.* **118**, 5502 (1996).
4. R. P. Hughes, D. C. Lindner, A. L. Rheingold, and G. A. P. Yap, *Inorg. Chem.* **36**, 1726 (1997).
5. E. J. Moore and R. Waymouth, *Chem. Eng. News* 6 (March 17, 1997).
6. D. L. Reger, T. D. Wright, C. A. Little, J. J. S. Lamba, and M. D. Smith, *Inorg. Chem.* **40**, 3810 (2001).

3. ANIONIC TRIS- AND BIS(DIPHENYLPHOSPHINOMETHYL)BORATES

Submitted by JONAS C. PETERS[*] and J. CHRISTOPHER THOMAS[*]
Checked by JULIE A. DUPONT[†] and CHARLES G. RIORDAN[†]

Chelating phosphines are perhaps more widely used than any other ligand class in synthetic inorganic and organic chemistry.[1] A wealth of literature exists describing the utility of neutral chelating phosphines for applications in catalytic and stoichiometric reaction chemistry. By comparison, the chemistry of anionic phosphine ligands remains relatively unexplored. Anionic (phosphino)borate ligands enable access to a wide range of charge neutral metal complexes whose reaction chemistry may be studied with respect to that of related discrete salt complexes supported by neutral phosphine chelates.[2] Herein we describe convenient procedures for the preparation of the tris(phosphino)borate ligand, $[PhB(CH_2PPh_2)_3]^-$,[3,4] and the bis(phosphino)borate ligand, $[Ph_2B(CH_2PPh_2)]^-$,[2b] as their tetra-*n*-butylammonium salts. The ammonium salts are

[*]Division of Chemistry and Chemical Engineering, California Institute of Technology, Pasadena, CA 91125.
[†]Department of Chemistry and Biochemistry, University of Delaware, Newark, DE 19716.

versatile reagents for the preparation of metal complexes of these (phosphino)-borate ligands, offering certain synthetic advantages by comparison to the lithium salt precursors. Additionally, the thallium complex $[PhB(CH_2PPh_2)_3]Tl$ provides a very useful transmetallation reagent for the tris(phosphino)borate ligand.[5] Its synthesis is also described.

A. TETRABUTYLAMMONIUM PHENYLTRIS (DIPHENYLPHOSPHINOMETHYL)BORATE, $[^nBu_4N][PhB(CH_2PPh_2)_3]$

$$3\,Li(TMEDA)CH_2PPh_2 + PhBCl_2 \longrightarrow [Li(TMEDA)][PhB(CH_2PPh_2)_3] + 2\,LiCl + 2\,TMEDA$$

$$[Li(TMEDA][PhB(CH_2PPh_2)_3] + [^nBu_4N][Cl] \longrightarrow [^nBu_4N][PhB(CH_2PPh_2)_3] + LiCl + TMEDA$$

■ **Caution.** *$PhBCl_2$ reacts vigorously with water and fumes in air. It should be handled under a nitrogen atmosphere, and care should be taken to minimize skin contact and respiratory exposure.*

Procedure

A 1-L round-bottomed Schlenk flask is equipped with a magnetic stirring bar and rubber septum and is charged with 33.3 g (103 mmol) of $Li(TMEDA)CH_2PPh_2$ (TMEDA = tetramethylethylenediamine),[6] suspended in 500 mL of dry and deoxygenated Et_2O, under a dinitrogen atmosphere. The stirring, yellow suspension is chilled to −78°C and, under a dinitrogen counterflow, the septum is replaced by a 150-mL dropping funnel charged with a solution of $PhBCl_2$* (5.475 g, 34.5 mmol) dissolved in toluene (70 mL). The phenyldichloroborane solution is added dropwise to the stirring suspension over a period of 20 min. The resulting yellow mixture is allowed to gradually warm to room temperature and stirring is continued for 20 h. A representative aliquot is analyzed by ^{31}P NMR spectroscopy to ensure complete consumption of $Li(TMEDA)CH_2PPh_2$ (addition of several drops of dry tetrahydrofuran to the sample aliquot affords a homogeneous solution that provides an accurate spectroscopic assay).

The remainder of the procedure is carried out under air and is executed relatively quickly to minimize ligand degradation. To precipitate the $[Li(TMEDA)][PhB(CH_2PPh_2)_3]$ product, the reaction mixture is cooled to −78°C and

*Available from Aldrich Chemical Company.

hexanes (400 mL) are added while stirring. The suspension is allowed to settle for at least 30 min at $-78°C$. The resulting white solid is collected by filtration on a coarse frit, and the filtrate is discarded. The off-white filtercake is washed with hexanes (2×300 mL) and extracted with methanol (75 mL), resulting in a somewhat cloudy suspension. This suspension is filtered through Celite on a medium-porosity sintered-glass frit. The methanol solution is added slowly to a solution of $[^{n}Bu_4N]Cl$ (19.2 g, 69 mmol), prepared by dissolving the $[^{n}Bu_4N]Cl$ in a mixture of 300 mL of deionized H_2O and 20 mL of diethyl ether. After the addition is complete, a sticky white solid results, which is collected on a medium-porosity frit and is repeatedly washed with copious amounts of water and diethyl ether (2×100 mL each). The resulting white powder is then dissolved in acetonitrile (100 mL), yielding a somewhat cloudy solution. The solution is then stirred over anhydrous sodium sulfate and subsequently filtered through Celite on a sintered-glass frit. The filtrate is thoroughly dried on a rotovap (rotary evaporator) and on a vacuum line, and then it is triturated several times with diethyl ether to ensure complete removal of acetonitrile. Finally, the resulting powder is suspended in diethyl ether (250 mL) and collected on a medium-porosity frit, affording 15.5 g (48.4%) of white $[^{n}Bu_4N][PhB(CH_2PPh_2)_3]$. This material is typically pure enough for further work (showing only trace acetonitrile impurities), but recrystallization from cold tetrahydrofuran/diethyl ether yields analytically pure material.

Properties

$[^{n}Bu_4N][PhB(CH_2PPh_2)_3]$ is a white, crystalline solid that is moderately air-stable and can be handled in nonacidic aqueous solutions for short periods without appreciable degradation. The ammonium salt is not soluble in hydrocarbons such as pentane and diethyl ether, nor aromatic solvents such as benzene and toluene, when in pure form. It is, however, appreciably soluble in toluene and diethyl ether prior to purification, partly accounting for the reduced yield. The salt is very soluble in tetrahydrofuran, acetone, acetonitrile, and methanol. The ^{1}H NMR spectrum (CD_3CN, 300 MHz) displays signals at δ 6.7–7.5 (m, 35H), 3.05 (m, 8H), 1.56 (m, 8H), 1.14 (br, 6H), 0.95 (t, 12H). The $^{31}P\{^{1}H\}$ NMR spectrum shows a single resonance at -10.9 ppm. Resonances in the ^{1}H and ^{31}P NMR are broadened when spectra are obtained in nonpolar solvents such as C_6D_6. Doping such solvents with small amounts of d_8-THF or d_6-acetone helps solubilize the ligand and affords sharper spectra. It is worth noting that the initial metathesis reaction of $Ph_2PCH_2Li(TMEDA)$ with $PhBCl_2$ proceeds in high yield according to ^{31}P NMR, and that the cation exchange also appears to be quantitative. The synthetic procedure described here ensures the isolation of a clean product free of LiCl, borane, and phosphine byproducts. Halogenated solvents such as chloroform and dichloromethane, even when dried and freshly distilled under dinitrogen, rapidly decompose both $[Li(TMEDA)][PhB(CH_2PPh_2)_3]$ and $[^nBu_4N][PhB(CH_2PPh_2)_3]$. The inability to use chlorinated solvents in the purification procedure partly accounts for the relatively tedious workup we recommend to obtain pure material. This problem is circumvented by preparation of the thallium complex of the ligand.

B. TETRABUTYLAMMONIUM DIPHENYLBIS (DIPHENYLPHOSPHINOMETHYL)BORATE, $[^nBu_4N][Ph_2B(CH_2PPh_2)_2]$

$$2\,Li(TMEDA)CH_2PPh_2 + Ph_2BCl \longrightarrow [Li(TMEDA)_2][Ph_2BCH_2PPh_2)_2] + LiCl$$

$$[Li(TMEDA)_2][Ph_2B(CH_2PPh_2)_2] + [^nBu_4N][Cl] \longrightarrow [^nBu_4N][Ph_2B(CH_2PPh_2)_2] + LiCl + 2\,TMEDA$$

■ **Caution.** *Ph_2BCl reacts vigorously with water and fumes in air. It should be handled under a nitrogen atmosphere, and care should be taken to minimize skin contact and respiratory exposure.*

Procedure

Under a dinitrogen atmosphere, a 500-mL round-bottomed Schlenk flask is charged with solid yellow $Ph_2PCH_2Li(TMEDA)$ (4.82 g, 15.0 mmol), and then diethyl ether (180 mL) to form a suspension. The reaction vessel is cooled to −78°C in a dry-ice/acetone bath. A toluene solution (10 mL) of Ph_2BCl* (1.514 g, 7.553 mmol) is added to the reaction vessel, dropwise, via syringe through a rubber septum under a constant dinitrogen purge. The reaction temperature is maintained at −78°C throughout the addition. The reaction mixture is stirred thoroughly and gradually allowed to warm to ambient temperature. Stirring is continued for an additional 14 h and affords a pale yellow precipitate. (*Note*: Analysis by ^{31}P NMR spectroscopy shows quantitative conversion to the desired product. The reaction volatiles are removed in vacuo, and the resulting solids are isolated by filtration onto a medium-porosity sintered-glass frit under a dinitrogen atmosphere. The white powder is then thoroughly washed with diethyl ether (5 × 10 mL). Drying this solid in vacuo provides light yellow, solid $[Li(TMEDA)_2]$ $[Ph_2B(CH_2PPh_2)_2]$ (5.67 g) and remaining LiCl salt. This product is completely freed from LiCl by extraction into warm toluene, filtering through Celite, and evaporation of the solvent. However, it is straightforward to

*Ph_2BCl is available from Organometallics, Inc., East Hampstead, NH.

perform the ammonium for lithium exchange without this final step. An ethanol solution (20 mL) of $[Li(TMEDA)_2][Ph_2B(CH_2PPh_2)_2]$ (3.51 g, 4.42 mmol) is first prepared. A solution of tetrabutylammonium chloride (1.424 g, 4.42 mmol) in dry ethanol (15 mL) is prepared separately. The ammonium chloride solution is then added dropwise to the ethanol solution of $[Li(TMEDA)_2][Ph_2B(CH_2PPh_2)_2]$ with stirring. The reaction mixture turns cloudy quickly, and after addition is complete, the mixture is allowed to settle for 30 min. The white solid precipitate is collected by filtration and is thoroughly washed with ethanol (30 mL) and then with diethyl ether (50 mL). Drying the powder in vacuo produces 3.13 g (88%) of the ammonium salt $[^nBu_4N][Ph_2B(CH_2PPh_2)_2]$.

Properties

$[^nBu_4N][Ph_2B(CH_2PPh_2)_2]$ is modestly air-stable, although less so than its tridentate derivative $[^nBu_4N][PhB(CH_2PPh_2)_3]$. Accordingly, it should be stored under nitrogen and handled with dry and deoxygenated solvents. Its ^1H NMR spectrum (d_6-acetone, 300 MHz) displays resonances at δ 7.29 (br s, 4H), 7.17 (m, 8H), 7.00 (m, 12H), 6.73 (m, 4H), 6.61 (t, 2H), 3.42 (m, 8H), 1.81 (m, 8H), 1.65 (br, 4H), 1.41 (sextet, 8H), 0.97 (t, 12H). The ^{31}P NMR spectrum (d_6-acetone, 121 MHz) shows a resonance at -8.8 ($^2J_{P-B} = 10\,Hz$) ppm. The ^{11}B NMR spectrum (d_6-acetone, 128.3 MHz) features one resonance at -12.6 ppm. In contrast to $[^nBu_4N][PhB(CH_2PPh_2)_3]$, $[^nBu_4N][Ph_2B(CH_2PPh_2)_2]$ decomposes rapidly in aqueous solution and water and therefore is not used in the workup procedure. The ammonium salt is very soluble in acetone, acetonitrile, and tetrahydrofuran and modestly soluble in warm toluene. Like the tris(phosphino)borate ligand, $[^nBu_4N][Ph_2B(CH_2PPh_2)_2]$ is unstable in chloroform and dichloromethane. A paper that generalizes this protocol to a family of $[R_2BPR'_2]^-$ ligands, and describes reliable procedures for the preparation of Ar_2BCl precursors, has been published.[7]

C. THALLIUM PHENYLTRIS(DIPHENYLPHOSPHINOMETHYL) BORATE, $[PhB(CH_2PPh_2)_3]Tl$

$$[Li(TMEDA)][PhB(CH_2PPh_2)_3] + TlPF_6 \longrightarrow [PhB(CH_2PPh_2)_3]Tl + LiPF_6 + TMEDA$$

■ **Caution.** *Thallium is a toxic heavy metal, and care should be taken to minimize skin and respiratory exposure to $TlPF_6$, and with respect to disposing of thallium waste products.*

Procedure

The following preparation is carried out on the benchtop under air. Solid [Li(TMEDA)][$PhB(CH_2PPh_2)_3$] (7.1 g, 8.6 mmol), generated as described above (LiCl impurities are not problematic as they are removed by the following procedure), is added to a 250-mL Erlenmeyer flask charged with a stirring bar and suspended in methanol (60 mL). To this stirring suspension is added an aqueous solution (30 mL) of $TlPF_6$ (3.00 g, 8.6 mmol) over a period of 5 min. A cloudy white suspension results, which is stirred for an additional 5 min, followed by extraction with dichloromethane (2 × 150 mL). The solution is filtered through Celite and thoroughly dried in vacuo to afford a light yellow powder. This powder is then washed with hexanes and diethyl ether (40 mL each), extracted into benzene (150 mL), stirred over $MgSO_4$, filtered, and then dried thoroughly in vacuo, affording the thallium complex as a fine yellow powder (5.00 g, 65%).

Properties

The thallium complex [$PhB(CH_2PPh_2)_3$]Tl is a yellow solid. ^1H NMR (C_6D_6, 300 MHz, 25°C): δ 8.13 (d, $J = 6.6$ Hz, 2H), 7.67 (m, $J = 7.5$ Hz, 2H), 7.42 (tt, $J = 6.6$, 1.2 Hz, 1H), 7.18–7.11 (m, 12H), 6.80–6.77 (m, 18H), 1.96 (br m, 6H). ^{31}P NMR (C_6D_6, 121.4 MHz, 25°C): 21.6 ppm (d, $^1J_{Tl-P} = 5214$ Hz for ^{205}Tl (70.5% abundance), $^1J_{Tl-P} = 5168$ Hz for ^{203}Tl (29.5% abundance). ^{13}C NMR (C_6D_6, 125.7 MHz, 25°C): δ 139.8, 132.5, 128.8–129.1 (overlapping resonances), 124.6, 17.0 (br). ^{11}B NMR (C_6D_6, 128.3 MHz, 25°C): −10.96 ppm. The compound is a versatile precursor to a wide range of transition metal complexes supported by the tris(phosphino)borate ligand. It is air- and water-stable for extended periods, and, unlike the lithium and ammonium salts of [$PhB(CH_2PPh_2)_3]^-$, it is both soluble and stable in chloroform and dichloromethane for days, making these useful solvents available for subsequent transmetallation chemistry.

Acknowledgments

The authors wish to thank those group members who aided in the development of these syntheses, especially Joseph Duimstra, Ian Shapiro, Connie Lu, and David Jenkins. The NSF (CHE-0132216), the DOE (PECASE), and the Dreyfus Foundation are acknowledged for their generous support. J. C. Peters also acknowledges Professor T. Don Tilley, Professor Daniel G. Nocera, and Dr. Alan A. Barney for contributions to the work described.

References

1. F. A Cotton and B. Hong, *Prog. Inorg. Chem.* **40**, 179 (1992).
2. (a) T. A. Betley and J. C Peters, *Angew. Chem., Int. Ed. Engl.* **42**, 2385 (2003); (b) J. C. Thomas and J. C. Peters, *J. Am. Chem. Soc.* **125**, 8870 (2003); (c) C. C. Lu and J. C. Peters, *J. Am. Chem. Soc.* **124**, 5272 (2002).
3. J. C. Peters, J. D. Feldman, and T. D. Tilley, *J. Am. Chem. Soc.* **121**, 9871 (1999).
4. A. A. Barney, A. F. Heyduk, and D. G. Nocera, *Chem. Commun.* 2379 (1999).
5. I. R. Shapiro, D. M. Jenkins, J. C. Thomas, M. W. Day, and J. C. Peters, *Chem. Commun.* 2152 (2001).
6. N. E. Schore, L. S. Benner, and B. E. Labelle, *Inorg. Chem.* **20**, 3200 (1981).
7. J. C. Thomas and J. C. Peters, *Inorg. Chem.* **42**, 5055 (2003).

4. SIX-COORDINATE ALUMINUM CATIONS BASED ON SALEN LIGANDS

Submitted by SAMEER SAHASRABUDHE,[*] BURL C. YEARWOOD,[*] and DAVID A. ATWOOD[*]
Checked by MICHAEL J. SCOTT[†]

Aluminum compounds with organic ligands are widely used as reagents in organic synthesis.[1] Transformations that can be achieved with these Lewis acid reagents include the reduction of ketones and aldehydes,[1] enantioselective Diels–Alder reactions,[2] and the "living" polymerization of oxiranes.[3,4] We have introduced a class of air-stable, cationic, six-coordinate aluminum complexes of general formula [(Schiff base)Al$(solv)_2$]X,[5–9] and have shown their utility as catalysts for the polymerization of propylene oxide.[8,9] Here we present detailed preparations for representative members of this class based on Salen (*N,N′*-bis(2-hydroxybenzylidene)ethylenediamine) and related ligands (see Fig. 1). These preparations may be applied with equal facility to related Schiff base ligands that have different connections between the two nitrogen atoms.

General Comments

All manipulations are conducted by using appropriate inert-atmosphere techniques. Toluene and THF are refluxed over sodium benzophenone and MeOH over Mg and then distilled prior to use. The ligands SalenH_2 and Salen^{t}BuH_2 were synthesized as described previously.[7] Dimethylaluminum chloride is available from Aldrich Chemical Co.

[*]Department of Chemistry, University of Kentucky, Lexington, KY 40506.
[†]Department of Chemistry, University of Florida, PO Box 117200, Gainesville FL 32611.

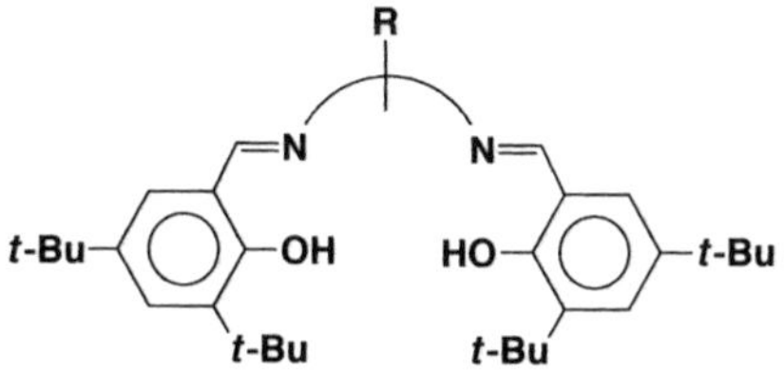

R	Ligand name
$(CH_2)_2$	Salen(^{t}Bu)
$(CH_2)_3$	Salpen(^{t}Bu)
$(CH_2)_4$	Salben(^{t}Bu)
C_6H_4	Salophen(^{t}Bu)
$C_6H_2Me_2$	Salomphen(^{t}Bu)

Figure 1. Naming scheme for Salen(^{t}Bu) class of ligands.

A. CHLORO-1,2-BIS(2-HYDROXYBENZYLIDENEIMINO) ETHANEALUMINUM, Al(Salen)Cl

$$\text{SalenH}_2 + \text{Me}_2\text{AlCl} \xrightarrow[25^\circ\text{C}]{\text{toluene}} \text{Al(Salen)Cl} + 2\,\text{CH}_4$$

■ **Caution.** *Dimethylaluminumchloride (Me_2AlCl) is a pyrophoric liquid. It must be handled under an inert atmosphere.*

Procedure

In a drybox, 5.0 g (18.6 mmol) of $SalenH_2$ is placed into a 250-mL side arm flask containing a 1-in. magnetic stirring bar. The ligand is dissolved in 125 mL of toluene, and then 1.72 g (18.6 mmol) of dimethylaluminumchloride is added slowly over the course of 2 min with stirring. A yellow solution and pale yellow solid results. The mixture is stirred for 2 h at 25°C, and then the solvent is removed by filtration. The stopcock of the sidearm flask is closed, a rubber septum placed in the neck, and the flask containing the precipitate is brought out of the drybox and placed onto a vacuum line equipped with a dry N_2 source. The side arm is evacuated, and then the flask is pressurized with nitrogen. The septum is replaced with a glass stopper, and the residue is dried under vacuum to yield 5.81 g (97%) of the desired product.

Anal. Calcd. for C, 58.46; H, 4.29. Found: C, 58.26; H, 4.54.

Properties

The product shows no appreciable solubility in Et_2O or THF; it will dissolve in protic solvents but in so doing forms the solvated salts. ^{1}H NMR (200 MHz, $CDCl_3$): δ 3.83 (m, 2H), 4.23 (m, 2H), 6.76–7.44 (m, 8H), 8.38 (s, 2H, CH=N).

IR (KBr) 3040 w, 2974 w, 1647 s, 1550 s, 1477 s, 1340 s, 756 s. The synthesis for AlLCl L = SalenCl, Acen, Salophen,[10] Salen^{t}Bu,[8] Salpen^{t}Bu, Salben^{t}Bu, Salophen^{t}Bu, Salomphen^{t}Bu is similar and leads to organic soluble compounds. The preparations for L = Salen, SalenCl, Acen, and Salophen are facilitated by the insolubility of the compounds, which precipitate essentially quantitatively from toluene.

B. DIAQUA-1,2-BIS(2-HYDROXYBENZYLIDENEIMINO) ETHANEALUMINUM CHLORIDE, [Al(Salen)(H_2O)$_2$]Cl

$$\mathrm{Al(Salen)} \xrightarrow[25^\circ\mathrm{C}]{\mathrm{H_2O}} [\mathrm{Al(Salen)(H_2O)_2}]\mathrm{Cl}$$

Procedure

In a drybox, a 100-mL flask containing a 1-in. magnetic stir bar is loaded with 1.0 g (3.04 mmol) of SalenAlCl. The stopcock is closed, a rubber septum is placed in the neck, and the flask is brought out of the drybox and placed onto a vacuum line equipped with a dry N_2 source. While stirring, 20 mL of distilled water is added by syringe over the course of a few seconds. The mixture is stirred for 2 h at 25°C, resulting in a pale yellow solution and a solid. Evaporation of the water over a period of 3 days results in quantitative yields of the product as a crystalline solid.

Anal. Calcd.: C, 52.69; H, 4.97. Found: C, 52.65; H, 5.02.

Properties

The compound [Al(Salen)(H_2O)$_2$]Cl is not air-sensitive. It has mp 162–165°C dec. The cation is soluble only in water. IR (KBr): 3057 s (br) 1639 s, 1560 m, 1473 s, 1296 s, 814 m, 760 s cm^{-1}. The synthesis and properties of the SalenCl and Acen derivatives are described in Ref. 7. Salophen, Salen(^{t}Bu), Salpen(^{t}Bu), Salben(^{t}Bu), Salophen^{t}Bu, and Salomphen^{t}Bu derivatives can also be prepared in a similar manner. In the solid state these compounds have been shown by X-ray crystallography to contain six-coordinate octahedral aluminum. The solvent molecules are trans to one another and the ligand coplanar with the aluminum atom. The chloride anions are connected by hydrogen bonds to the solvents on the aluminum as well as additional solvent in the crystal lattice.

C. DIMETHANOL-1,2 BIS-(2-HYDROXY-3,5-BIS(*tert*-BUTYL)BENZYLIDENEIMINO)ETHANEALUMINUM CHLORIDE, [Al (Salen^{t}Bu)(MeOH)$_2$]Cl

$$\text{Al(Salen}^t\text{Bu)Cl} \xrightarrow[25^\circ\text{C}]{\text{MeOH}} [\text{Al(Salen}^t\text{Bu)(MeOH)}_2]\text{Cl}$$

Procedure

In a drybox a 100-mL sidearm flask containing a 1-in. magnetic stirring bar is loaded with 0.350 g (0.633 mmol) of (Salen^{t}Bu)AlCl. The stopcock is closed, a rubber septum placed on the neck, and the flask brought out of the drybox and placed onto a vacuum line equipped with a dry N_2 source. The flask is pressurized with N_2, and with stirring, 7 mL of MeOH is added by syringe over the course of a few seconds. The product turns white and dissolves over 5 min. Slow evaporation of the solvent in air leads to a nearly quantitative yield of pale yellow crystals. The solid can also be prepared by stirring Salen(^{t}Bu)AlCl in MeOH for 3 h and then removing the solvent under vacuum.

Anal. Calcd.: C, 66.16; H, 8.82. Found: C, 65.78; H, 8.68.

Properties

The compound (mp > 260 dec.) dissolves in organic solvents such as CH_2Cl_2 and $CHCl_3$. In the solid state the compound has been shown by X-ray crystallography to contain six-coordinate, octahedral aluminum.[7] As with the water cation, the solvent molecules are trans to one another and the ligand is coplanar with the aluminum atom. ^{1}H NMR (400 MHz, $CDCl_3$): δ 1.29 (s, 18H, ^{t}Bu), 1.53 (s, 18H, ^{t}Bu), 3.46 (s, 6H, CH_3OH), 3.79 (s, 2H, CH_2CH_2), 4.12 (s, 2H, CH_2CH_2), 7.04 (d, 2H, PhH), 7.55 (d, 2H, PhH), 8.38 (s, 2H, HC=N). IR (KBr): 3406 w, 2955 s, 1639 s, 1554 m, 1442 m, 1276 m, 875 m, 756 m cm^{-1}. The Salen, Acen, Salophen, Salen^{t}Bu, Salen^{t}Bu, Salophen^{t}Bu, and Salomphen^{t}Bu derivatives can be prepared in a similar manner.

D. DIMETHANOL-1,2-BIS(2-HYDROXY-3,5-BIS(*tert*-BUTYL)BENZYLIDENEIMINO)ETHANEALUMINUM TETRAPHENYLBORATE, [Al(Salen^{t}Bu)(MeOH)$_2$][BPh$_4$]

$$\text{Al(}^t\text{BuSalen)Cl} + \text{Na[BPh}_4] \xrightarrow[25^\circ\text{C}]{\text{MeOH}} [\text{Al(Salen}^t\text{Bu)(MeOH)}_2[\text{BPh}_4] + \text{NaCl}$$

Procedure

In a drybox a 250-mL sidearm flask containing a 1-in. magnetic stirring bar is loaded with a mixture of 1.50 g (2.71 mmol) of (Salen^{t}Bu)AlCl and 0.928 g (2.71 mmol) of sodium tetraphenylborate ($NaBPh_4$). As solids, they do not react. The stopcock of the sidearm flask is closed, a rubber septum placed on the neck, and the flask brought out of the drybox and placed onto a vacuum line equipped with a dry N_2 source. Then 20 mL of MeOH is added to the reactants. A yellow solid appears and then goes back into solution. The yellow solution is stirred for 18 h and then allowed to stand for 24 h. During this time a small amount of white precipitate settles out of solution. The solution is filtered, and the filtrate concentrated to 15 mL and then cooled to −30°C. After several days, pale yellow plates are formed. These are isolated by filtration under nitrogen and dried under vacuum to give the product in essentially quantitative yield.

Anal. Calcd. for C, 77.32; H, 8.28. Found: C, 77.32; H, 8.42.

Properties

The compound (mp 118–121°C) is soluble in solvents such as $CHCl_3$, $CHCl_2$, and THF. It is moderately sensitive to ambient conditions and forms an uncharacterized hydrate on exposure to air. ^1H NMR (400 MHz, $CDCl_3$) δ 1.31 (s, 18H, ^{t}Bu), 1.46 (s, 18H, ^{t}Bu), 2.93 (s, 6H, CH_3OH), 3.14 (s, 4H, CH_2CH_2), 6.71–7.66 (m, 26H, PhH, and CH=N)). IR (KBr) 3437 m, 3057 m, 1626 s, 1543 m, 1473 s, 1255 s, 850 m, 705 s cm^{-1}. The Salen, Acen, SalenCl,[6] Salophen, Salpen^{t}Bu, Salben^{t}Bu, Salophen^{t}Bu, and Salomphen^{t}Bu derivatives can be prepared in a similar manner.

E. DIMETHANOL-1,2-BIS(2-HYDROXY-3,5-BIS(*tert*-BUTYL) BENZYLIDENEIMINO)ETHANEALUMINUM *para*-TOLUENESULFONATE, [Al(^{t}BuSalen)(MeOH)$_2$][OTs]

$$[\text{Al}(\text{Salen}^t\text{Bu}]\text{Cl} \xrightarrow[\text{HOTs}]{\text{MeOH}} [\text{Al}(\text{Salen}^t\text{Bu})(\text{MeOH})_2][\text{OTs}] + \text{HCl}$$

In a drybox, a 250-mL sidearm flask containing a 1-in. magnetic stirring bar is loaded with a mixture of 1.00 g (1.81 mmol) of (Salen^{t}Bu)AlCl and 0.351 g (1.81 mmol) of sodium *para*-toluenesulfonate (NaOTs). As solids, they do not react. The stopcock of the sidearm flask is closed, a rubber septum placed on the neck, and the flask brought out of the drybox and placed onto a vacuum line equipped with a dry N_2 source. Then 20 mL of MeOH is added to the reactants. The mixture is stirred, and after a few minutes the solids dissolve to form a yellow solution. The mixture is further stirred for 24 h, and then the small amount of white solid formed is removed by filtration. The resulting solution is cooled to −30°C, and after several days, opaque yellow crystals form that are isolated by filtration and dried under vacuum (0.624 g, 41%).

Anal. Calcd. for C, 65.40; H, 8.17. Found: C, 65.37; H, 7.95

Properties

The compound (mp 135–138°C) is soluble in toluene, chloroform, and THF. It is moderately air-stable. X-ray analysis shows it to contain a six-coordinate, distorted octahedral aluminum with the Salen^{t}Bu ligand occupying the four equatorial positions and the two methanol molecules in the axial positions. ^{1}H NMR (400 MHz, $CDCl_3$): δ 1.20 (s, 18H, ^{t}Bu), 1.36 (s, 18H, ^{t}Bu), 3.03 (m, 3H, CH_3OH), 3.70 [s (br), 4H, CH_2CH_2], 6.77 (d, 2H, OTs PhH), 6.82 (d, 2H ligand PhH), 7.15 (d, 2H, OTs PhH), 7.37 (d, 2H ligand PhH), 7.96 (s, 2H, CH=N). IR (KBr): 3142 m, 2955 s, 1637 s, 1550 m, 1444 m, 1257 s, 1172 s, 860 m, 682 m cm^{-1}.

F. BIS(TETRAHYDROFURAN)-1,2-BIS(2-HYDROXY-3,5-BIS (*tert*-BUTYL)BENZYLIDENEIMINO)ETHANEALUMINUM TETRAPHENYLBORATE, [Al(^{t}BuSalen)(THF)$_2$][BPh$_4$]

$$\text{Al(Salen}^t\text{Bu)Cl} + \text{NA[BPh}_4] \xrightarrow{\text{THF}} [\text{Al(}^t\text{(BuSalen)(THF)}_2][\text{BPh}_4] + \text{NaCl}$$

Procedure

In a drybox, a 250-mL sidearm flask containing a 1-in. magnetic stirring bar is loaded with a mixture of 4.00 g (7.23 mmol) of (Salen^{t}Bu)AlCl and 2.51 g (7.35 mmol) of sodium tetraphenylborate ($NaBPh_4$). As solids, they do not react. The stopcock of the sidearm flask is closed, a rubber septum placed on the neck, and the flask brought out of the drybox and placed onto a vacuum line equipped with a dry N_2 source. Then 50 mL of degassed THF is added to the reactants. The pale yellow suspension is refluxed for 12 h. The solution is filtered under N_2. The THF is removed under vacuum, yielding 5.93 g (84%) of the compound as a pale yellow solid.

Anal. Calcd. for C, 78.35; H, 8.42. Found: C, 78.44; H, 8.29.

Properties

The compound is soluble in toluene, chloroform, and THF. It is moderately air-stable. ^{1}H NMR (400 MHz, $CDCl_3$): δ 1.28 (s, 18H, ^{t}Bu), 1.33 (s, 18H, ^{t}Bu), 1.83 (m, 4H, THF), 2.74 (s, 2H, CH_2CH_2), 3.18 (s, 2H, CH_2CH_2), 3.71 (m, 8H, THF), 6.90–7.61 [m, 26H, PhH, and CH=N)]. IR (KBr) 3055 m, 2957 s, 1629 s, 1556 m, 1446 m, 1276 m, 1047 s, 862 m, 734 m cm^{-1}. The Salpen-^{t}Bu, Salben^{t}Bu, Salophen^{t}Bu, and Salomphen^{t}Bu derivatives can be prepared in a similar manner. The Salpen^{t}Bu derivative was shown by x-ray studies to contain a six-coordinate, octahedral aluminum with the Salpen^{t}Bu ligand occupying the four equatorial positions and the two THF molecules in the axial positions.

References

1. (a) D. Schinzer, ed., *Selectivities in Lewis Acid Promoted Reactions*, Kluwer Academic Publishers, Dordrecht, 1989; (b) H. Yamamoto, in *Organometallics in Synthesis*, M. Schlosser, ed., Wiley, Chichester, UK, 1994, Chapter 7.
2. Y. Hayashi, J. J. Rohde, and E. J. Corey, *J. Am. Chem. Soc.* **118**, 5502 (1996).
3. (a) K. Maruoka, S. Nagahara, and H. Yamamoto, *J. Am. Chem. Soc.* **112**, 6115 5475 (1990); (b) K. Maruoka, S. Nagahara, and H. Yamamoto, *Tetrahedron Lett.* **31**, 5475 (1990); (c) K. Maruoka, A. B. Concepcion, N. Murase, M. Oishi, N. Hirayama, and H. Yamamoto, *J. Am. Chem. Soc.* **115**, 3943 (1993).
4. (a) E. J. Vandenberg, *J. Polym. Sci., A-1*, **7**, 525 (1969); (b) C. L. Jun, A. Le Borgne, and N. Spassky, *J. Polym. Sci., Polym. Symp.* **74**, 31 (1986); (c) V. Vincens, A. Le Borgne, and N. Spassky, *Makromol. Chem. Rapid Commun.* **10**, 623 (1989); (d) A. Le Borgne, V. Vincens, M. Jouglard, and N. Spassky, *Makromol. Chem. Macromol. Symp.* **73**, 37 (1993); (e) H. Sugimoto, C. Kawamura, M. Kuroki, T. Aida, and S. Inoue, *Macromolecules* **27**, 2013 (1994); (f) T. Aida, R. Mizutta, Y. Yoshida, and S. Inoue, *Makromol. Chem.* **182**, 1073 (1981); (g) K. Shimasaki, T. Aida, and S. Inoue, *Macromolecules* **20**, 3076 (1987); (h) S. Asano, T. Aida, and S. Inoue, *J. Chem. Soc. Chem.Comm.* 1148 (1985).
5. D. A. Atwood, J. A. Jegier, and D. Rutherford, *Inorg. Chem.* **35**, 63 (1996).
6. D. Rutherford and D. A. Atwood, *Organometallics* **15**, 4417 (1995).
7. D. A. Atwood, J. A. Jegier, and D. Rutherford, *J. Am. Chem. Soc.* **117**, 6779 (1995); D. A. Atwood and M. J. Harvey, *Chem. Rev.*, **101**, 37 (2001).
8. J. A. Jegier, M. A. Munoz-Hernandez, and D. A. Atwood, *J. Chem. Soc., Dalton Trans.* 2583 (1999).
9. M. A. Munoz-Hernandez, M. L. McKee, T. S. Keizer, B. C. Yearwood, and D. A. Atwood, *J. Chem Soc., Dalton Trans.* 410 (2002).
10. R. H. Holm, G. W. Everett, and A. Chakravorty, *Prog. Inorg. Chem.* **7**, 83 (1966).
11. P. L. Gurian, L. K. Cheatham, J. W. Ziller, and A. R. Barron, *J. Chem. Soc., Dalton Trans.* 1449 (1991).

5. TABULAR α-ALUMINA

Submitted by RICHARD F. HILL,[*] G. LYNN WOOD,[*†]
R. DANZER,[‡] and ROBERT T. PAINE,[§]
Checked by TIMOTHY BOYLE[¶]

The chemical and physical properties of alumina make it one of the most industrially significant bulk electronic and structural ceramic materials produced worldwide.[1] The material has a wide structural diversity, adopting at least 15 crystallographic phases. The most prominent, α-Al_2O_3 and δ-Al_2O_3, are made commercially in tonnage quantities by using well-developed chemical processes. Ceramic-grade, calcined α-Al_2O_3 powders are usually obtained as agglomerates that are dry ground, sorted, and classified into specific product lines. The resulting irregularly shaped particles are suitable for many uses; however, there are increasing fundamental scientific and practical needs for higher-quality materials with well-defined morphologies. In particular, spherical and tabular morphologies are in demand. Tabular aluminas are typically obtained by grinding and sintering powders calcined in the temperature region 1600–1850°C, and the resulting high-density, recrystallized particles have an elongated (50–400-μm), flat, platelike morphology. Unfortunately, these powders have a wide size distribution and contain large amounts of fines that reduce their utility for adsorbent, catalyst support and ceramic filler applications. For fundamental laboratory studies in these application fields it would be useful to have available more well-characterized tabular α-Al_2O_3 powders. Patent literature suggests that fluoride may be used to promote alumina crystal growth,[2,3] and we have reported the use of boehmite/hydrofluoric acid (HF_{aq}) gel slurries as a starting material for tabular α-Al_2O_3 synthesis.[4] Here we provide a simplified procedure for obtaining α-Al_2O_3 platelets with a narrow diameter and size distribution (average size 8 μm distribution 5–12 μm).

Equipment

All manipulations, including sample heatings, are performed in a well-ventilated fume hood. A small, handheld stirrer is used initially to mix the reaction components.

[*]Center for Micro-Engineered Materials, University of New Mexico, Albuquerque, NM 87131.
[†]Department of Chemistry, Valdosta State University, Valdosta, GA 31698.
[‡]Institute of Structural and Functional Ceramics, University of Leoben, A8700 Leoben, Austria.
[§]Department of Chemistry, University of New Mexico, Albuquerque, NM 87131.
[¶]Sandia National Laboratory, Albuquerque, NM 87106.

Materials

Boehmite alumina powder is obtained from Condea-Vista as plural-grade 200 (Inorganic Specialties Division, Condea Vista, 900 Threadneedle, #100, Houston, TX 70079) and aqueous HF (50 wt%) is purchased from J. T. Baker.

Procedure

■ **Caution.** *Hydrofluoric acid causes severe burns and all appropriate safety literature describing precautions should be consulted before use. Lab coat, face-shield, and heavy vinyl gloves should be worn when preparing, stirring, and transferring aqueous HF solutions. The fume hood should be approved for use with hydrofluoric acid. The high-temperature oven should also be placed in the hood since aluminum fluoride and HF are evolved in the heat treatment.*

A solution of aqueous hydrofluoric acid (1.7 wt%) is prepared by diluting commercial, 50 wt%, HF with deionized water in a polypropylene bottle. A 100 g portion of the diluted HF was transferred to a Nalgene plastic beaker and boehmite alumina powder (20 g) is added over 5 min with manual stirring using a stainless-steel spatula. After approximately 5 min, the mixture has the appearance of a moderately thick, white paste. An additional quantity (100 g) of water is added, and the mixture is stirred for 3 min. Additional boehmite powder (100 g) is added with continuous stirring. The resulting slurry forms a thick gel. This gel is stirred for 5 min at high speed using a standard laboratory paddle mixer* until the slurry takes the consistency of plumber's putty. (*Note*: This is an important variable, since the paste should not separate into solid and liquid components during subsequent processing. If a powerful stirrer is not available, the final mixing can be accomplished by hand mixing using a spatula. The checker found hand mixing to be adequate.) The resulting paste is dried at 140°C overnight in a standard laboratory oven.† This produces an agglomerated solid mass that is easily ground with a large mortar and pestle to give a powder (maximum agglomerate size 5 mm). A portion of this solid (25.6 g) is transferred to a tapered 50-mL [$5 \times 4.8 \times 3.0$-cm (height × top × bottom)] alumina crucible fitted with a flat, smooth-fitting alumina lid. The covered crucible is placed in a furnace‡

*Cole-Parmer Instrument Co. Chicago, IL. Alternatively, a variable-speed shop drill may be used with a standard stirrer shaft.

†Since the inside of the oven will be exposed to corrosive HF and AlF_3 vapors, an old drying oven is preferred that is vented through the top to a fume hood. Alternatively, a small vacuum oven may be used, but the vacuum applied should be minimal.

‡This may be any high-temperature tube or muffle furnace. The furnace should be placed in a fume hood.

with an air atmosphere and is heated from 20 to 1075°C at a rate of 10°C/min. This temperature is maintained for 180 min, then the temperature is increased to 1200°C at 5°C/min. This temperature is maintained for 180 min. At the end of this time, the covered crucible is briefly withdrawn from the furnace with appropriate tongs, the lid is removed, and the uncovered crucible is returned to the furnace. Heating at 1200°C is continued for another 300 min. During the heating of the covered crucible at 1200°C, AlF_3 sublimes to the top of the crucible, forming a seal that prevents most of the AlF_3 vapors from escaping. This results in intimate Al_2O_3/AlF_3 contact that enhances the platelet growth. When the crucible lid is removed and the crucible reheated, AlF_3 sublimes from the sample, as indicated by formation of white vapor. At the end of the heating cycle, the crucible and contents are cooled to room temperature, and the resulting powder (20.3 g) is removed from the crucible. The powder can be deagglomerated with a mortar and pestle, using very light pressure, and then is screened through a 325-mesh sieve.

Anal. Calcd. for Al, 52.92; O, 47.08. Found: Al, 52.88; O, 47.05. C, H, $N < 0.5\%$, $F < 380$ ppm.

Figure 1. Scanning electron micrograph of tabular α-Al_2O_3.

Properties

The compound is a free-flowing white solid. The powder X-ray diffraction (XRD) is in complete agreement with that for α-Al_2O_3 (corundum JCPDS) card file 46-1212: *XRD:* (d, Å; I,%): 3.487 (33), 2.555 (73); 2.382 (25), 2.088 (100); 1.741 (35); 1.603 (86); 1.512 (6); 1.405 (22); 1.374 (40); 1.240 (12). A scanning electron micrograph (SEM) (Fig. 1) shows the particles have the desired platelike morphology with average particle diameter of 8 μm. If fines appear in the SEM, this indicates that the last heating stage was not of sufficient length. The surface area for typical samples is 0.1 m^2/g. There is some variability found in average particle size. The variables include size of the crucible and the amount of powder calcined in the crucible. Larger platelets are obtained when the crucible is nearly filled to the top with the ground agglomerate.

References

1. M. Howe-Grant, ed., *Encyclopedia of Chemical Technology*, 4th ed., Wiley, New York,
2. J. Black and J. W. K. Lau, U.S. Patent 5,137,959 (1992).
3. W. W. Weber and J. A. Herbst, U.S. Patent 4,379,134 (1983).
4. R. F. Hill, R. Danzer, and R. T. Paine, *J. Am. Ceram. Soc.* **84**, 514 (2001).

6. PARA-SUBSTITUTED ARYL ISOCYANIDES

Submitted by NICOLE L. WAGNER,[*] KRISTEN L. MURPHY,[*] DANIEL T. HAWORTH,[*] and DENNIS W. BENNETT[*]
Checked by MATTHIAS ZELLER,[†] NATHAN TAKAS,[†] STEVEN J. DiMUZIO,[†] and ALLEN D. HUNTER[†]

Isocyanides behave as unique ligands in transition metal complexes, capable of strong coordination to both low- and high-valence metal centers. Subtle changes in the substituent para to the isocyanide moiety of these ligands result in modification of their π acidity, which can subsequently alter the geometry of the transition metal complexes dramatically.[1,2] In the course of studying these complexes, the general synthetic routes to representative isocyanides have been modified, and these procedures are detailed in this contribution.

[*]Department of Chemistry, University of Wisconsin, Milwaukee, WI 53211.
[†]Department of Chemistry, Youngstown State University, Youngstown, OH 44555.

Starting Materials

All starting materials are commercial samples used without further purification. The following reagents are available from Acros-Fisher: *p*-phenylenediamine (99%), *p*-anisidine (99%), trisphosgene (99%), and tungsten(VI) chloride (99%). 2,5-Dimethyl,1,4-phenyldiamine is obtained from Aldrich Chemical Co. Phosgene (99%) is obtained from Matheson. Reagent-grade solvents (99.5%) are obtained from VWR, with the exception of THF(Fisher). Solvents for the synthesis of the isocyanides are used without further purification. Solvents used for the organometallic synthesis are purified as described in procedure D.

■ **Caution.** *Isocyanides are toxic and have a distinctive odor that is very offensive. Care should be taken to avoid inhalation, particularly when handling liquid isocyanides. All reactions should be carried out in a well-ventilated hood. Phosgene gas is a relatively common synthetic reagent and can be safely used with care, but it is extremely toxic and should be handled only by the most experienced personnel in the laboratory. The use of a self-contained breathing apparatus is strongly recommended when phosgene is used in these reactions. Although phosgene is less costly, liquid diphosgene (trichloromethylchloroformate) or solid triphosgene [bis(trichloromethyl)carbonate] can be used as safer alternatives to phosgene in the synthetic approaches described here. It is, nevertheless, important to note that all three phosgenation reagents are extremely irritating to the eyes and skin, and if inhaled, either directly or via their decomposition products, can cause severe respiratory damage.*

A. 1,4-DIISOCYANOBENZENE, $p\text{-}CNC_6H_4NC$

$$p\text{-}H_2N(C_6H_4)NH_2 + 2\,HCOOH \longrightarrow p\text{-}OHCHN(C_6H_4)NHCHO + 2\,H_2O$$

$$p\text{-}OHCHN(C_6H_4)NHCHO + 2\,COCl_2 + 4\,Et_3N \longrightarrow$$
$$p\text{-}CN(C_6H_4)NC + 4\,[Et_3NH]Cl + 2\,CO_2$$

Procedure

This is a modification of a synthetic method initially reported by Ugi.[3]

■ **Caution.** *Phosgene is a highly toxic gas that reacts with water to form HCl and has a characteristic odor of freshly mown hay. It is necessary to carry out this portion of the reaction using standard Schlenk techniques and in a well-ventilated fume hood. Unreacted phosgene can be neutralized with aqueous base.*

A 250-mL round-bottomed flask equipped with a water collector (Dean–Stark trap), condenser, and magnetic stirrer is charged with *p*-$H_2N(C_6H_4)NH_2$ (36 g, 0.33 mol), formic acid (69 mL, 1.8 mol, 3 : 1 excess), and toluene (80 mL, reagent grade). The mixture is heated to reflux until the water is collected (~3 h). The flask is cooled, and formamide is collected by filtration, and rinsed with copious amounts of ethanol and diethylether until formic acid is washed away. Color and percent yield of formamide (~70%) is dependent on the purity of the amine. The reaction can be scaled up, depending on collector size.

A 1-L three-necked round-bottomed flask equipped with a mechanical stirrer and dry-ice/propanol-cooled cold fingers is charged with *p*-$OHCHN(C_6H_4)NH$-CHO (48.2 g, 0.294 mol) dissolved in CH_2Cl_2 (500 mL, reagent grade), and triethylamine (125 mL, 1.7 mol). The mixture is cooled in an ice/salt bath and flushed with nitrogen. Phosgene* (32 mL, 0.587 mol), collected into a liquid nitrogen cooled collector under reduced pressure, is allowed to warm slowly and drip into the reaction mixture (~3 h) under a slow stream of nitrogen. Excess phosgene is vented through a series of flasks filled with air, oil, and saturated NaOH, respectively. The mixture is stirred for an additional 2 h and allowed to warm to room temperature. $NaHCO_3$ (10%, 200 mL) is added to neutralize any remaining phosgene. The mixture is washed with water (200 mL), HCl (0.1 M, 200 mL) and saturated NaCl (150 mL), and the separated CH_2Cl_2 layer is dried over Na_2SO_4. The solvent is removed under vacuum, and the solid isocyanide is sublimed under vacuum at 30–60°C. Yield 50–80%.

Anal. Calcd. for $C_8H_4N_2$: C, 74.99; H, 3.15; N, 21.85. Found: C, 74.75; H, 3.08; N, 21.78.

The analogous compounds *p*-$CN(C_6H_4)CN$, *p*-$CN(C_6H_{10})NC$, *p*-$CN(C_6(CH_3)_4)NC$, and *p*-$CN(C_6H_4)R$ where R = NO_2, SH, COOH, and Cl can be prepared in a similar manner.

B. 1,4-DIISOCYANO-2,5-DIMETHYLBENZENE, *p*-$CN(C_6H_2(CH_3)_2)NC$

$$p\text{-}H_2N(C_6H_2(CH_3)_2)N_2H + HCOOH \longrightarrow p\text{-}OHCHN(C_6H_2(CH_3)_2)NHCHO + H_2O$$

$$3\,p\text{-}OHCHN(C_6H_2(CH_3)_2)NHCHO + 2\,CO(COCl_3)_2 + 12\,Et_3N \longrightarrow 3\,p\text{-}CN(C_6H_2(CH_3)_2)NC + 12\,[Et_3NH]Cl + 6\,CO_2$$

*Checkers used liquid Diphosgene (Aldrich) with no loss in yield.

Procedure

This is a modification of a synthetic method reported by Sahu.[4] This method can be adapted for the synthesis of the isocyanides prepared in Section 6.A. Substitution of triphosgene for phosgene provides a safer alternative for the synthesis of isocyanides, in general. On the other hand, phosgene is less expensive and the phosgene reaction is easier to scale up. *p*-$H_2N(C_6H_2(CH_3)_2)NH_2$ is converted to the corresponding formamide by the procedure outlined in Section 6.A.

A 250-mL two- or three-necked round-bottomed flask equipped with magnetic stirrer, condenser, and dropping funnel is charged with *p*-$OHCHN(C_6H_2$-$(CH_3)_2)NHCHO$ (0.75 g, 3.9 mmol), triphosgene (0.77 g, 2.6 mmol), and $CHCl_3$ (40 mL, reagent grade) to dissolve the solids. Triethylamine (2.2 mL, 15.6 mmol) in $CHCl_3$ (10 mL) is added dropwise to the stirring mixture. The mixture is gradually warmed to 50°C and stirred for 3 h. The solution is washed with water (10 mL), HCl (0.1 M, 20 mL), $NaHCO_3$ (10%, 20 mL), and saturated NaCl (15 mL), then the separated organic layer is dried over Na_2SO_4. After filtration the solvent is removed under vacuum. The product is purified by sublimation at 30–60°C.

Anal. Calcd. for $C_{10}H_8N_2$: C, 76.90; H, 5.17; N, 17.93. Found: C, 76.96; H, 5.17; N, 17.95.

C. 1-ISOCYANO-4-METHOXYBENZENE, *p*-$CN(C_6H_4)OCH_3$

$$p\text{-}H_2N(C_6H_4)OCH_3 + CHCl_3 \longrightarrow p\text{-}CN(C_6H_4)OCH_3 + 3\,HCl$$

Procedure

This is a modification of a synthetic method reported by Weber.[5] This single-step phase transfer synthesis provides obvious advantages over the methods described in Sections 6.A and 6.B. It is especially useful for the preparation of liquid isocyanides, since they are stabilized in the presence of the unreacted amine that remains in the reaction system until final purification. Unfortunately, the method does not appear to be general. In addition to *p*-$H_2N(C_6H_4)OCH_3$, *p*-$CN(C_6H_4)CF_3$, *p*-$CN(C_6H_4)F$ and *p*-$CN(C_6H_4)CH_3$ can be synthesized by this method.

A 125-mL round-bottomed flask equipped with a condenser and magnetic stirrer is charged with *p*-$H_3CO(C_6H_4)NH_2$ (12.3 g, 0.10 mol), $CHCl_3$ (10.5 mL, reagent grade), benzyltriethylammoniumchloride (BTEAC, 0.32 g, 1.4 mmol), and CH_2Cl_2 (33 mL, reagent grade). NaOH (50%, 35 mL) is added to the mixture after cooling to room temperature. The mixture is warmed to refluxing temperature and allowed to reflux for 3–5 h. Then the reaction mixture is diluted with cold water (100 mL) and extracted with portions of CH_2Cl_2 (100 mL). The CH_2Cl_2 washings are combined and washed with water (25 mL) and saturated NaCl (25 mL) and then dried over $MgSO_4$. The solvent is removed under vacuum. Any unreacted amine is separated from the isocyanide using a silicagel column with CH_2Cl_2 as the eluant. The isocyanide elutes first, and remaining amine is washed from the column using THF.

Anal. Calcd. for C_8H_7NO: C, 72.15; H, 5.31; N, 10.52. Found: C, 72.34; H, 5.20; N, 10.33.

D. BIS(1,4-DIISOCYANOBENZENE)BIS(1,2-BIS (DIPHENYLPHOSPHINO)ETHANE)TUNGSTEN(0), W(dppe)$_2$(CNC$_6$H$_4$NC)$_2$

$$\mathrm{W(dppe)_2(N_2)_2} + 2\,p\text{-}\mathrm{CN(C_6H_4)NC} \longrightarrow cis\text{-}\mathrm{W(dppe)_2(CNC_6H_4NC)_2} + 2\,\mathrm{N_2}$$

This synthesis has been reported by Bennett and co-workers.[1,2] A 100-mL round-bottomed flask equipped with sidearm and magnetic stirrer is charged with *trans*-$(N_2)_2$W(dppe)$_2$ (0.31 g, 0.3 mmol, synthesized according to literature[7]), *p*-CN(C_6H_4)NC (1.9 g, 1.5 mmol, 2.5 : 1 excess), and THF (80 mL, distilled with Na[Ph_2CO]) and refluxed for 6 h. (*Note*: alternatively the reaction may be stirred under irradiation from a 75-W incandescent tungsten filament source for 48 h.) Hexanes (50 mL, distilled from P_2O_5) are added, and the flask is allowed to sit for 3–5 days to grow crystals, or the reaction solvent volume is reduced to 20–30 mL and hexanes (40 mL) are added to precipitate the product. The red product is filtered, rinsed with 4 : 1 hexanes/toluene (distilled with Na[Ph_2CO]), and dried under vacuum.

Properties of Intermediates and Products

Both formamides and solid isocyanides are moderately stable in air but tend to oxidize over time. Consequently, they should be stored under nitrogen at or below 4°C. Accurate percent yields for the isocyanide compounds are not given, since, as previously observed,[6] they are stabilized in the presence of excess starting materials and are generally purified only as needed for further synthesis.

Characterization of product is confirmed via ^{1}H NMR, IR, and mass spectra. Mass spectra of all these isocyanides exhibit a parent ion of maximum intensity and characteristic fragmentation peaks. Mass spectroscopy is also a useful tool for detecting unreacted amine or formamide. Isocyanides decompose in acid and are soluble in hexane, THF, toluene, CH_2Cl_2, CH_3CN, and alcohols.

1,4-Diisocyanobenzene is a colorless crystalline solid. The IR spectrum exhibits the strong characteristic ν_{CN} stretching frequency at 2127 cm^{-1} (KBr). ^{1}H NMR: aromatic singlet at 7.45 ppm. The sublimed solid exhibits a melting point of 152°C.

1,4-Diisocyano-2,5-dimethylbenzene is a colorless crystalline solid. IR: $\nu_{CN} = 2120$ cm^{-1} (KBr). ^{1}H NMR: aromatic singlet at δ 7.3; methyl proton peak at δ 2.4. The sublimed solid exhibits a melting point of 160°C.

1-Isocyano-4-methoxybenzene is a colorless liquid. IR: $\nu_{CN} = 2122$ cm^{-1}; ^{1}H NMR: pair of doublets for the nonequivalent aromatic protons at δ 7.3 and 6.9; methoxy singlet at δ 3.8. The purified solid exhibits a melting point of 32°C.

Bis(1,4-diisocyanobenzene)bis(1,2–bis(diphenylphosphino)ethane)tungsten(0) is a deep red crystalline solid. It is remarkably air-stable in the solid state. IR: ν_{CN} (sym) = 1761 cm^{-1} and ν_{CN} (asym) = 1662 cm^{-1} for the coordinated isocyanide stretching frequencies, and ν(CN) = 2115 cm^{-1} for the uncoordinated end of the ligand. ^{31}P NMR: 32.1 ppm ($J_{WP} = 196$ Hz); 47.5 ppm ($J_{WP} = 280$ Hz).

Acknowledgments

The submitters wish to provide a special note of thanks to the checkers, Professor Hunter and his students. Their useful and constructive comments were of great assistance in the final preparation of this manuscript. The checkers would also like to acknowledge the contributions of Cynthia Perrine, James Updegraff, Robert Wilcox, Les McSparrin, Lisa Walther, Floyd Snuder, and Monique Peace for checking one or more variants of these reactions.

References

1. C. Hu, W. C. Hodgeman, and D. W. Bennett, *Inorg. Chem.* **35**, 1621 (1996).
2. N. L. Wagner, F. E. Laib, and D. W. Bennett, *J. Am. Chem. Soc.* **122**, 10856 (2000).
3. I. Ugi, U. Fetzer, U. Eholzer, H. Knupfer, and K. Offermann, *Angew. Chem., Int. Ed. Eng.* **4**, 472 (1965).
4. D. P. Sahu, *Indian J. Chem.* **32B**, 385 (1993).
5. W. P. Weber and G. W. Gokel, *Tetrahedron Lett.* **17**, 1637 (1972).
6. L. Malatesta, F. Bonati, *Isocyanide Complexes of Metals*, Wiley, Chichester, UK, 1969.
7. J. R. Dilworth and R. L. Richards, *Inorg. Synth.* 119 (1980).

7. UNSYMMETRIC TRIPOD LIGANDS RELATED TO TRIS(PYRAZOL-1-YL)METHANE

Submitted by PETER K. BYERS,* ALLAN J. CANTY,† and R. THOMAS HONEYMAN†
Checked by JAMES R. GARDINIER‡ and DANIEL L. REGER‡

Tripodal ligands with a bridgehead carbon atom have assumed an important role in coordination and organometallic chemistry. Of particular interest are tris(pyrazol-1-yl)methanes,[1] isoelectronic with widely used tris(pyrazol-1-yl)borates. Unsymmetric ligands closely related to tris(pyrazol-1-yl)methane have been reported,[2,3] and their lower symmetry is useful in a range of applications,[2–6] including spectroscopic studies,[3,4] and studies of isomerism[2,5] and cycloplatination reactions.[2b,5d] Symmetric tris(pyridin-2-yl)methane as a ligand has also been widely studied,[7] and we describe here the facile syntheses of an unsymmetric mixed pyrazole/pyridine tripod ligand bis(pyrazol-1-yl)(pyridin-2-yl)methane, $(pz)_2(py)CH$, and closely related bis(pyrazol-1-yl)(*N*-methylimidazol-2-yl)methane, $(pz)_2(mim)CH$ (see Fig. 1). These ligands contain heterocycles with closely related but different donor properties,[4a,6,8] and subtle differences in donor group geometries presented to the metal center in complexes.[4a] The synthetic strategy developed by Peterson is employed, relying on the condensation of bis(pyrazol-1-yl)methanone, $(pz)_2C{=}O$, with aldehydes, $R(H)C{=}O$, to give $(pz)_2(R)CH$ and carbon dioxide.[9] The syntheses of the precursor reagents $(pz)_2C{=}O$[9c] and *N*-methylimidazole-2-carbaldehyde {(mim)(H)C=O},[10] developed from reported syntheses, are given as procedures 6.A and 6.B, respectively.

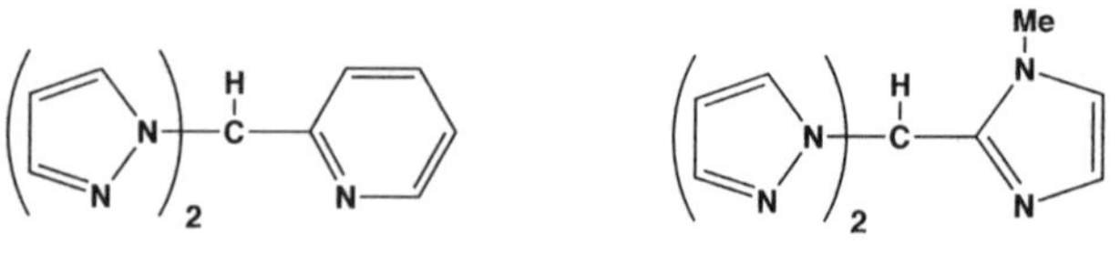

Figure 1. Tripod ligands.

*School of Chemistry, University of Birmingham, Edgbaston, Birmingham B15 2TT, Great Britain.
†School of Chemistry, University of Tasmania, Hobart, Tasmania 7001, Australia.
‡Department of Chemistry and Biochemistry, University of South Carolina, Columbia, SC 29208.

General Comment

Anaerobic techniques and dry solvents and reagents are used for all preparations.

A. BIS(PYRAZOL-1-YL)METHANONE, $(pz)_2C{=}O$

$$2\,pzH + 2\,Et_3N + Cl_2C{=}O \longrightarrow (pz)_2C{=}O + 2\,[Et_3NH]Cl$$

■ **Caution.** *Because phosgene is volatile (bp 8°C) and highly toxic by inhalation or skin contact, is a severe lachrymator and is moisture-sensitive, this reaction should be performed in a well-ventilated hood. See general comments for choosing phosgene or related reagents elsewhere in this volume.*

Procedure

Pyrazole (5.0 g, 73.5 mmol, Aldrich) and triethylamine (10.25 mL, 73.5 mmol) are added to diethyl ether (200 mL) in a three-necked 500-mL round-bottomed flask fitted with an overhead stirrer and nitrogen inlet. The mixture is stirred for 5 min using an overhead stirrer.* Phosgene (19 mL, 1.93 M in toluene, as received from Fluka) is added in two portions using a pressure equalized dropping funnel. Stirring is continued for 15 min, the precipitate removed by filtration under vacuum, and the volume of the filtrate reduced by rotary evaporation† to give a pale yellow oil. Hexane (10 mL) is added and colorless crystals of the product form over 3 h at room temperature or 1 h when cooled in ice. The product is collected by filtration and dried under high vacuum. The yield is 5.65 g (95%).‡

Properties

The solid product is unstable at ambient temperature and should be stored at −20°C; mp lit.[10a] 61.5–62.5°C. ^{1}H NMR in $CDCl_3$: δ 8.72 (1H, dd, $J = 3.1$ Hz, H_5), 7.89 (1H, d, $J = 1$ Hz, H_3), 6.54 (1H, dd, $J = 3.1$ Hz, H_4).

B. *N*-METHYLIMIDAZOLE-2-CARBALDEHYDE, (mim)CHO

$$Li(mim) + (Me_2N)(H)C{=}O + 2\,HCl \longrightarrow (mim)(H)C{=}O + [Me_2NH_2]Cl + LiCl$$

*The checkers used magnetic stirring with intermittent shaking as needed.

†Checkers used cannula filtration and vacuum distillation of solvent.

‡The checkers report a yield of 76%, m.p. 59–60°C.

Procedure

Butyllithium (140 mL of 1.2 M solution in diethyl ether, 168 mmol, prepared as reported[11]) is added dropwise to a stirred solution of *N*-methylimidazole (13.4 mL, 168 mmol, Aldrich) in diethyl ether (50 mL) at $-80°C$ in a three-necked 500-mL round-bottomed flask fitted with a nitrogen inlet. The solution is allowed to warm slowly to 0°C, then cooled to $-80°C$, and added dropwise at this temperature via a jacketed dropping funnel to a mixture of *N,N*-dimethylformamide (25 mL, Aldrich) and diethyl ether (30 mL) stirred with an overhead stirrer at $-80°C$ in a three-necked 500-mL round-bottomed flask fitted with a nitrogen inlet.* The white suspension is allowed to warm to room temperature, and stirred for 6 h, and aqueous HCl (100 mL, 5 M) added dropwise over a few minutes. The organic layer is removed using a separating funnel and washed with 5 M aqueous HCl (2×20 mL). The combined acid extracts are made slightly alkaline with Na_2CO_3, extracted with dichloromethane (2×40 mL), and the combined extracts are dried over $MgSO_4$. After filtration, removal of solvent, and vacuum distillation (60–65°C, 1 torr), the product is obtained as a clear oil that crystallizes on standing (12.8 g, 69%).†

Properties

The product should be stored at $-20°C$. ^1H NMR in $CDCl_3$: δ 9.82 (s, 1H, CHO), 7.28 [s, 1H, H(4 or 5)], 7.12 [s, 1H, H(5 or 4)], 4.03 (s, 3H, NMe).

C. BIS(PYRAZOL-1-YL)(PYRIDIN-2-YL)METHANE, $(pz)_2(py)CH$

$$(pz)_2C{=}O + (py)(H)C{=}O \longrightarrow (pz)_2(py)CH + CO_2$$

Procedure

Bis(pyrazol-1-yl)methanone (0.98 g, 6.3 mmol) and pyridine-2-aldehyde (0.60 mL, 6.3 mmol, Merck) are added to a 50-mL Schlenk tube under nitrogen. A catalytic amount of anhydrous cobalt(II) chloride (0.01 g) is added, although the reaction does proceed satisfactorily without the catalyst. The mixture is gently warmed to ~40°C via an external water bath until evolution of carbon dioxide is observed, and the mixture is then cooled and set aside until the reaction has subsided. Water (5 mL) is added and the mixture extracted with dichloromethane

*The checkers used a 1.6 M solution of butyllithium in hexanes (Aldrich) and cannula transfer of DMF/Et_2O solution to the Li(mim) suspension.

†The checkers report a yield of 50% at half-scale, crystals at 0°C melting at ~23°C.

(2 × 20 mL). The combined extracts are dried over $MgSO_4$ and filtered, and the solvent is removed in a vacuum.* The product is recrystallized from hot hexane (0.64 g, 45%).

Properties

^{1}H NMR in $CDCl_3$: δ 8.61 [ddd, 1H, H(6)], 7.88 (s, 1H, CH), ~7.85 [m, 3H, $H(4)_{py}$ and $H(5)_{pz}$], 7.42 [ddd, 1H, $H(5)_{py}$], 7.15 [d, 1H, $H(3)_{py}$], 6.36 [dd, 2H, H(4)].

D. BIS(PYRAZOL-1-YL)(*N*-METHYLIMIDAZOL-2-YL)METHANE, $(pz)_2(mim)CH$

$$(pz)_2C{=}O + (mim)(H)C{=}O \longrightarrow (pz)_2(mim)CH + CO_2$$

Procedure

Bis(pyrazol-1-yl)methanone (0.98 g, 6.3 mmol) and *N*-methylimidazole-2-aldehyde (0.60 mL, 6.3 mmol) are added to a 50-mL Schlenk tube under nitrogen in an open system. A vigorous reaction commences immediately to produce a red-brown tar and carbon dioxide. The material thus obtained is dissolved in dichloromethane and chromatographed on a column (~2 × 15 cm; Merck 60, 230/240-mesh silicagel; 1 atm or medium pressure of nitrogen applied) with dichloromethane used as the eluent. A colorless solution of the product is obtained, as other product(s) are either not eluted or have low R_f values.† Addition of hexane (20 mL) to the dichloromethane eluent, followed by slow removal of dichloromethane under a vacuum at room temperature gives $(pz)_2(mim)CH$ as white crystals (1.7 g, 49%).‡

Properties

^{1}H NMR in $CDCl_3$: δ 8.00 (s, 1H, CH), 7.93 [d, 2H, H(5)], 7.52 [s. 2H. H(3)]. 7.16 [d, 1H, H(4 or $5)_{mim}$], 6.87 [d, 1H, H(5 or $4)_{mim}$], 3.56 [s, 3H, NMe].

*The checkers monitored purity by thin-layer chromatography (TLC) (silica/Et_2O, product R_f ~0.4). Hot hexane extract was evaporated, residue extracted with Et_2O, and solution passed over a silica column. Evaporation gave a pale yellow oil that crystallized on standing. Recrystallization from hot hexane cooled to −20°C overnight gave colorless needles in a yield of 30%, mp 78.5–79.0°C.

†The checkers found pyrazole in the first fraction eluting.

‡The checkers obtained 0.84 g (58% yield, mp 119–120°C) using 13 mg $CoCl_2$ as catalyst with $(pz)_2CO$ in 2 mL of THF, and adding the solution of (mim)(H)C=O in 2 mL THF by cannula.

References

1. (a) S. Trofimenko, *Progr. Inorg. Chem.* **34**, 115 (1986); (b) D. L. Reger, J. E. Collins, D. L. Jameson, and R. K. Castellano, *Inorg. Synth.* **32**, 63 (1998).
2. (a) P. K. Byers, A. J. Canty, and R. T. Honeyman, *J. Organomet. Chem.* **385**, 417 (1990); (b) A. J. Canty and R. T. Honeyman, *J. Organomet. Chem.* **387**, 247 1990).
3. (a) T. C. Higgs and C. J. Carrano, *Inorg. Chem.* **36**, 291 (1997); (b) T. C. Higgs and C. J. Carrano, *Inorg. Chem.* **36**, 298 (1997).
4. T. Astley, A. J. Canty, M. A. Hitchman, G. L. Rowbottom, B. W. Skelton, and A. H. White, *J. Chem. Soc., Dalton Trans.* 1981 (1991); (b) A. J. Canty, P. R. Traill, R. Colton, and I. M. Thomas, *Inorg. Chim. Acta* **210**, 91 (1993).
5. (a) P. K. Byers and A. J. Canty, *Organometallics* **9**, 210 (1990); (b) P. K. Byers, A. J. Canty, B. W. Skelton, and A. H. White, *Organometallics* **9**, 826 (1990); (c) A. J. Canty, R. T. Honeyman, B. W. Skelton, and A. H. White, *J. Organomet. Chem.* **396**, 105 (1990); (d) P. K. Byers, A. J. Canty, and R. T. Honeyman, *Adv. Organomet. Chem.* **34**, 1 (1991); (e) A. J. Canty, R. T. Honeyman, B. W. Skelton, and A. H. White, *J. Organomet. Chem.* **424**, 381 (1992).
6. T. Astley, M. A. Hitchman, B. W. Skelton, and A. H. White, *Aust. J. Chem.* **50**, 145 (1997); (b) P. A. Anderson, T. Astley, M. A. Hitchman, F. R. Keene, B. Moubaraki, K. S. Murray, B. W. Skelton, E. R. Tiekink, H. Toftlund, and A. H. White, *J. Chem. Soc., Dalton Trans.* 3512 (2000).
7. For convenient preparations of tris(pyridin-2-yl)methane from bis(pyridin-2-yl)methanone, see (a) G. R. Newkome, V. K. Gupta, H. C. R. Taylor, and F. R. Fronczek, *Organometallics* **3**, 1549 (1984); (b) A. J. Canty and N. J. Minchin, *Aust. J. Chem.* **39**, 1063 (1986); (c) A. J. Canty, N. Chaichit, B. M. Gatehouse, E. E. George, and G. Hayhurst, *Inorg. Chem.* **20**, 2414 (1981); (d) F. R. Keene, M. R. Snow, P. J. Stephenson, and E. R. T. Tiekink, *Inorg. Chem.* **27**, 2040 (1988).
8. A. J. Canty and C. V. Lee, *Organometallics* **1**, 1063 (1982).
9. (a) K. I. Thé and L. K. Peterson, *Can. J. Chem.* **51**, 422 (1973); (b) K. I. Thé, L. K. Peterson, and E. Kiehlmann, *Can. J. Chem.* **51**, 2448 (1973); (c) L. K. Peterson, E. Kiehlmann, A. R. Sanger, and K. I. Thé, *Can J. Chem.* **52**, 2367 (1974).
10. P. E. Iversen and H. Lund, *Acta Chem, Scand.* **20**, 2649 (1966).
11. R. G. Jones and H. Gilman, *Org. React.* **6**, 339 (1951).

8. DIFLUOROTRIPHENYLARSENIC(V), AsF_2Ph_3

Submitted by FRANCISCO J. ARNÁIZ* and MARIANO J. MIRANDA*
Checked by JOHN R. SHAPLEY†

$$AsPh_3 + H_2O_2 + 2\,HF \longrightarrow AsPh_3F_2 + 2\,H_2O$$

The pentavalent triphenylarsenic dihalides are of interest because they present structures that are dependent on the halogen[1–3] and are convenient sources for the preparation of arsenic ylides.[4,5] The chemistry of triphenylarsenic difluoride,

*Laboratorio de Quimica Inorganica, Universidad de Burgos, 09001 Burgos, Spain.
†Department of Chemistry, University of Illinois at Urbana—Champaign, Urbana, IL 61801.

by far the most air-stable of the halide derivatives, has not been much explored. This is probably because of the synthetic procedures currently available: (1) prolonged heating of triphenylarsine with sulfur tetrafluoride[6] or (2) halogen exchange between triphenylarsenic dichloride and silver fluoride.[7] However, we have found that triphenylarsenic difluoride precipitates in good yield when a saturated solution of triphenylarsine oxide in methanol is treated with concentrated hydrofluoric acid. Here we describe a simple one-pot synthesis based on triphenylarsine, that is especially useful when the oxide is not available. It is based in the facile oxidation of triphenylarsine with hydrogen peroxide and the immediate conversion of the oxide into the fluoride.

Procedure

■ **Caution.** *Arsenic compounds are poisonous. Concentrated hydrogen peroxide can cause severe burns. Hydrofluoric acid is neurotoxic and corrosive, and causes severe burns. All manipulations should be conducted in an efficient fume hood, and gloves and goggles should be worn.*

In a 100-mL Teflon* beaker immersed in a bath at 80–100°C are placed 10 g (32.7 mmol) of $AsPh_3$,† 10 g of glacial acetic acid and a magnetic stirring bar. Then 10 g (88.2 mol) of 30% H_2O_2 is added dropwise to the abovementioned mixture with stirring. Some boiling or frothing of the solution may occur, since the reaction is exothermic and oxygen is released. Then 10 g (240 mmol) of 48% HF is added with stirring to the warm solution. During the addition, a white precipitate forms. The beaker is immersed in an ice-cold water bath, and the mixture is stirred for 10 min. Then, 20 g of cold distilled water is added to the mixture and the stirring is continued for further 10 min. The solid product is collected by vacuum filtration, washed with three 5-mL portions of methanol, and dried in air for 10 min. The solid is transferred to a Teflon vial, which is introduced into a round-bottomed flask and maintained under vacuum at room temperature for 2 h. Yield: 10 g (90%).

Anal. Calcd. for $AsPh_3F_2$: C, 62.81; H, 4.39; F, 11.04. Found: C, 62.6; H, 4.3; F, 11.3.

Properties

White AsF_2Ph_3 melts at 137–139°C. The IR spectrum, taken as a Nujol mull, has the characteristic band $\nu_{as}(AsF_2)$ at 517 cm^{-1}. The ^{19}F NMR spectrum in $CDCl_3$

*Similar results are obtained by using a variety of plastic vessels.
†The procedure was checked at half-scale (yield 93%).

shows only a singlet at −87.78 ppm respect to $CFCl_3$ (−86.01 in acetone-d_6). It is insoluble in water, slightly soluble in diethyl ether and cold methanol, and soluble in acetone, ethanol, dichloromethane, chloroform, acetonitrile, dimethyl formamide, and dimethyl sulfoxide, solvents from which it can be recrystallized. It reacts with KOH in aqueous solution to yield Ph_3AsO, but it resists the action of cold water for hours. It is very stable at room temperature and can be manipulated in air without special precautions. A very pure product (mp 139–140°C) is obtained by recrystallization from ethanol.

References

1. A. Augustine, G. Ferguson, and F. C. March, *Can. J. Chem.* **53**, 1647 (1975).
2. N. Bricklebank, S. M. Godfrey, H. P. Lane, C. A. McAuliffe, R. G. Pritchard, and J. M. Moreno, *J. Chem. Soc., Dalton Trans.* 3873 (1995) and refs. therein.
3. B. Neumüller, S. Chitsatz, and K. Dehnicke, *Z. Naturforsch.* **54b**, 1611 (1999).
4. G. O. Doak and L. D. Freedman, *Organometallic Compounds of Arsenic, Antimony and Bismuth*, Wiley-Interscience, New York, 1970.
5. S. Chitsatz, B. Neumüller, and K. Dehnicke, *Z. Anorg. Allg. Chem.* **626**, 634 (2000) and refs. therein.
6. W. C. Smith, *J. Am. Chem. Soc.* **82**, 6176 (1960).
7. M. H. O'Brien, G. O. Doak, and G. G. Long, *Inorg. Chim. Acta.* **1**, 34 (1967).

9. TETRAMETHYLAMMONIUM SALTS OF SUPEROXIDE AND PEROXYNITRITE

Submitted by D. SCOTT BOHLE[*†] and ELISABETH S. SAGAN[*]
Checked by WILLEM H. KOPPENOL[‡] and REINHARD KISSNER[‡]

Peroxynitrite is an important cytotoxin that results in vivo from the direct reaction of superoxide and nitric oxide. There are several methods to synthesize peroxynitrite, including the solid-state photolysis of KNO_3,[1] the ozonation of sodium azide,[2] or the alkaline reaction of nitrite esters with hydrogen peroxide.[3] Perhaps the most common peroxynitrite synthesis involves treating an acidified solution of hydrogen peroxide with nitrite, followed by an immediate quench with base.[4] Although this synthesis is facile, quick, and affordable, it results in

*Department of Chemistry, University of Wyoming, Laramie, WY 82071.
†Current address: Department of Chemistry, McGill University, Montreal, H3A 2K6 Quebec, Canada.
‡Laboratorium für Anorganische Chemie, ETH Zurich, CH-8092 Zurich, Switzerland.

contamination with nitrite, nitrate, and/or hydrogen peroxide. Hydrogen peroxide may be removed from these solutions by treatment with manganese dioxide, but the peroxynitrite from this method has not yet been isolated free from the nitrite or nitrate contaminants. The presence of nitrite or nitrate may be inconsequential for many applications of peroxynitrite, in which case this synthesis is quite adequate. However, for mechanistic studies, either for accurate determination of product stoichiometry or for isotopic exchange experiments, such as with ^{18}O, the abovementioned methods cannot be used. The method described here involves two stages: a 3-day solid-state synthesis/mixing reaction that is followed by a pair of liquid ammonia extraction/lyophilization steps to produce pure tetramethylammonium superoxide; the second stage requires less than 2 h to dissolve the tetramethylammonium superoxide, nitrosylate it, and then lyophilize the product to give pure tetramethylammonium peroxynitrite. Several synthetic routes to tetramethylammonium superoxide have been reported over the years.[5,6] However, some of these methods involve cumbersome separations. The synthesis of tetramethylammonium superoxide from a solid-state metathesis reaction between potassium superoxide and tetramethylammonium hydroxide pentahydrate has proved to be the most reliable with the highest purity. The synthesis of the tetramethylammonium peroxynitrite salt mimics the biosynthesis of peroxynitrite, in which nitric oxide reacts directly with superoxide. The improved synthesis has ensured 100% purity of the peroxynitrite salt with no contamination of starting material, nitrate, or nitrite. Nitric oxide is generated by the reduction of nitrite with iron sulfate under acidic conditions, which guarantees that the gas is readily dried and free of nitrous oxide or nitrogen dioxide, as is found in commercially available nitric oxide gas cylinders. In addition, the exact control of stoichiometry means over nitrosylation is avoided, and ^{15}N can be efficiently incorporated into the product by the use of commonly available ^{15}N nitrite salts.

A. TETRAMETHYLAMMONIUM SUPEROXIDE, $[N(CH_3)_4]O_2$

$$11\,KO_2 + [N(CH_3)_4]OH \cdot 5\,H_2O \xrightarrow{NH_3(l)} [N(CH_3)_4]O_2 + 11\,KOH + 7.5\,O_2$$

■ **Caution.** *The use of predried tetramethylammonium hydroxide pentahydrate*[7] *has resulted in a serious explosion during the solid-phase metathesis of potassium superoxide with tetramethylammonium hydroxide pentahydrate. This can be avoided by using the pentahydrate as supplied without drying and by employing an increased amount of potassium superoxide. The latter can easily be separated during the liquid ammonia extraction.*

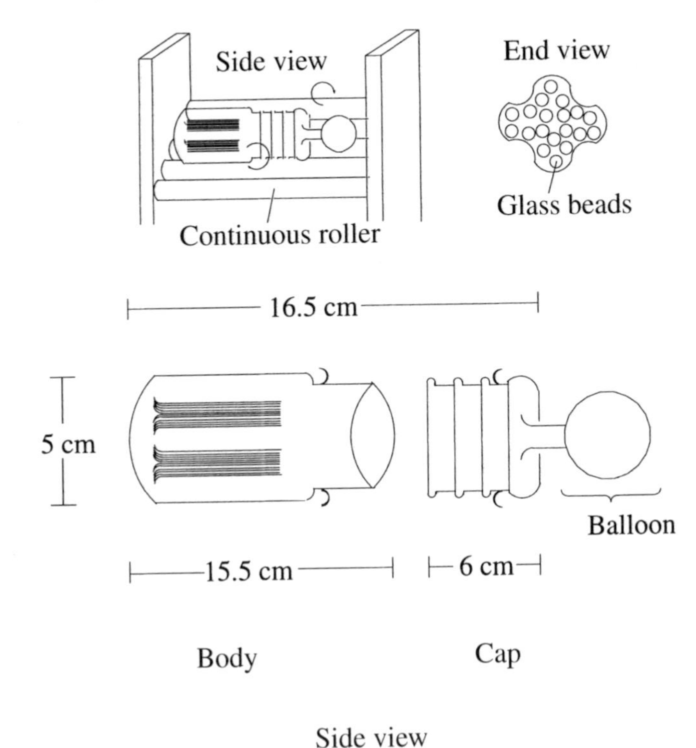

Figure 1. Solid-state reaction setup.

Procedure

Commercially available potassium superoxide and tetramethylammonium hydroxide pentahydrate from Aldrich were used as supplied without further purification. Industrial-grade nitrogen gas does not require any further purification, but the ammonia gas, 99.999%, should be at a minimum purged through Drierite before use. All work is done under a nitrogen flow unless otherwise stated.

The reaction apparatus is shown in Fig. 1 and consists of a beveled flask fit with male and female T45/50 ground-glass joints. The female top has a tapered joint such that a balloon, used to maintain positive pressures throughout the reaction, fastened with tape, fits easily. The female top is 6 cm in length, while the male bottom is 15.5 cm in length. The entire flask is 5 cm in height. The continuous roller is a Star Manufacturing, Inc. model 12 silvertone roller. The roller bed is 19.3 cm in length. Each roller has a circumference of 9.2 cm, with a spacing of 1.2 cm between each roller. The reaction vessel has a rotation rate of 2.5 revolutions per minute (rpm).

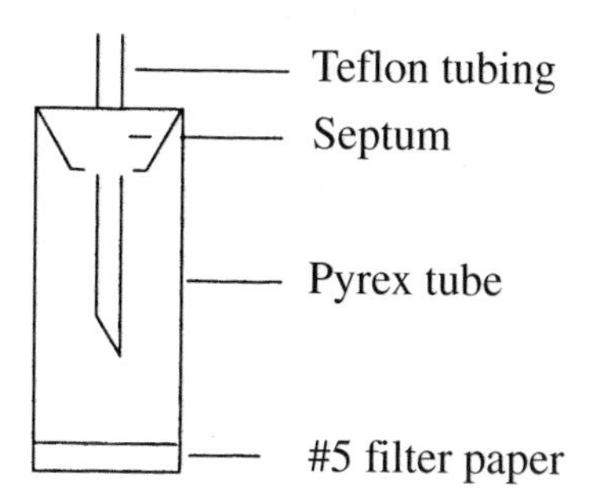

Figure 2. Canula filter. The Teflon tubing is inserted through the NMR tube septum. The filter paper is taped onto the pyrex tubing using Teflon tape.

In a drybox, 10.0 g (141 mmol) of potassium superoxide, 1.2 g (6.6 mmol) of tetramethylammonium hydroxide pentahydrate, and 15 mL of small glass beads are added to the reaction flask. The solid yellow mixture is continuously rotated slowly for 3 days. The reaction vessel is taken into the drybox daily to scrape any caked-up product from the sides of the flask and to ensure optimum mixing. For the first day, it is sufficient to tap the flask walls to obtain free-flowing contents. After 3 days, the mixture is taken into the drybox and transferred to a T14/20 100-mL round-bottomed flask fit with the canula filter (Fig. 2). The round-bottomed flask is connected to a receiving flask of the same type via Teflon tubing. Once taken out of the drybox, the reaction flask is immersed in a dry-ice/ethanol bath. A medium flow of nitrogen is purged through the flask. Liquid ammonia, 40 mL, is condensed into the reaction flask using a Teflon tube. The flow of nitrogen and ammonia is monitored by allowing it to escape into the hood through an oil bubbler filled with paraffin oil. After the ammonia has condensed, shaking the flask will produce a pale yellow slurry of dissolved tetramethylammonium superoxide to be filtered from the insoluble potassium hydroxide. The dry-ice/ethanol bath is switched to the receiving flask, and the ammonia/superoxide solution is filtered to the receiving flask with a forced nitrogen stream. After filtration the liquid ammonia is boiled off with a nitrogen flow to leave a pale yellow solid of tetramethylammonium superoxide. This extraction procedure is conducted 3 times to give 0.341 g, 49% yield of pure product.

Properties

The product is a very pale yellow to cream-colored solid. It is hygroscopic and is soluble in liquid ammonia, with limited solubility and stability in acetonitrile, dimethyl sulfoxide, and dimethylformamide.[6] Water or moisture from the air leads to disproportionation, and so the product must be stored under anhydrous conditions. It is characterized in the solid state by Raman spectroscopy and has a single peak at 1118 cm^{-1}. The UV–vis, magnetic susceptibility, elemental analysis, and electrochemistry have been reported elsewhere.[5,6]

B. TETRAMETHYLAMMONIUM PEROXYNITRITE, $[N(CH_3)_4][ONOO]$

$$NaNO_2 + FeSO_4 \cdot 7\,H_2O \xrightarrow{H_2O} NO + Fe^{+3} + SO_4^{2-} + 6H_2O + Na^+ + 2OH^-$$

$$[N(CH_3)_4]O_2 + NO(g) \xrightarrow{NH_3(1)} [N(CH_3)_4][ONOO]$$

Procedure

Sodium nitrite, 99.99%, ferrous sulfate heptahydrate, sulfuric acid, sodium hydroxide, phosphorus pentoxide, and potassium hydroxide are reagent grade.

The experimental setup is shown in Fig. 3. The NO generator consists of a 100-mL T14/20 three-necked flask fitted with septa on the two outside necks. The central neck is connected to a dry-ice condenser that is filled with dry ice/ethanol during the synthesis. The condenser is connected to the NO line via clear tubing. The NO line consists of three drying tubes connected together in a line by

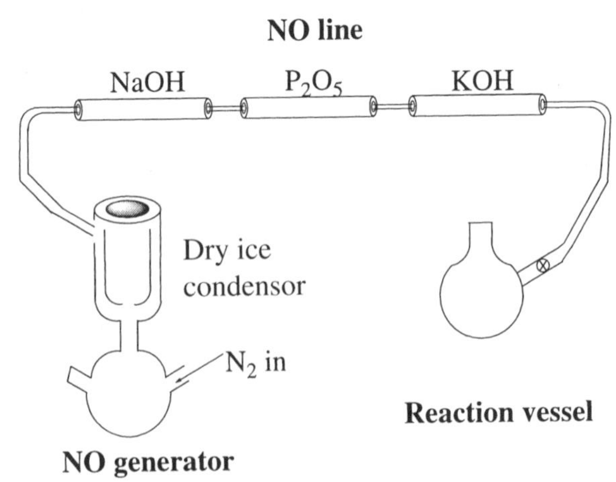

Figure 3. Experimental design for $ONOO^-$ synthesis. The dry-ice condensor of the "NO generator" is used to trap any water that may evolve after addition of iron sulfate. The three drying tubes of the "NO line" trap any remaining water before it reaches the reaction vessel.

clear tubing. The first tube contains sodium hydroxide pellets. The second and third tubes contain phosphorus pentoxide powder and potassium hydroxide pellets, respectively. The inside ends of each tube are packed with glass wool. The NO line is connected to a glass pipette fit with teflon tubing. The Teflon tube is inserted into the reaction vessel and under the liquid surface to introduce the NO.

For the nitric oxide gas generation, a degassed solution of sodium nitrite is prepared by charging a T14/20 100-mL three-necked flask with 0.200 g sodium nitrite (3 mmol) and adding 4 mL distilled water outside the hood. The solution is purged with nitrogen for 15 min. The purge rate is monitored through the exit oil bubbler. The nitrite solution flask is connected to a dry-ice condenser filled with a dry-ice/ethanol bath to capture any water that may escape. An acidified solution of iron sulfate is prepared in air by charging a 100-mL round-bottomed flask with 5 g ferrous sulfate heptahydrate (18 mmol), 20 mL distilled water, and 1 mL concentrated sulfuric acid. The light green solution is purged with nitrogen for 15 mins. Although the iron sulfate solution is stable for up to one week, fresh solutions are to be used when possible. In the drybox, a T14/20 100-mL Schlenk flask is charged with 120 mg (1.13 mmol) of tetramethylammonium superoxide and is fitted with a septum. The flask is removed from the drybox and is purged with nitrogen for 3 min. The flask is immersed in a dry-ice/ethanol bath. Liquid ammonia, 40 mL, is condensed into the flask by using a Teflon tube inlet. The nitric oxide scrubbing line is purged with nitrogen for 10 min. The Teflon tubing used to introduce the ammonia gas is replaced with the tubing from the nitric oxide line. Dry nitric oxide gas is generated by adding the iron sulfate solution dropwise to the nitrite solution. The clear, colorless nitrite solution turns from a green-brown sludge to a milky light brown to clear dark brown, indicating the completion of nitric oxide gas generation. The total volume of iron sulfate solution (0.856 M) used ranges from 1.25 to 2.0 mL. As the nitric oxide dissolves into the superoxide solution, the initially clear, colorless solution changes to light yellow to orange, indicating the formation of tetramethylammonium peroxynitrite. The dry-ice/ethanol bath and Teflon tubing are removed. Roughly 50% of the liquid ammonia is evaporated under a nitrogen flow. The remaining peroxynitrite/ammonia solution is lyopholized by freezing the solution with liquid nitrogen, placing it under a dynamic high vacuum, and removing the liquid nitrogen bath. After lyophilization, the fluffy orange solid, tetramethylammonium peroxynitrite, is brought into the drybox (0.162 g, 97% yield).

Properties

The product is a yellow-orange to orange, moisture-sensitive solid. It is stable for a minimum of 4 weeks if stored in a good inert-atmosphere box. It is stable in solution only in liquid ammonia and basic water. The product is soluble, but slowly reacts with dimethylsulfoxide, acetonitrile, alcohols, and halocarbons. Characterization of this product is performed by Raman spectroscopy, which provides an accurate indication of purity. The solid-state peroxynitrite stretching frequencies are as follows: δ_{ONOO}, 591 cm^{-1}; δ_{ONOO^-}, 634 cm^{-1}; $\nu_{O\text{–}O}$, 803 cm^{-1}; $\nu_{N\text{–}O}$, 926 cm^{-1}; and $\nu_{N=O}$, 1459 cm^{-1}. Tetramethylammonium peroxynitrite begins to decompose even on dilution in 1 M NaOH, resulting in an inaccurate measurement of concentration and purity of the solid. Other methods of tetramethylammonium peroxynitrite characterization include UV–visible spectroscopy and ^{15}N NMR spectroscopy, which are reported in detail elsewhere.[8,9]

References

1. P. A. King, V. E. Anderson, J. O. Edwards, G. Gustafson, R. C. Plumb, and J. W. Suggs, *Am. Chem. Soc.* **114**, 5430–5432 (1992).
2. W. A. Pryor, R. Cueto, X. Jin, W. H. Koppenol, M. Ngu-Schwemlein, G. L. Squadrito, P. L. Uppu, and R. M. Uppu, *Free Rad. Biol. Med.* **18**, 75–83 (1995).
3. J. R. Luis, M. E. Peña, and A. Ríos, *J. Chem. Soc., Chem. Commun.* 1298–1299 (1993).
4. A. Saha, S. Goldstein, D. Cabelli, and G. Czapski, *Free Rad. Biol. Med.* **24**, 653–659 (1998).
5. A. D. McElroyand and J. S. Hashman, *Inorg. Chem.* **3**, 1798–1799 (1964).
6. D. T. Sawyer, T. S. Calderwood, K. Yamaguchi, and C. T. Angelis, *Inorg. Chem.*, **22**, 2577–2583 (1983).
7. K. Yamaguchi, T. S. Calderwood, and D. T. Sawyer, *Inorg. Chem.* **25**, 1289–1290 (1986).
8. D. S. Bohle, B. Hansert, S. C. Paulson, and B. D. Smith, *J. Am. Chem. Soc.* **116**, 7423–7424 (1994).
9. D. S. Bohle, P. A. Glassbrenner, and B. Hansert, *Meth. Enzymol.* **269**, 302–311 (1996).

10. TELLURIUM–NITROGEN COMPOUNDS

Submitted by TRISTRAM CHIVERS,* NICOLE SANDBLOM,* and GABRIELE SCHATTE*
Checked by DEAN M. GIOLANDO†

The first structural characterization of a tellurium diimide RNTe(μ-N^{t}Bu)$_2$TeNR (R = PPh$_2$NSiMe$_3$) was reported in 1994.[1] Subsequently, the synthesis of the symmetric derivative, R = ^{t}Bu, was achieved by the reaction of Li[HN^{t}Bu] with TeCl$_4$ in a 4 : 1 molar ratio in toluene.[2] Several LiCl-containing byproducts are also obtained in this solvent.[3] In THF, however, the tellurium diimide, ^{t}BuN-Te(μ-N^{t}Bu)$_2$TeN^{t}Bu, is obtained in 90% yield and is readily purified.[4] Unlike sulfur or selenium diimides, the tellurium diimides are dimeric and form either cis or

*Department of Chemistry, University of Calgary, Calgary, Alberta, Canada, T2N 1N4.
†Department of Chemistry, University of Toledo, Toledo OH 43606.

trans isomers in the solid state.[1,2] In *cis*-^{t}BuNTe(μ-N^{t}Bu)$_2$TeN^{t}Bu the terminal ^{t}Bu groups are in the endo positions with respect to the Te_2N_2 ring.[2]

This new reagent has proved useful for the development of TeN chemistry. For example, reactions of ^{t}BuNTe(μ-N^{t}Bu)$_2$TeN^{t}Bu with Li[HN^{t}Bu] or K[O^{t}Bu] generate the anions $[Te(N^{t}Bu)_3]^{2-}$ and $[Te(N^{t}Bu)_2(O^{t}Bu)]^{-}$, respectively.[5,6] The dimer ^{t}BuNTe(μ-N^{t}Bu)$_2$TeN^{t}Bu may act as a bridging or chelating ligand toward Cu^{+} and Ag.[7] The redistribution of ^{t}BuNTe(μ-N^{t}Bu)$_2$TeN^{t}Bu with tellurium tetrahalides yields the imidotellurium(IV) dihalides $(^{t}BuNTeX_2)_n$ (X = Cl, Br).[4] The cycloaddition reaction of ^{t}BuNTe(μ-N^{t}Bu)$_2$TeN^{t}Bu with ^{t}BuNCO generates the ureatotelluroxide $\{OC(\mu\text{-}N^{t}Bu)_2TeO\}_2$.[8] The dimeric compound $\{Li_2[Te(N^{t}Bu)_3]\}_2$ is a useful reagent for incorporating other elements, such as B,[5] P,[5] Sb,[9] Bi,[9] or In[10] into the TeN ring system. The stannatellurone $\{^{t}BuNSn(\mu\text{-}N^{t}Bu)_2TeN^{t}Bu\}(\mu_3\text{-}SnTe)$ is obtained from the reaction of $\{Li_2[Te(N^{t}Bu)_3]\}_2$ with Sn(II) salts.[11]

General Procedures

■ **Caution.** *$TeCl_4$ is highly toxic as well as hygroscopic. Consequently, reactions should be carried out in an efficient fume hood. Reaction vessels that come in contact with tellurium compounds are rinsed with water first (fume hood), the washings are combined in a beaker, and concentrated nitric acid is added the next day. The water-rinsed reaction vessels are placed into a concentrated nitric acid bath for 24 h before final cleaning.*

Because of the oxygen and moisture sensitivity of the starting materials and products, all operations are carried out under an atmosphere of dry argon using Schlenk or inert-atmosphere drybox techniques. Familiarity with solvent transfer using cannula techniques, filtration under inert-gas atmosphere, and vacuum techniques is required.[12] All glassware is dried in an oven at ≈120°C. The solvents *n*-hexane, tetrahydrofuran, and diethyl ether are twice dried over sodium/benzophenone and then distilled under argon. Solvents are stored over activated molecular sieves (3 Å) in 500-mL glass vessels equipped with PTFE piston valves (Chemglass) prior to use. Tellurium tetrachloride (ÆSAR, 99%) and *n*-BuLi (Aldrich, 2.5 M solution in hexanes) are used without further purification and are stored under argon.

A. LITHIUM *tert*-BUTYLAMIDE, Li[HN^{t}Bu]

$$^{t}BuNH_2 + {}^{n}BuLi \rightarrow Li[HN^{t}Bu] + C_4H_{10}$$

Procedure

The following procedure is suitable for the synthesis of Li[HN^{t}Bu] in large quantities. Neat ^{t}BuNH$_2$ (65 mL, 0.61 mol) and *n*-hexane (170 mL) are added to a 1-L round-bottomed flask equipped with a gas inlet sidearm and containing a magnetic stirring bar. (*Note*: ^{t}BuNH$_2$ is used in excess to ensure that all *n*-BuLi is used up during the reaction.) A calibrated 500-mL dropping funnel with pressure equalizer and Teflon key is connected to the flask. The reaction vessel is cooled to -10°C (ice/NaCl). The dropping funnel is charged with the yellow solution of *n*-BuLi in hexanes (2.5 M, 200 mL, 0.5 mol), and this solution is added dropwise (1 h) to the stirred solution of ^{t}BuNH$_2$ in *n*-hexane.

■ **Caution.** *The reaction is exothermic and n-butane gas is produced. Therefore, sufficient cooling and a slow addition of the n-BuLi/hexanes solution are required. n-BuLi can ignite on contact with water or air. During the addition of n-BuLi solutions the storage bottle should be kept in the metal container provided by the supplier, and leather gloves should be worn because disposable gloves react immediately with n-BuLi.*

The colorless solution becomes pale yellow, and a white precipitate of Li[HN^{t}Bu] is formed. The reaction mixture is stirred for an additional hour at -10°C and finally allowed to reach room temperature. After 0.5 h the volatile materials are removed under vacuum. The white solid (38.50 g, 0.487 mol, 97.4%) is pumped to dryness under vacuum by placing a warm water bath underneath the flask ($T = 30$°C) and stored in a 150-mL Nalgene container in the drybox. (*Note*: The solid readily adsorbs *n*-hexane and appears dry. If the solid forms big clumps during the process of removing the solvent, it is necessary to move the flask into a drybox and break up the clumps. Afterwards, the remaining solvent is removed under vacuum. This procedure ensures the complete removal of *n*-hexane.)

■ **Caution.** *Li[HN^{t}Bu] reacts violently with water. After recovering the product, the remaining Li[HN^{t}Bu] in the reaction vessel must be destroyed carefully. The closed flask (under an argon atmosphere) is taken out of the drybox, placed into an ice bath, and 200 mL of isopropanol is added carefully. After a few minutes 200 mL of ice water is added to the iso-propanol solution. The mixture is stirred until no solid remains, and then the contents treated as normal organic waste.*

Properties

Lithium *tert*-butyl amide is an air- and moisture-sensitive flammable solid that can be stored for up to 4 months in the drybox. In the solid state it has an octameric, cyclic ladder structure.[13] ^{1}H NMR (23°C): in C_7D_8 or C_6D_6, δ 1.37 (9 H, ^{t}Bu); in d_8-THF, δ 1.07 (9 H, ^{t}Bu) and -1.55 (1 H, NH). ^{13}C NMR (23°C): in C_6D_6 (from Ref. 13), δ 51.84 ($C(CH_3)_3$) and 38.17 ($C(CH_3)_3$); in d_8-THF, δ 51.65 ($C(CH_3)_3$) and 38.82 ($C(CH_3)_3$). ^{7}Li NMR (23°C): in d_8-THF, 3.22 ppm ($\delta\nu_{1/2} = 13.4$ Hz).

B. BIS(μ-*tert*-BUTYLIMIDO)-BIS(*tert*-BUTYLIMIDO)DITELLURIUM, ^{t}BuNTe(μ-N^{t}Bu)$_2$TeN^{t}Bu (1)

$$2\,TeCl_4 + 8\,Li[HN^tBu] \longrightarrow {}^tBuNTe(\mu\text{-}N^tBu)_2TeN^tBu + 8\,LiCl + 4\,{}^tBuNH_2$$

Procedure

Li[HN^{t}Bu] (finely powdered; 11.90 g, 148.6 mmol) and $TeCl_4$ (finely powdered; 10.00 g, 37.1 mmol) are loaded in the drybox into two separate round-bottomed sidearm flasks (500 and 1000 mL, respectively). The two reaction vessels are removed from the drybox and connected to a standard argon gas/vacuum line. Dry THF (100 and 150 mL) is added to the flasks containing the $TeCl_4$ and Li[HN^{t}Bu], respectively. The yellow $TeCl_4$/THF solution is cooled immediately to -78°C (ethanol/dry ice), and then the white suspension of Li[HN^{t}Bu] in THF is transferred slowly via a double-tipped, stainless-steel cannula (16 gauge, length 2 ft). When 50 mL of the Li[HN^{t}Bu]/THF suspension have been added, an orange slurry is observed. After 100 mL, the slurry darkens to yellow-brown, and, on completion of the addition, the color of the reaction mixture is dark brown. The Li[HN^{t}Bu] vessel is rinsed with THF (20 mL), and the washings are added to the reaction mixture. The reaction mixture is stirred for 3 h at -78°C. (*Note*: For optimum yields, it is important to keep the cold bath at -78°C by addition of dry ice.) The cold bath is then allowed to reach 23°C slowly, whereupon the color of the reaction mixture changes to red-orange. The color intensifies to dark red after the solution is stirred for an additional 48 h at 23°C. Volatile materials are removed under vacuum, and the solid orange residue is extracted with diethyl ether (~200 mL) during vigorous stirring. The precipitate is allowed to settle, and the dark red solution is then filtered by using a cannula through a fine-porosity sintered-glass frit (4–8 μm) into a second 500-mL Schlenk flask under slightly reduced pressure. The remaining solid is washed twice with Et_2O (~30 mL), and the washings are filtered into the solution of the first extraction. Removal of solvent under vacuum affords ^{t}BuNTe(μ-N^{t}Bu)$_2$TeN^{t}Bu as an orange microcrystalline solid (9.03 g, 90% yield based on $TeCl_4$). This product contains small amounts of LiCl.

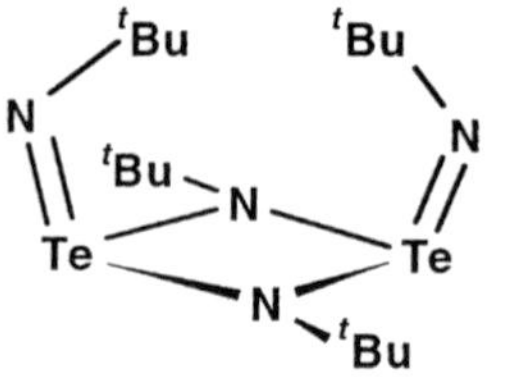

1

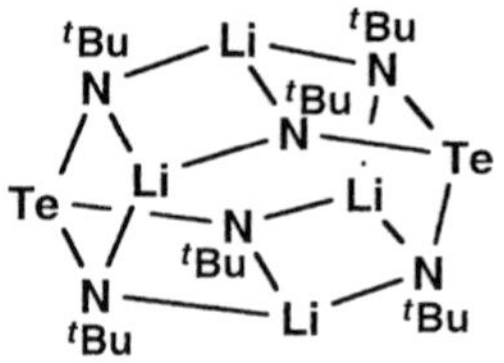

2

Anal. Calcd. for $Te_2C_{16}H_{36}N_4$: C, 35.61; H, 6.72; N, 10.38. Found: C, 35.11; H, 6.58; N, 9.98; Cl, 0.16.

LiCl-free ^{t}BuNTe(μ-N^{t}Bu)$_2$TeN^{t}Bu is obtained via vacuum sublimation. In a typical experiment a sublimation apparatus is charged with 1.770 g of the crude product. Sublimation is conducted at 90–95°C/10^{-3} mbar with the cold finger at 18°C. Over 24 h pure ^{t}BuNTe(μ-N^{t}Bu)$_2$TeN^{t}Bu sublimes onto the cold finger as a yellow-orange solid (1.239 g).

Properties

^{t}BuNTe(μ-N^{t}Bu)$_2$TeN^{t}Bu is an air- and moisture-sensitive orange solid that can be stored in glass vials for up to 6 months in the drybox. It is extremely soluble in toluene, *n*-hexane, *n*-pentane, diethyl ether, and tetrahydrofuran, but it decomposes in chlorinated solvents (CH_2Cl_2, $CHCl_3$), acetonitrile or carbon disulfide. The dimer ^{t}BuNTe(μ-N^{t}Bu)$_2$TeN^{t}Bu melts without decomposition at 100–102°C. Solutions of ^{t}BuNTe(μ-N^{t}Bu)$_2$TeN^{t}Bu show the presence of two isomers, A and B (^{1}H and ^{13}C NMR; ratio A/B $= 4:1$; see Fig. 1). However, both isomers yield the same products in further reactions. ^{1}H NMR (23°C): in C_6D_6 (see Fig. 1), δ 1.66 (9H, ^{t}Bu), 1.25 (9H, ^{t}Bu), A; δ 1.58, 1.27, (9H, ^{t}Bu), B; in C_7D_8, δ 1 (9H, ^{t}Bu), 1.27 (9H, ^{t}Bu) A; 1.55, 1.29, (9H, ^{t}Bu), B; in d_8-THF, 1.43, A; 1.45, B. ^{13}C NMR (23°C): in C_7D_8, δ 63.68, 57.92 (*C*CH$_3$), and 37.00, 36.28 (C*C*H$_3$), A, δ 64.56, 58.21 (*C*CH$_3$) and 36.50, 35.34 (C*C*H$_3$), B. ^{125}Te NMR (23°C): in C_7D_8, δ 1476 ($\Delta\nu_{1/2} \approx 3000\,\mathrm{Hz}$).

C. BIS{DILITHIUM[TRIS(*tert*-BUTYLIMIDO)TELLURITE]}, {Li$_2$[Te(N^{t}Bu)$_3$]}$_2$ (2)

$$^t\mathrm{BuNTe}(\mu\text{-N}^t\mathrm{Bu})_2\mathrm{TeN}^t\mathrm{Bu} + 4\,\mathrm{Li[HN}^t\mathrm{Bu]} \longrightarrow \{\mathrm{Li_2[Te(N}^t\mathrm{Bu)_3]}\}_2 + 2\,^t\mathrm{BuNH_2}$$

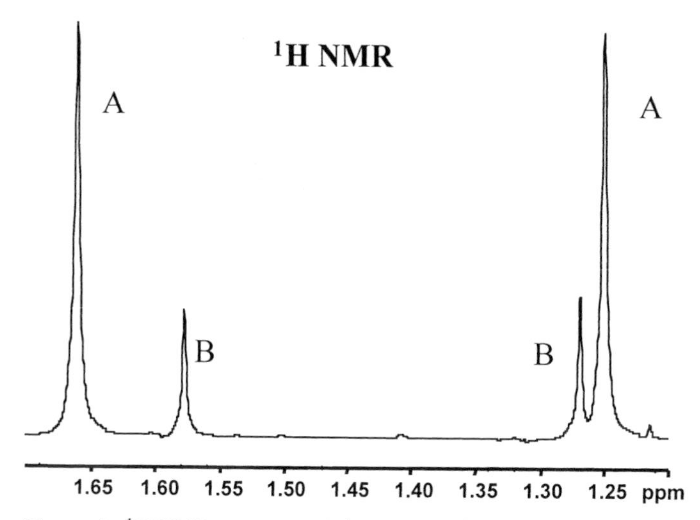

Figure 1. ^1H NMR spectrum of [^{t}BuNTe(μ-N^{t}Bu)$_2$TeN^{t}Bu] in C_6D_6.

Procedure

Li[HN^{t}Bu] (finely powdered; 0.720 g, 8.98 mmol) and [^{t}BuNTe(μ-N^{t}Bu)$_2$TeN^{t}Bu] (1.200 g, 2.22 mmol) are loaded in the drybox into two separate pear-shaped Schlenk flasks (150 mL). The two reaction vessels are connected to a standard argon gas/vacuum line. Dried toluene (10 mL) is added to each flask. The colorless Li[HN^{t}Bu]/toluene solution is added dropwise by cannula, with stirring, to the red ^{t}BuNTe(μ-N^{t}Bu)$_2$TeN^{t}Bu/toluene solution cooled to −78°C. No immediate color changes are observed. The reaction mixture is warmed slowly to 23°C after 15 min at −78°C. The color of the reaction mixture changes slowly from red to golden brown-yellow, and the volatile materials are removed immediately under dynamic vacuum (after ~1.5 h). Then ~3 mL of toluene is added to the pale yellow-brown solid, resulting in a white precipitate and a brown-yellow solution. The solution is cooled to −20°C to produce more {Li$_2$[Te(N^{t}Bu)$_3$]}$_2$. The brown-yellow solution is transferred via a cannula into another Schlenk vessel. This purification process is repeated twice to give {Li$_2$[Te(N^{t}Bu)$_3$]}$_2$ (1.082 g, 3.05 mmol; 69%) as a pale yellow solid.

Anal. Calcd. for $Li_4Te_2C_{24}H_{54}N_6$: C, 40.62; H, 7.67; N, 11.84. Found: C, 40.22; H, 7.10; N, 12.19.

Properties

{Li$_2$[Te(N^{t}Bu)$_3$]}$_2$ is an air- and moisture-sensitive solid that can be stored for several months in the drybox. This dimeric cluster adopts a distorted hexagonal prismatic (or cyclic ladder) structure.[5] ^{1}H NMR (23°C): in C_7D_8 or C_6D_6, δ 1.34 (^{t}Bu); in d_8-THF, δ 1.30 (^{t}Bu). ^{13}C NMR (23°C): in C_7D_8, δ 56.20 ($C(CH_3)_3$) and 37.85 ($C(CH_3)_3$); in d_8-THF, δ 56.09 ($C(CH_3)_3$) and 38.83 ($C(CH_3)_3$). ^{7}Li NMR (23°C): in C_7D_8, δ 0.77; in d_8-THF, δ 3.91.

References

1. (a) T. Chivers, X. Gao, and M. Parvez, *J. Chem. Soc., Chem. Commun.* 2149 (1994); (b) T. Chivers, X. Gao, and M. Parvez, *Inorg. Chem.* **35**, 9 (1996).
2. T. Chivers, X. Gao, and M. Parvez, *J. Am. Chem. Soc.* **117**, 2359 (1995).
3. T. Chivers, X. Gao, and M. Parvez, *Inorg. Chem.* **35**, 4336 (1996).
4. T. Chivers, G. D. Enright, N. Sandblom, G. Schatte, and M. Parvez, *Inorg. Chem.* **38**, 5431 (1999).
5. T. Chivers, X. Gao, and M. Parvez, *Angew. Chem., Int. Ed. Engl.* **34**, 2549 (1995).
6. T. Chivers, X. Gao, and M. Parvez, *Inorg. Chem.* **35**, 553 (1996).
7. T. Chivers, M. Parvez, and G. Schatte, *Angew. Chem., Int. Ed. Engl.* **38**, 2217 (1999).
8. T. Chivers, C. Jaska, N. Sandblom, and G. Schatte, *Chem. Commun.* 1657 (2000).
9. T. Chivers, M. Parvez, and G. Schatte, *Inorg. Chem.* **38**, 1380 (1999).
10. T. Chivers, and G. Schatte, *Eur. J. Inorg. Chem.* 2266 (2002).
11. T. Chivers and G. Schatte, *Chem. Commun.* 2264 (2001).
12. D. F. Shriver and M. A. Drezden, *The Manipulation of Air-Sensitive Compounds,* 2nd ed., Wiley-Interscience, New York, 1986.
13. N. D. R. Barnett, W. Clegg, L. Horsburgh, D. M. Lindsay, Qi-Y. Liu, F. M. Mackenzie, R. E. Mulvey, and P. G. Williard, *Chem. Commun.* 2321 (1996).

Chapter Two

ORGANOMETALLIC AND COORDINATION COMPLEXES

11. ADDITION COMPOUNDS OF DIBROMODIOXOMOLYBDENUM(VI)

Submitted by FRANCISCO J. ARNÁIZ,[*] MARÍA R. PEDROSA,[*] and RAFAEL AGUADO[*]
Checked by HUNG KAY LEE[†]

Addition compounds of MoO_2Br_2 are useful precursors for the synthesis of species that are relevant in catalytic chemical and biochemical oxotransfer processes.[1] Also, molybdenum bromides are on the borderline of stability for +5 and +6 oxidation states of the metal, which makes them potentially useful for redox studies. These compounds are usually prepared by treating MoO_2Br_2 with the appropriate ligand in aprotic solvents,[2] and in a few cases by addition of the ligand to a solution of $H_2MoO_4 \cdot H_2O$ in HBr.[3] However, the presence of $MoO_2Br_2(H_2O)_2$ in solutions of alkali molybdates in concentrated hydrobromic acid[4] and the extractability of this species with diethyl ether suggest that a number of addition compounds could be readily accessible from commercial sodium molybdate. In this manner a number of adducts with O-donor ligands can be isolated.[5] Here we describe the preparation of the outer-sphere, hydrogen-bonded complex $MoO_2Br_2(H_2O)_2 \cdot$diglyme as well as those of the adducts $MoO_2Br_2(L)_2$,

[*]Laboratorio de Química Inorgánica, Universidad de Burgos, 09001 Burgos, Spain.
[†]Department of Chemistry, The Chinese University of Hong Kong, Shatin, New Territories, Hong Kong, China.

Inorganic Syntheses, Volume 34, edited by John R. Shapley
ISBN 0-471-64750-0

L = dimethyl formamide (DMF), dimethyl sulfoxide (DMSO), and triphenylphosphine oxide ($OPPh_3$). The former illustrates how simple polyethers may stabilize the hydrated dioxomolybdenum(VI) bromide core. The others are representative examples of addition compounds with ligands that are able to displace water from the coordination sphere of molybdenum but at the same time are not sufficiently basic toward the proton to cause formation of hydrobromides and molybdates.

■ **Caution.** *Concentrated hydrobromic acid is corrosive, volatile (fumes in air), and toxic. Diethyl ether, diglyme, dimethyl formamide, and dimethyl sulfoxide are flammable and toxic. Triphenylphosphine oxide is toxic. Use safety gloves and goggles, avoid inhalations of vapors and dust, and conduct all operations in a well-ventilated hood far from ignition sources.*

A. STOCK SOLUTION OF $MoO_2Br_2(H_2O)_2$ IN DIETHYL ETHER

$$Na_2MoO_4 \cdot 2H_2O \xrightarrow[Et_2O]{HBr} MoO_2Br_2(H_2O)_2 + 2NaBr$$

Procedure

To a 50-mL Erlenmeyer flask charged with 25 g (145.2 mmol) of concentrated ($\sim$47%) hydrobromic acid and a stirring bar and immersed in an ice-cold water bath, 5.0 g (20.7 mmol) of powdered $Na_2MoO_4 \cdot 2H_2O$ is added with stirring. After 5 min of stirring the molybdate dissolves and some sodium bromide precipitates. The flask is removed from the bath, and 25 mL of dry, freshly distilled diethyl ether is added at room temperature. The mixture is vigorously stirred for 5 min or shaken for 1 min in a separatory funnel. The yellow (upper) diethyl ether phase is collected in a 100-mL Erlenmeyer flask, and the extraction is repeated (2 $\times$ 25 mL). The combined ether extract is stirred for 5 min with 5 g of anhydrous magnesium sulfate. The solution is collected by filtration through a glass frit into a 250-mL round-bottomed flask, and the magnesium sulfate is washed with three 10-mL portions of dry diethyl ether. Approximately 90–100 mL of a yellow solution, which contains more than 95% of the initial molybdenum as $MoO_2Br_2(H_2O)_2$, is obtained. It can be stored for weeks at room temperature in a tightly closed container, and for months in the refrigerator, where yellow $MoO_2Br_2(H_2O)_2 \cdot 2Et_2O$ eventually crystallizes.

B. DIAQUADIBROMODIOXOMOLYBDENUM(VI)DIGLYME COMPLEX, $MoO_2Br_2(H_2O)_2 \cdot$DIGLYME

$$MoO_2Br_2(H_2O)_2 + \text{diglyme} \longrightarrow MoO_2Br_2(H_2O)_2 \cdot \text{diglyme}$$

Procedure

Freshly distilled diglyme (5 g, 37.3 mmol) is added to a diethyl ether stock solution prepared as described in Section 11.A. The resulting solution is concentrated on a rotary evaporator at room temperature to a final volume of $\sim$25 mL. A copious mass of yellow crystals forms in the process. The solid is collected by vacuum filtration, washed with two 5-mL portions of diethyl ether, and dried under mild vacuum. Typical yield, with regard to the initial molybdate, is 7.0 g (74%). A second crop of crystals is obtained by concentrating the combined mother liquor and washings, followed by cooling at $-30°C$ for 2 h. Overall yield: 8.5 g (90%).

Anal. Calcd. for $MoO_2Br_2(H_2O)_2.C_6H_{14}O_3$: Mo, 20.95; C, 15.74; H, 3.96. Found: Mo, 20.80; C, 15.81; H, 4.03.

Properties

Yellow $MoO_2Br_2(H_2O)_2 \cdot$diglyme melts at 54–55°C with decomposition. The product is insoluble in hexane and soluble in most polar organic solvents. The IR spectrum, taken as Nujol or KBr dispersions, has characteristic bands for ν_{MoO_2} at 913 and 951 cm^{-1}. The ^1H NMR spectrum (80 MHz, acetone-d_6) exhibits signals at δ 3.29 (s, 6H, CH_3), 3.51 (s, 8H, CH_2) and 4.96 (s, 4H, H_2O). The compound is sufficiently stable to be manipulated in air at room temperature without special precautions for short periods. It develops a light green coloration, although the degree of decomposition is not substantial, after weeks in a tightly closed container in the dark at room temperature.

C. BIS(DIMETHYLFORMAMIDE) DIBROMODIOXOMOLYBDENUM(VI), $MoO_2Br_2(DMF)_2$

$$MoO_2Br_2(H_2O)_2 + 2DMF \longrightarrow MoO_2Br_2(DMF)_2 + 2H_2O$$

Procedure

To a diethyl ether solution of $MoO_2Br_2(H_2O)_2$ prepared as described in Section 11.A, a solution of 3.0 g (41.0 mmol) of freshly distilled DMF in 15 mL of Et_2O is added with stirring. The pale yellow precipitate is filtered, washed with three 10-mL portions of Et_2O, and dried under vacuum. Yield 8.4 g (93%).

Anal. Calcd. for $MoO_2Br_2(C_3H_7NO)_2$: Mo, 22.11; C, 16.61; H, 3.25; N, 6.46. Found: Mo, 22.05; C, 16.64; H, 3.30; N, 6.39.

Properties

Solid $MoO_2Br_2(DMF)_2$ melts at 139–141°C with decomposition. The IR spectrum, taken as a KBr dispersion, has characteristic bands for ν_{MoO_2} at 903 and 940 cm^{-1}. The ^1H NMR spectrum in acetone-d_6 exhibits signals at δ 3.03 (s, 3H, CH_3), 3.22 (s, 3H, CH_3), 8.26 (s, 1H, CH). The complex is insoluble in hexane and diethyl ether and is soluble in methanol, ethanol, dichloromethane, chloroform, acetone, dimethyl formamide, and dimethyl sulfoxide. It is stable in air at room temperature and can be manipulated without special care. This product is specially useful for the synthesis of a number of adducts with pyridine and related bases, since the dimethyl formamide displaced can be readily removed by washing with most common organic solvents.

D. BIS(DIMETHYLSULFOXIDE) DIBROMODIOXOMOLYBDENUM(VI), $MoO_2Br_2(DMSO)_2$

$$MoO_2Br_2(H_2O)_2 + 2DMSO \longrightarrow MoO_2Br_2(DMSO)_2 + 2H_2O$$

Procedure

To a diethyl ether solution of $MoO_2Br_2(H_2O)_2$ prepared as described in Section 11.A, a solution of 3.2 g (41.0 mmol) of DMSO in 25 mL of acetone is added with stirring. The resulting pale yellow precipitate is filtered, washed with three 10-mL portions of acetone, and dried under vacuum. Yield 8.6 g (93%).

Anal. Calcd. for $MoO_2Br_2(C_2H_6OS)_2$: Mo, 21.61; C, 10.82; H, 2.72. Found: Mo, 21.48; C, 10.89; H, 2.78.

Properties

Solid $MoO_2Br_2(DMSO)_2$ melts at 143–144°C with decomposition. The IR spectrum, taken as a KBr dispersion, has characteristic bands for ν_{MoO_2} at 891 and 923 cm^{-1}. The ^1H NMR spectrum in acetone-d_6 exhibits a signal at δ 2.95 (s, CH_3). The compound is insoluble in hexane and diethyl ether; is slightly soluble in dichloromethane, chloroform, methanol, ethanol and acetone; and is soluble in dimethyl formamide and dimethyl sulfoxide. It is stable in air at room temperature and can be stored for weeks in the dark.

E. BIS(TRIPHENYLPHOSPHINEOXIDE) DIBROMODIOXOMOLYBDENUM(VI), $MoO_2Br_2(OPPh_3)_2$

Procedure

To a diethyl ether solution of $MoO_2Br_2(H_2O)_2$ prepared as described in Section 11.A (at-half-scale), a solution of 5.4 g (19.4 mmol) of $OPPh_3$ in 60 mL of acetone is added with stirring. The yellow precipitate is filtered, washed with three 10-mL portions of acetone, and dried under vacuum. Yield, with regard to $OPPh_3$, 7.9 g (96%).

Anal. Calcd. for $MoO_2Br_2(C_{18}H_{15}OP)_2$: Mo, 11.36; C, 51.21; H, 3.58. Found: Mo, 11.28; C, 51.18; H, 3.52.

Properties

$MoO_2Br_2(OPPh_3)_2$ decomposes before melting. The IR spectrum, taken as KBr dispersion, has characteristic bands for ν_{MoO_2} at 902 and 945 cm^{-1}. The ^1H NMR spectrum in acetone-d_6 exhibits a signal at δ 7.61 (m, CH). The compound is insoluble in hexane and diethyl ether; is slightly soluble in methanol, ethanol, acetone, dichloromethane, and chloroform; and is soluble in dimethyl formamide and dimethyl sulfoxide. It is stable at room temperature and can be manipulated in air without special precautions.

References

1. H. K. Lee, Y. L. Wong, Z. Y. Zhou, Z. Y. Zhang, D. K. P. Ng, and T. C. W. Mak, *J. Chem. Soc., Dalton Trans.* 539 (2000) and refs. therein.
2. (a) T. V. Iorns and F. E. Stafford, *J. Am. Chem. Soc.* **88**, 4819 (1966); (b) F. E. Kühn, E. Herdweck, J. J. Haider, W. A. Herrman, I. S. Goncalves, A. D. Lopes, and C. C. Romao, *J. Organomet. Chem.* **583**, 3 (1999).
3. R. J. Butcher, H. P. Ganz, R. G. A. R. Maclagan, H. K. J. Powell, C. J. Wilkins, and S. H. Yong, *J. Chem. Soc., Dalton Trans.* 1223 (1975).
4. J. M. Coddington and M. J. Taylor, *J. Chem. Soc. Dalton Trans.* 41 (1990).
5. (a) F. J. Arnáiz, R. Aguado, M. R. Pedrosa, J. Mahía, and M. A. Maestro, *Polyhedron* **20**, 2781 (2001); (b) F. J. Arnáiz, R. Aguado, M. R. Pedrosa, J. Mahía, and M. A. Maestro, *Polyhedron* **21**, 1635 (2002).

12. OXORHENIUM(V) OXAZOLINE COMPLEXES FOR OXYGEN ATOM TRANSFER

Submitted by LEE D. MCPHERSON,[*] VIRGINIE M. BéREAU,[*] and MAHDI M. ABU-OMAR[*]
Checked by VESELA UGRINOVA,[†] LIHUNG PU,[†] SCHEROI D. TAYLOR, and SETH N. BROWN[†]

Several rhenium complexes in various oxidation states have been shown to be effective catalysts for oxygen atom transfer (OAT) reactions.[1–11] In particular, $Re(O)(hoz)_2Cl$ (Hhoz = [2-(2′ hydroxyphenyl)-2-oxazoline]), was demonstrated to be a useful OAT catalyst between sulfides and sulfoxides.[12] In addition, this oxorhenium(V) complex is an efficient catalyst for the reduction of perchlorate by oxygen atom transfer in the presence of sulfides under mild conditions.[13] Given that perchlorate salts are environmental contaminants,[14,15] and given their biological toxicity[16] and kinetic inertness in solution, $Re(O)(hoz)_2Cl$ and its derivatives are potentially interesting for environmental applications. These complexes are air-stable and water-tolerant, and have a bench life of more than one year. These properties are extremely attractive because they reduce the amount of time spent on catalyst preparation and provide the researcher with greater flexibility in tailoring and exploring their catalytic chemistry.

Improved syntheses of $Re(O)(hoz)_2Cl$, $Re(O)(hoz)_2(OTf)$, and $Re(O)(hoz)Cl_2(PPh_3)$ are presented herein. These syntheses have the advantage of providing consistent results, higher yields, and an avoidance of sideproducts. The starting rhenium complexes, $Re(O)Cl_3(PPh_3)_2$[17] and $Re(O)Cl_3(OPPh_3)(SMe_2)$,[18,19] are prepared according to literature methods; however, the procedure for $Re(O)Cl_3(OPPh_3)(SMe_2)$ is included in Section 12.B. Preparation of the ligand, Hhoz, following previous literature methods,[20] is included in the following section.

A. 2-(2′-HYDROXYPHENYL)-2-OXAZOLINE (Hhoz)

OH O ... O + H_2N ... OH ⟶ OH O ... N(H) ... OH —(1. $SOCl_2$; 2. $NaHCO_3$)⟶ OH O ... N

■ **Caution.** *Thionyl chloride is corrosive and emits HCl gas. Handle with caution in a well-ventilated area.*

[*]Department of Chemistry and Biochemistry, University of California, Los Angeles, CA 90024.
[†]Department of Chemistry and Biochemistry, University of Notre Dame, Notre Dame, IN 46556.

2-Aminoethanol (Sigma; 4.1 mL, 0.068 mol) is added to a 100-mL flask with a stirring-bar. Ethylsalicylate (Aldrich; 10 mL, 0.068 mol) is added, and a reflux condenser is fitted to the flask. The mixture is refluxed neat for 2.5 h, and then the product solvent (EtOH) is completely removed under reduced pressure. The crude, white solid is washed with two 10-mL aliquots of cold water. It is dissolved in 25 mL of CH_2Cl_2 and dried with $MgSO_4$. The filtered solution is evaporated under reduced pressure, and the solid residue is dried under vacuum to yield 8.5 g (69%) of white solid, 2-hydroxy-*N*-(2′-hydroxyethyl)benzamide.

2-Hydroxy-*N*-(2′-hydroxyethyl)benzamide (6.2 g, 0.034 mol) is added to a 250-mL flask with a stirring-bar under argon. Dichloromethane (150 mL) is added to the flask, and the flask is cooled to 0°C with an ice bath. Freshly distilled thionyl chloride (Aldrich; 8.1 mL, 0.068 mol) is added to the solution slowly via syringe. The mixture is stirred for 18 h at room temperature. The solid precipitate is collected by filtration with a large, fine-mesh glass frit. It is then dissolved in 100 mL, of water, and 25 mL saturated aqueous sodium bicarbonate solution is slowly added. The resulting pale pink solid is collected by filtration using a coarse-mesh glass frit. The solid is extracted with five 25-mL aliquots of diethyl ether, which are then combined. The diethyl ether is removed under reduced pressure to yield 3.0 g (54%) of pale pink crystals.

Anal. Calcd. for $C_9H_9NO_2$: C, 66.23; H, 5.56; N, 5.56. Found: C, 66.35; H, 5.49; N, 5.56.

Properties

Hhoz is a pale pink, crystalline substance and is stable in air. It is soluble in ethanol, acetonitrile, dichloromethane, diethyl ether, and pentane. Its IR spectrum shows an imine peak at 1640 cm^{-1} and an aromatic stretch at 1620 cm^{-1} in a KBr pellet. ^1H NMR ($CDCl_3$): δ 12.14 (s, 1H), 7.67 (d, 1H), 7.37 (t, 1H), 6.94 (m, 2H), 4.43 (t, 2H), 4.09 (t, 2H).

B. TRICHLOROOXO(DIMETHYLSULFIDE)(TRIPHENYLPHOSPHINEOXIDE)RHENIUM(V), $Re(O)Cl_3(OPPh_3)(SMe_2)$

$$Re(O)Cl_3(PPh_3)_2 + 2\ H_3C{-}S(=O){-}CH_3 \longrightarrow Re(O)Cl_3(OPPh_3)(SMe_2) + OPPh_3 + SMe_2$$

■ **Caution.** *Concentrated HCl is extremely corrosive and reacts violently with base. Handle with caution. Dimethylsulfide is toxic and has an unpleasant odor. The synthesis must be carried out in a well-ventilated hood.*

$Re(O)Cl_3(PPh_3)_2$[17] (10 g, 0.011 mol) is placed in a 500-mL round-bottomed flask in benzene (250 mL) with a magnetic stirring bar. Concentrated hydrochloric acid (Fisher; 44 mL, 12 M) is added slowly to the suspension followed by dimethylsulfoxide (Fisher; 10 g, 0.13 mol). The suspension is stirred for 5 days closed to the atmosphere. The resulting light green solid is collected by suction filtration, and then washed with (2 × 30 mL) acetone and then with cold dichloromethane (2 × 30 mL). After drying in air, the yield is 7.5 g (95%).

Properties

$Re(O)Cl_3(OPPh_3)(SMe_2)$ is a lime green powder. It is stable in air and slightly soluble in ethanol. Its IR spectrum in Nujol shows a strong Re=O stretch at 981 cm^{-1} along with stretches at 1130 and 1138 cm^{-1}. ^1H NMR (CD_2Cl_2): δ 7.6 (m, 15H), 2.70 (s, 6H). ^{31}P NMR (CD_2Cl_2): 48 ppm (s, Re—$OPPh_3$). There is no visible decomposition of the product when stored for six months as a solid in a desiccator.

C. CHLOROBIS(2-(2′-HYDROXYPHENYL)-2-OXAZOLINE) OXORHENIUM(V), $ReOCl(hoz)_2$

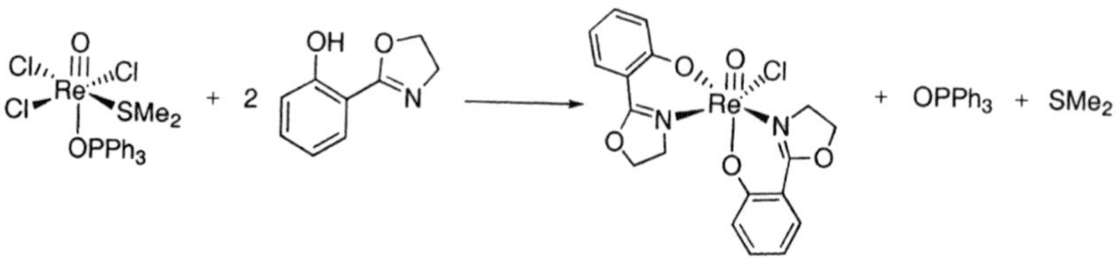

■ **Caution.** *Dimethylsulfide is toxic and has an unpleasant odor. The synthesis must be carried out in a well-ventilated hood.*

$Re(O)Cl_3(OPPh_3)(SMe_2)$ (510 mg, 0.786 mmol) is placed in a 50-mL round-bottomed flask with 2.2 equiv of the ligand, Hhoz (280 mg, 1.72 mmol). Ethanol (30 mL, 200-proof) is added to the flask and is followed with 2,6-lutidine (Aldrich; 500 μL, 4.30 mmol). Finally, a magnetic stirring bar is added, and a reflux condenser is fitted to the flask. The light green suspension is heated under reflux for 3 h under ambient atmosphere, and then is cooled to room temperature. The dark green precipitate is collected by filtration through a fine-mesh glass frit, and it is thoroughly washed with diethyl ether. After drying in air, the yield is 351 mg (80%).

Anal. Calcd. for $C_{18}ClH_{16}N_2O_5Re$: C, 38.5; H, 2.9; N, 5.0. Found: C, 38.2; H, 2.8; N, 5.0.

Properties

$Re(O)(hoz)_2Cl$ is a forest green powder. It is stable in air and is soluble in dichloromethane. It is partially soluble in acetonitrile and chloroform. ^1H NMR (CD_2Cl_2): δ 7.93 (d, 1H), 7.68 (d, 1H), 7.45 (t, 1H), 7.20 (t, 1H), 6.96 (t, 1H), 6.87 (d, 1H), 6.81 (t, 1H), 6.75 (d, 1H), 5.05 (m, 1H), 4.97 (t, 2H), 4.85 (m, 1H), 4.72 (t, 2 H), 4.20 (m, 2H). Its IR spectrum shows a strong Re=O stretch at 973 cm^{-1} in a KBr pellet. It decomposes at 268°C. There is no visible decomposition of the product when stored for 6 months as a solid or a solution.

D. BIS(2-(2′-HYDROXYPHENYL)-2-OXAZOLINE)OXORHENIUM(V) TRIFLUOROMETHANESULFONATE, $[ReO(hoz)_2][OTf]$

+ 2 AgOTf

L = OH_2 or CH_3CN

$ReO(hoz)_2Cl$ (203 mg, 0.362 mmol) is added to a dry 50-mL Schlenk flask with a magnetic stirring bar under an argon atmosphere. Acetonitrile is added to the flask, and the mixture is allowed to stir at 50°C for 30 min or until all the solid is dissolved. Silver trifluoromethanesulfonate (Lancaster; 102 mg, 0.397 mmol) is added, and an oven-dried reflux condenser is fitted to the flask. The solution is heated under reflux for 1 h and then is allowed to cool. The solution is filtered through a fine-mesh glass frit under ambient atmosphere. The solvent is removed under reduced pressure, and the crude dark green solid is redissolved in a minimum of dichloromethane. The solution is stored at −10°C in a freezer for 2 h to facilitate further precipitation of impurities, which are then filtered out using a fine-mesh glass frit. The filtrate is evaporated, and the dark green solid residue is dried under vacuum. The yield is 233 mg (95%).

Anal. Calcd. for $C_{19}F_3H_{16}N_2O_8ReS$: C, 33.8; H, 2.4; N, 4.1; S, 4.7. Found: C, 32.6; H, 2.4; N, 4.2; S, 4.5.

Properties

$[Re(O)(hoz)_2][OTf]$ is a dark green solid that is very soluble in acetonitrile and dichloromethane. ^1H NMR (CD_2Cl_2): δ 7.98 (d, 2H), 7.60 (t, 2H), 7.19 (d, 2H), 6.95 (t, 2H), 4.95 (m, 6H), 4.38 (m, 2H). It catalyzes the reduction of perchlorate at a rate equal to that for $Re(O)(hoz)_2Cl$. It is stable in air and does not visibly decompose in solid or solution form for 6 months. The oxazoline ligands are C_2-symmetric around the metal, leaving the oxo ligand at one axial position and a trans solvent molecule occupying the other axial position.[12]

E. DICHLORO(2-(2′-HYDROXYPHENYL)-2-OXAZOLINE)-(TRIPHENYLPHOSPHINE)OXORHENIUM(V), $ReO(hoz)Cl_2(PPh_3)$

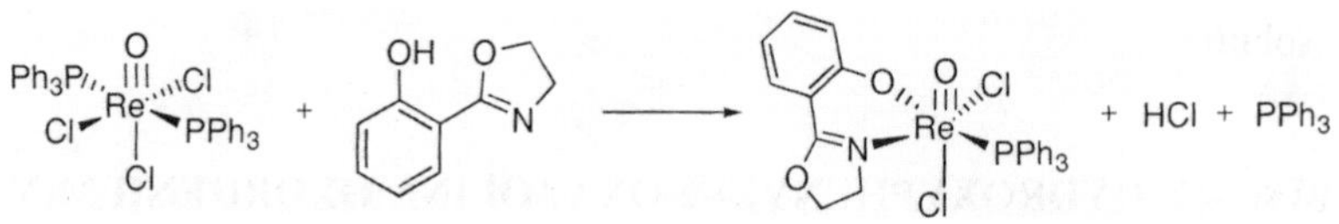

$Re(O)Cl_3(PPh_3)_2$ (500 mg, 0.60 mmol) is placed in a 100-mL round-bottomed flask with 1 equiv of the ligand, Hhoz (98 mg, 0.60 mmol). Ethanol (50 mL, 200-proof) and a magnetic stirring bar are added and a reflux condenser is fitted to the flask. The yellow suspension is heated under reflux for 1 h under ambient atmosphere. The mixture is filtered while hot through a fine-mesh glass frit. The crude green solid is washed with 50 mL of hot ethanol and 10 mL of ether (yield 272 mg 65%). The solid is dissolved in a minimum of dichloromethane and layered with *n*-pentane.* The solution is stored at $-10°C$ overnight, and the resulting green precipitate is collected on fine-mesh glass frit. The yield is 105 mg (25% overall).

Anal. Calcd. for $C_{27}H_{23}Cl_2NO_3PRe$: C, 46.5; H, 3.32; N, 2.01. Found: C, 46.7; H, 3.25; N, 2.09.

Properties

$Re(O)(hoz)Cl_2(PPh_3)$ is a light green compound that is inactive as a catalyst for perchlorate reduction. It is soluble in dichloromethane. It is partially soluble in ethanol and acetonitrile. ^1H NMR (CD_2Cl_2): δ 7.30 (m, 16H), 6.93 (t, 1H), 6.77 (t, 1H), 6.43 (d, 1H), 4.65 (q, 1H), 4.21 (q, 1H), 3.99 (q, 1H), 3.26 (m, 1H). The compound is stable in air and shows no visible decomposition after 6 months as a solid. Its IR spectrum features a Re=O stretch of 964 cm^{-1}.

Acknowledgments

The submitters acknowledge financial support provided by the National Science Foundation and the Arnold and Mabel Beckman Foundation.

*The checkers have noted that column chromatography on silicagel (CH_2Cl_2:pentane, 10 : 1) of the crude product facilitated removal of triphenylphosphine and improved the overall yield to 35%.

References

1. R. R. Conry and J. M. Mayer, *Inorg. Chem.* **29**, 4862 (1990).
2. J. Takacs, M. R. Cook, P. Kiprof, J. G. Kuchler, and W. A. Herrmann, *Organometallics* **10**, 316 (1991).
3. Z. Zhu and J. H. Espenson, *J. Mol. Catal. A* **103**, 87 (1995).
4. M. M. Abu-Omar, E. H. Appleman, and J. H. Espenson, *Inorg. Chem.* **35**, 7751 (1996).
5. S. N. Brown and J. M. Mayer, *J. Am. Chem. Soc.* **118**, 12119 (1996).
6. J. B. Arterburn, M. C. Perry, S. L. Nelson, B. R. Dibble, and M. S. Holguin, *J. Am. Chem. Soc.* **119**, 9309 (1997).
7. C. C. Romao, F. E. Kuhn, and W. A. Herrmann, *Chem. Rev.* **97**, 3197 (1997).
8. K. P. Gable, F. A. Zhuravlev, and A. F. T. Yokochi, *Chem. Commun.* **799** (1998).
9. J. H. Espenson, *Chem. Commun.* 479 (1999).
10. K. P. Gable and E. C. Brown, *Organometallics* **19**, 944 (2000).
11. G. S. Owens, J. Arias, and M. M. Abu-Omar, *Catal. Today* **55**, 317 (2000).
12. J. Arias, C. R. Newlands, and M. M. Abu-Omar, *Inorg. Chem.* **40**, 2185 (2001).
13. M. M. Abu-Omar, L. D. McPherson, J. Arias, and V. M. Béreau, *Angew. Chem., Int. Ed.* **39**, 4310 (2000).
14. E. T. Urbansky and M. R. Schock, *J. Environ. Manage.* **56**, 79 (1999).
15. B. Gu, G. M. Brown, L. Maya, M. J. Lance, and B. A. Moyer, *Environ. Sci. Technol.* **35**, 3363 (2001).
16. R. V. Burg, *J. Appl. Toxicol.* **15**, 237 (1995).
17. G. W. Parshall, *Inorg. Synth.* **17**, 110 (1977).
18. D. E. Grove and G. J. Wilkinson, *J. Chem. Soc. A* 1224 (1966).
19. M. M. Abu-Omar and S. I. Khan, *Inorg. Chem.* **37**, 4979 (1998).
20. H. R. Hoveyda, S. J. Rettig, and C. Orvig, *Inorg. Chem.* **31**, 5408 (1992).

13. ALLYL AND DIENYL COMPLEXES OF RUTHENIUM

Submitted by ALBRECHT SALZER,[*] ANDRÉ BAUER,[*] STEFAN GEYSER,[*] and FRANK PODEWILS[*]
Checked by GREGORY C. TURPIN[†] and RICHARD D. ERNST[†]

Complexes of ruthenium with π-bonded ligands are becoming increasingly important in organometallic chemistry and homogeneous catalysis. There is, however, still a lack of reliable and simple methods to prepare allyl and pentadienyl complexes of ruthenium by a general route. Previous methods include the "Fischer–Müller synthesis," where $RuCl_3 \cdot nH_2O$ is treated with an isopropyl Grignard reagent and irradiated in the presence of a diolefin,[1] and the "Vitulli method," where $RuCl_3 \cdot nH_2O$ is treated with zinc powder in the presence of diolefins.[2,3] Both methods are sometimes difficult to reproduce. Commercial

[*]Institut für Anorganische Chemie, RWTH Aachen, D 52056 Aachen, Germany.
[†]Department of Chemistry, University of Utah, Salt Lake City, UT 84112.

$RuCl_3 \cdot nH_2O$ is a nonstoichiometric material that may differ from one supplier to the other. The complex $[Ru(H_2O)_6]^{2+}$ is in some cases an alternative precursor,[4] but is difficult to prepare. We wish to present an improved synthesis of di-μ-chloro(dichloro)bis(η^3:η^3-2,7-dimethylocta-2,6-diene-1,8-diyl)diruthenium[5–7] and its application as a versatile precursor for Ru(II) complexes with η^5-dienyl ligands.[8,9]

General Remarks

All compounds are routinely handled and stored under nitrogen in Schlenk flasks, although most solids can be handled safely in air. Solvents are dried and degassed before use.

A. DI-μ-CHLORO(DICHLORO)BIS(η^3:η^3-2,7-DIMETHYLOCTA-2,6-DIENE-1,8-DIYL) DIRUTHENIUM, $\{Ru(\eta^3\text{: }\eta^3\text{-}C_{10}H_{16})Cl_2\}_2$

$$2RuCl_3nH_2O + 4C_5H_8 \rightarrow \{Ru(\eta^3 : \eta^3\text{-}C_{10}H_{16})Cl_2\}_2$$

Procedure

A solution of 18 g (68 mmol) of commercial ruthenium trichloride hydrate (Johnson Matthey, 40–43% ruthenium) in a mixture of isoprene (680 mL) and 2-methoxyethanol (280 mL) is heated at reflux for 10 days under inert gas (nitrogen or argon). The purple crystalline product is collected in a medium-porosity sintered-glass funnel, washed with diethyl ether, and dried in vacuo; yield 19.9 g (95%).

Anal. Calcd. for $C_{20}H_{32}Cl_4Ru_2$: C, 38.97 H, 5.23. Found: C, 39.00, H, 5.19.

Properties

Di-μ-chloro(dichloro)bis[η^3-η^3-2,7-dimethylocta-2,6-diene-1,8-diyl)diruthenium (compound A) is a purple crystalline compound, that is indefinitely stable at room temperature under a nitrogen atmosphere. For short periods it can also be handled in air. MS (EI) *m/z*: 616 (M^+). In the solid state the product shows symmetry C_1,[10] but it exists as a mixture of two diastereomers in solution.[7] It reacts with various Lewis bases L to form the monomers $Ru(\eta^3\text{:}\eta^3\text{-}C_{10}H_{16})Cl_2L$[11,12] and it shows catalytic activity in ROMP polymerization.[13]

B. BIS(η^5-2,4-DIMETHYLPENTADIENYL)RUTHENIUM ("OPEN RUTHENOCENE"), $Ru(\eta^5\text{-}C_5H_5Me_2)_2$

$$\frac{1}{2}(Ru(\eta^3:\eta^3\text{—}C_{10}H_{16})Cl_2]_2 + 2C_7H_{12} + Li_2CO_3 \rightarrow$$
$$Ru(C_7H_{11})_2 + 2LiCl + C_{10}H_{16} + CO_2 + H_2O$$

Procedure

A solution of product A (0.31 g, 0.5 mmol) and Li_2CO_3 (0.15 g, 2.0 mmol) in a mixture of 10 mL ethanol, 26 μL (0.5 mmol) acetonitrile, and 2,4-dimethylpenta-1,3-diene (Aldrich) (0.38 g, 4.0 mmol) is heated at reflux for 4 h in a 100-mL round-bottomed flask with a standard tapered joint.* The solvent is evaporated under reduced pressure, and the flask is connected to a short, straight, water-cooled reflux condenser (Fig. 1). The top of the reflux condenser is attached to a high-vacuum pump, and the brown crude product is purified by sublimation at 80°C and 10^{-3} bar. The yellow solid product is collected from the lower end of the reflux condenser. On a larger scale, it is more convenient to extract the dry brown crude product with diethyl ether, filter the solution through a pad of alumina, evaporate the solvent, and then sublime the product. Yield 95%.

Anal. Calcd. for $C_{14}H_{22}Ru$: C, 57.7; H, 7.6. Found: C, 57.99; H, 7.82.

Properties

The pale yellow crystalline material is soluble in nonpolar solvents such as hexane, diethyl ether, or benzene. The solutions are air-sensitive. ^{1}H NMR (200 MHz, C_6D_6): δ 4.63 (s, 1H), 2.71 (s, 2H, $J = 2$ Hz), 1.73 (s, 6H), 0.86 (s, 2H, $J = 2$ Hz).^{13}C NMR (50 MHz, C_6D_6): δ 100.3 (C-2), 97.8 (C-3), 46.8 (C-1), 26.3 (C-4).

C. BIS(η^5-CYCLOPENTADIENYL)RUTHENIUM ("RUTHENOCENE"), $Ru(\eta^5\text{-}C_5H_5)_2$

$$RuCl_3 \cdot nH_2O + 2C_5H_6 + Li_2CO_3 \rightarrow Ru(C_5H_5)_2 + 2LiCl + CO_2 + H_2O$$

*The checkers report that the reactions leading to dienyl complexes proceed best when heating is slowly commenced; alternatively, the reactions can be carried out by stirring the mixtures at room temperature for 4–20 h. The time is determined primarily by particle size and the amount of excess carbonate that is used, indicating a heterogeneous reaction. As a result, very finely ground sodium or potassium carbonates may be used successfully.

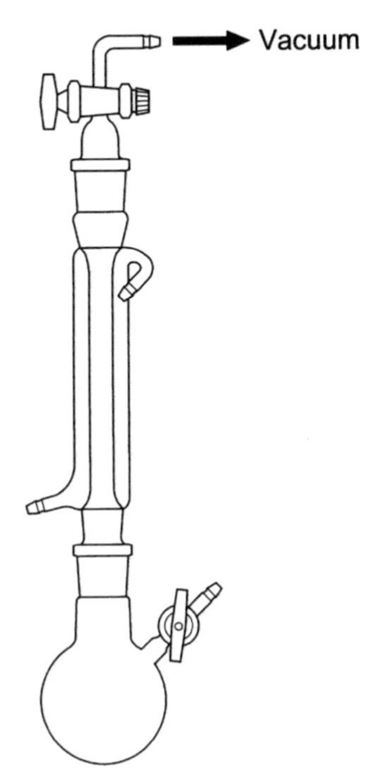

Figure 1. Apparatus for preparation of "open ruthenocene" and "ruthenocene."

Procedure

A mixture of 1.0 g (27 mmol) of $RuCl_3 \cdot nH_2O$, 2.0 g (27 mmol) of lithium carbonate, and 0.66 g (10 mmol) of freshly distilled cyclopentadiene is heated at reflux in 50 mL of isopropanol for 16 h in a 100-mL round-bottomed flask with a standard tapered joint. The solvent is evaporated under reduced pressure, and the flask is connected to a short straight water-cooled reflux condenser (Fig. 1). The compound is sublimed at 80°C and 10^{-3} bar and collected from the bottom of the condenser. Yield: 0.88 g (95%) of off-white crystals.

Alternatively, the complex can be made in identical yields following the procedure outlined in Section 13.B starting from $(Ru(\eta^3{:}\eta^3\text{-}C_{10}H_{16})Cl_2)_2$.

Anal. Calcd. for $C_{10}H_{10}Ru$: C, 51.94, H, 4.36. Found: C, 50.99, H, 4.41.

Properties

Ruthenocene is an air-stable, off-white crystalline material, soluble in all nonpolar as well as most polar organic solvents without decomposition. ^1H NMR

(200 MHz, C_6D_6): δ 4.46 (s, Cp). ^{13}C NMR (50 MHz, C_6D_6): δ 70.4 (Cp). MS (EI) *m/z*: 232 (M^{+}, 100%), 169 (M^{+}-Cp, 20%).

D. (η^6-BENZENE(η^5-2,4-DIMETHYLPENTADIENYL)RUTHENIUM TETRAFLUOROBORATE, $[Ru(\eta^6-C_6H_6)(\eta^5-C_5H_5Me_2)][BF_4]$

■ **Caution.** *The reaction should be carried out in a well-ventilated hood. Benzene is a known carcinogen, so contact with the liquid or vapor should be avoided.*

$$\frac{1}{2}(Ru(\eta^3:\eta^3\text{-}C_{10}H_{16})Cl_2]_2 + C_7H_{12} + C_6H_6 + 2AgBF_4 \rightarrow$$
$$[Ru(C_7H_{11})(C_6H_6)]BF_4 + 2AgCl + C_{10}H_{16}$$

Procedure

To a solution of silver tetrafluoroborate (0.77 g, 4.0 mmol) in 20 mL of ethanol, 2,4-dimethylpenta-1,3-diene (0.62 g, 6.5 mmol), and 2 mL of benzene are added. This solution is added dropwise to a vigorously stirred solution of product A (0.61 g, 2.0 mmol), in 15 mL of dichloromethane. Silver chloride is separated by centrifugation and decanting the supernatant liquids and then it is washed with 5 mL of dichloromethane. The combined dichloromethane solutions are concentrated to ~5 mL, and the orange product is precipitated by addition of an equal volume of diethyl ether. Yield 75%.

Anal. Calcd for $C_{13}H_{17}RuBF_4$:C, 43.24; H, 4.74. Found: C, 43.29; H, 4.67.

Properties

The orange crystalline compound is insoluble in hexane or diethyl ether, but it is soluble in polar solvents such as dichloromethane, acetone, and nitromethane. ^1H NMR (300 MHz, CD_2Cl_2): δ 6.36 (s, 6H,C_6H_6), 3.53 (d, 2H, $J = 4$ Hz, CH_2 exo), 2.16 (s, 6H, CH_3), 1.15 (d, 2H, $J = 4$ Hz, CH_2 endo); ^{13}C NMR (75 MHz, CD_2Cl_2): 106.3 (2 CCH_3), 97.9 (1 CH), 92.7 (6 CH), 51.6 (2 CH_2), 25.7 (2 CH_3).

E. BIS(η^5-CYCLOHEPTADIENYL)RUTHENIUM, $Ru(\eta^5-C_7H_9)_2$

$$\frac{1}{2}(Ru(\eta^3:\eta^3\text{—}C_{10}H_{16})Cl_2)_2 + 2C_7H_{10} + Li_2CO_3 \rightarrow$$
$$Ru(C_7H_9)_2 + 2LiCl + C_{10}H_{16} + CO_2 + H_2O$$

Procedure

This compound is prepared as described in Section 13.B, but substituting 1,3-cycloheptadiene (Aldrich). Yield 80%.

Anal. Calcd. for $C_{14}H_{18}Ru$: C, 58.52; H, 6.31. Found: C. 57.99; H, 6.21.

Properties

The yellow crystalline material is soluble in nonpolar solvents such as hexane, diethyl ether, or benzene. The solutions are air-sensitive. ^{1}H NMR (300 MHz, C_6D_6): δ 5.01 (t, 2H, H-1), 4.37 (dd, 4H, H-2). 3.88 (m, 4H, H-3), 2.15, 1.66 (m, 8H, H-4). ^{13}C NMR (75 MHz, C_6D_6): δ 94.56 (C-1); 88.21 (C-2); 64.88 (C-3); 34.95 (C-4).

F. BIS(η^5-INDENYL)RUTHENIUM, $Ru(C_9H_7)_2$

$$\frac{1}{2}[Ru(\eta^3:\eta^3\text{-}C_{10}H_{16})Cl_2]_2 + 2C_9H_8 + Li_2CO_3 \rightarrow$$
$$Ru(C_9H_7)_2 + 2LiCl + C_{10}H_{16} + CO_2 + H_2O$$

Procedure

A solution of product A (0.24 g, 0.4 mmol) and Li_2CO_3 (0.29 g, 4.0 mmol) in a mixture of 10 mL of ethanol, 26 μL (0.5 mmol) of acetonitrile, and freshly distilled indene (Aldrich) (0.36 g, 3.1 mmol) is heated under reflux for 4 h. After evaporation of the solvent, the orange residue is extracted with diethyl ether. The solution is filtered through a pad of alumina and then cooled to −80°C. Orange crystals separate, which are isolated by filtration, and dried under vacuum. Yield 46%.

Anal. Calcd. for $C_{18}H_{14}Ru$: C, 65.25; H, 4.26. Found: C, 65.02; H, 4.19.

Properties

The product is an orange crystalline compound, soluble in diethyl ether and aromatic solvents. ^{1}H NMR (200 MHz, C_6D_6): δ 6.52–6.69 (m, 8H, H-1,2); 4.80 (d, $^{3}J = 2.4$ Hz, 4H, H-3); 4.46(t, $^{3}J = 2.4$ Hz, 2H, H-4). MS (EI) *m/z*: 332 (M^+, 100%), 217 (M^+ −indene, 20%).

G. BIS(η^5-PENTAMETHYLCYCLOPENTADIENYL)RUTHENIUM, $Ru(\eta^5-C_5Me_5)_2$

$$\frac{1}{2}[Ru(\eta^3:\eta^3\text{—}C_{10}H_{16})C_{12}]_2 + 2C_{10}H_{16} + Li_2CO_3 \rightarrow$$
$$Ru(C_{10}H_{15})_2 + 2LiCl + C_{10}H_{16} + CO_2 + H_2O$$

Procedure

The off-white crystalline compound is prepared as described in Section 13.F, using pentamethylcyclopentadiene (Strem Chemicals) instead of indene. Yield 47%.

Anal. Calcd. for $C_{20}H_{30}Ru$: C, 64.66; H, 8.15. Found: C, 65.01; H, 8.19.

Properties

The product forms air-stable off-white crystals, soluble in all nonpolar solvents as well as in dichloromethane, acetone, or ethanol. ^{1}H NMR (200 MHz, C_6D_6): δ 1.64 (s, Me). ^{13}C NMR (50 MHz, C_6D_6): δ 82.9 (CCH_3), 10.5 (CCH_3) MS (EI) *m/z*: 372 ($M,^+$ 100%).

References

1. J. Müller, C. G. Kreiter, B. Mertschenk, and S. Schmitt, *Chem. Ber.* **108**, 273 (1975).
2. P. Partici, G. Vitulli, M. Paci, and L. Porri, *J. Chem. Soc., Dalton Trans.* 1961 (1980).
3. R. D. Ernst and L. Stahl, *Organometallics* **2**, 1229 (1983).
4. M. Stebler-Roethlisberger, A. Salzer, H. B. Bürgi, and A. Ludi, *Organometallics* **5**, 298 (1986).
5. L. Porri, M. C. Gallazzi, A. Colombo, and G. Allegra, *Tetrahedron Lett.* **47**, 4187 (1965).
6. R. Roulet and D. N. Cox, *Inorg. Chem.* **29**, 1360 (1990).
7. J. G. Toerien and P. H. van Rooyen, *J. Chem. Soc. Dalton Trans.* 1963 (1991).
8. A. Bauer, U. Englert, S. Geyser, F. Podewils, and A. Salzer, *Organometallics* **19**, 5471 (2000).
9. A. Salzer, S. Geyser, and F. Podewils, Eur. Patent Appl. 110429.8-2110.
10. A. Colombo and G. Allegra, *Acta Cryst. B* **27**, 1653 (1971).
11. R. Roulet, D. N. Cox, and R. W. H. Small, *J. Chem. Soc., Dalton Trans.* 2013 (1991).
12. .R. A. Head, J. F. Nixon, J. R. Swain, and C. M. Woodard, *J. Organomet. Chem.* **76**, 393 (1974).
13. W. A. Herrmann, N. C. Schattenmann, O. Nuyken, and S. C Glader, *Angew. Chem., Int. Ed. Engl.* **35**, 1087 (1996).

14. TRIS(4,7-DIPHENYL-1,10-PHENANTHROLINE) RUTHENIUM(II) CHLORIDE, $[Ru(4,7\text{-}Ph_2Phen)_3]Cl_2$

Submitted by GLEN W. WALKER* and DANIEL G. NOCERA*
Checked by SHAWN SWAVEY† and KAREN J. BREWER†

$$1/n\{Ru(1,5\text{-COD})Cl_2\}n + 3L \rightarrow [RuL_3]Cl_2 + 1,5\text{-COD}$$
$$(L = 4,7\text{-diphenyl-1,10-phenanthroline})$$

The title complex is an important luminescent probe for oxygen, which has found use for a variety of analytical applications.[1] Measurements of oxygen concentration rely on the quenching of the intense orange emission from the long-lived[3] MLCT excited state of the ruthenium complex by an energy transfer reaction with ground-state triplet oxygen. The concentration of oxygen in solution or thin films is determined from a Stern–Volmer analysis of the decrease in luminescence intensity from the probe complex.[1] Optical fiber arrays[2] and thin-film gas sensors[3] have been designed on the basis of this Stern–Volmer quenching scheme. The lumophore is also used in the aerospace industry, where specialty coatings containing luminescent oxygen probes are used to create optical maps of the pressure variations across aerodynamic surfaces in wind tunnels.[4,5]

The established syntheses of the title complex involve reactions of 4,7-diphenyl-1,10-phenanthroline with potassium aquopentachlororuthenate(III) in aqueous *N,N*-dimethylformamide[6] or with ruthenium trichloride in aqueous ethanol.[7] The yields obtained by these routes are typically low ($\sim$5%), and the purification of the complex may require extensive chromatography. More recent approaches to the synthesis of ruthenium(II) polypyridyl complexes involve the reaction of the polymeric ruthenium(II) compound, $\{Ru(CO)_2Cl_2\}_n$, with the appropriate diimine ligand in refluxing 2-methoxyethanol.[8] These syntheses are generally rapid and proceed in good to high yields. We now report a related approach for the synthesis of the title complex in which the organometallic polymerdichloro(1,5-cyclooctadiene) ruthenium(II), $\{Ru(1,5\text{-}(COD)Cl_2\}_n$, is the source of ruthenium(II) in refluxing 2-methoxyethanol. This new synthetic method provides a straightforward and efficient route to the complex that can be completed in 2 days. The related dimine ligands, 1,10-phenanthroline and 2,2′-bipyridine, may be treated with $\{Ru(COD)Cl_2\}_n$ under similar conditions to give the corresponding ruthenium(II) tris(α-diimine chelate) complexes.

*Department of Chemistry, Massachusetts Institute of Technology, Cambridge, MA 02139.
†Department of Chemistry, VPI & State University, Blacksburg, VA 24061.

Procedure

■ **Caution.** *2-Methoxyethanol is a teratogen. Appropriate protection for hands, eyes, and clothing should be worn.*

A suspension of dichloro(1,5-cyclooctadiene)ruthenium(II) polymer (0.281 g, 1.00 mmol of Ru, Strem) and 4,7-diphenyl-1,10-phenanthroline (1.01 g, 3.01 mmol, Aldrich) in nitrogen-purged 2-methoxyethanol (100 mL, Aldrich) is heated to reflux under nitrogen in a 250-mL round-bottomed flask equipped with a reflux condenser. After 6 h, the clear dark red-orange solution is cooled to room temperature under nitrogen, filtered, and then concentrated to a dark red oil by rotary evaporation at 65–70°C. The red oil is dissolved in 12 mL of CH_2Cl_2. Then diethyl ether is added slowly until the point of crystallization (~16 mL). The mixture is chilled in a refrigerator overnight at −4°C, and the bright red-orange crystalline product is isolated by suction filtration and washed with diethyl ether. A second crop of red-orange crystals is obtained from the filtrate. The combined product is reprecipitated from CH_2Cl_2 (12–15 mL) with diethyl ether (20 mL) to remove traces of the starting ligand and a purple impurity. The microcrystalline orange powder is isolated by suction filtration, washed with diethyl ether, and dried under vacuum at 50–55°C overnight. Combined yield: 0.90 g (71%).

Anal. Calcd. for $C_{72}H_{48}N_6Cl_2Ru \cdot 6H_2O$: C, 67.71; H, 4.73; N, 6.58; Cl, 5.55. Found: C, 67.60; H, 4.52; N, 6.59; Cl, 5.81.

Properties

The product complex is a bright red-orange crystalline solid that is very soluble in methanol, soluble in dichloromethane, sparingly soluble in acetone, and insoluble in water. ^{1}H NMR ($CDCl_3$, 300 MHz): δ 8.82 (d, 6 H, $J_{HH} = 5.5$ Hz), 8.19 (s, 6 H), 7.92 (d, 6 H, $J_{HH} = 5.5$ Hz), 7.66 (m, 12 H), 7.55 (m, 18 H). ^{13}C NMR ($CDCl_3$, 75.4 MHz): δ 154.0, 148.7, 148.3, 135.8, 130.1, 129.7, 129.1, 128.7, 127.5, 126.1. The complex forms an intense red-orange solution with a bright orange luminescence under irradiation with blue light. The electronic absorption spectrum of the complex in methanol shows peaks at 461 nm ($\varepsilon = 3.60 \times 10^4$ M^{-1} cm^{-1}) and 436 nm ($\varepsilon = 3.55 \times 10^4$ M^{-1} cm^{-1}), and a shoulder at 311 nm ($\varepsilon = 4.6 \times 10^4$ M^{-1} cm^{-1}). The luminescence spectrum of the complex in methanol at room temperature reveals an unsymmetric emission band with a maximum at 613 ± 2 nm ($\lambda_{ex} = 460$ nm). The excited-state lifetimes and quenching kinetics with oxygen for this complex have been determined.[9] The Ru(III)/Ru(II) reduction potential of the complex has been established by cyclic voltammetry.[6]

References

1. J. N. Demas and B. A. DeGraff, *J. Chem. Educ.* **74**, 690 (1997).
2. Y. Zhao, A. Richman, C. Storey, N. B. Radford, and P. Pantano, *Anal. Chem.* **71**, 3887 (1999).
3. J. R. Bacon and J. N. Demas, *Anal. Chem.* **59**, 2780 (1987).
4. K. S. Schanze, B. F. Carroll, and S. Korotkevitch, *AIAA* (American Institute of Aeronautics and Astronautics) *J.* **35**, 306 (1997).
5. M. Gouterman, *J. Chem. Educ.* **74**, 697 (1997).
6. C.-T. Lin, W. Böttcher, M. Chou, C. Creutz, and N. Sutin, *J. Am. Chem. Soc.* **98**, 6536 (1976).
7. R. J. Watts and G. A. Crosby, *J. Am. Chem. Soc.* **93**, 3184 (1971).
8. N. C. Thomas and G. B. Deacon, *Inorg. Synth.* **25**, 107 (1989).
9. D. García-Fresnadillo, Y. Georgiadou, G. Orellana, A. M. Braun, and E. Oliveros, *Helv. Chim. Acta* **79**, 1222 (1996).

15. METAL COMPLEXES OF A DIIMINO (TETRACYANO)PYRROLIZINIDE LIGAND DERIVED FROM TETRACYANOETHYLENE

Submitted by ALBERTO FLAMINI*
Checked by PRASANA GHALSASI,† TRENT D. SELBY,† and JOEL L. MILLER†

The reductive condensation of two molecules of tetracyanoethylene provides an anionic 2*H*-pyrrolyl compound L′ ($C_{11}H_2N_7^-$) by a high-yield, gram-scale reaction.[1] The sodium salt of L′ is soluble in water, whereas the tetraphenylarsonium salt is soluble in organic solvents.[2] Protonation of L′ induces intramolecular cyclization to form an aminoiminopyrrolizine compound HL.[3] Similarly, the interaction of metal ions with L′ (or with HL accompanied by deprotanation) provides metal complexes of the analogous diiminopyrrolizine ligand.[4,5,6] (see Scheme 1). The compounds ML_2 in particular are highly colored tetraaza complexes with interesting comparisons to porphyrin and phthalocyanine complexes.[5] In addition, kinetic and thermodynamic parameters concerning the acid–base equilibria between HL and L′ in aqueous solution in the pH range 0–13 have been determined.[7] Substitution of one hydrogen atom of the amino group in L′ by an aliphatic group R gives the corresponding amino-substituted-L′ derivatives (L′R),[8] and subsequent protonation gives the analogous neutral imino-substituted derivatives (HLR).[9]

*CNR, Istituto di Struttura della Materia, Area della Ricerca di Roma, 00016 Monterotondo Stazione, Roma, Italy.
†Department of Chemistry, University of Utah, Salt Lake City, UT 84112.

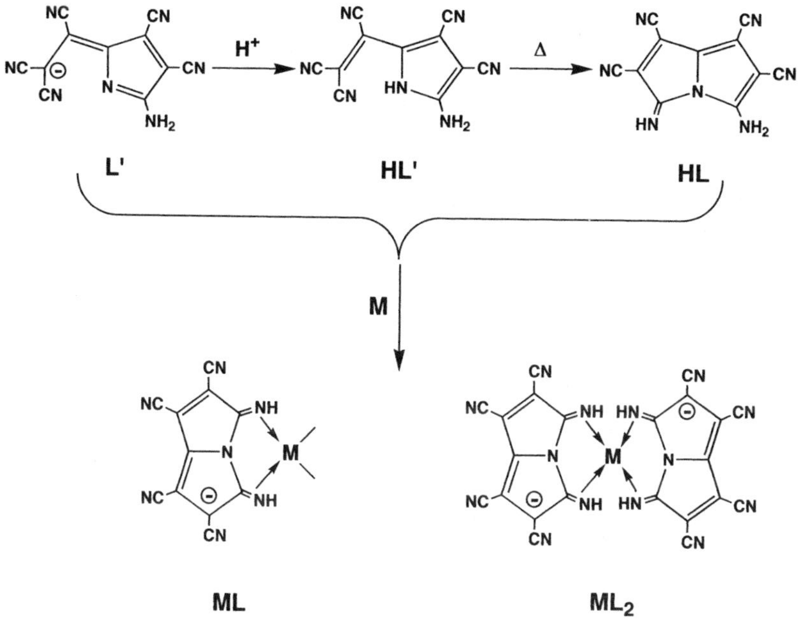

Scheme 1. Coordination reactions of L′ and HL.

The following procedures detail the preparation of L′ as its tetraphenylarsonium and sodium salts, the conversion of L′ to HL, and the synthesis of Ni(II) and Cu(II) complexes of L by using either L′ or HL.

■ **Caution.** *All procedures should be performed in a well-ventilated hood, and protective gloves should be worn.*

A. TETRAPHENYLARSONIUM SALT OF 2-(5-AMINO-3,4-DICYANO-2*H*-PYRROL-2-YLIDENE)-1,1,2-TRICYANOETHANIDE, $[AsPh_4]L'$

$$2Na_2C_6N_4 \xrightarrow[2.\ H_2O]{1.\ SiMe_3Cl} 2NaCl + NaCN + 2SiMe_3OH + C_{11}N_7H_2Na(NaL')$$

$$NaL' + AsPh_4Cl \xrightarrow{H_2O} [AsPh_4]L' + NaCl$$

Procedure

■ **Caution.** *The aqueous filtrate, separated after addition of [$AsPh_4$]Cl, contains NaCN. Appropriate precautions must be taken before discarding it.*

Na_2TCNE·2DME (5.3 g, 15 mmol),[10] weighed in a small Schlenk tube under N_2, is transferred to a 250-mL, two-necked, round-bottomed flask under N_2. Then THF (30 mL, freshly dried and deoxygenated by distilling over sodium benzophenoneketyl under N_2), is added. The flask is equipped with a magnetic stirring bar and a wide neck with a septum rubber stopper, and the solution is then cooled to $-10°C$ under N_2. Me_3SiCl (5.7 mL, 45 mmol,* distilled at atmospheric pressure under N_2 and stored under N_2) is added slowly with stirring by a syringe through the rubber stopper and over a period of ~3 min in order to avoid any increase of temperature (the reaction is mildly exothermic). The suspension is maintained at low temperature for 30 min under N_2 with stirring, left to warm to room temperature, kept at room temperature for 1 h, and finally heated to 80°C for 3 h to assure completion of the reaction. During this time the color of the suspension turns gradually from brown to deep violet, and a precipitate of sodium chloride is formed.

The solvent is removed under vacuum (10^{-1} torr). At this point of the procedure, operating under an inert atmosphere is no longer necessary, and air is allowed to enter the reaction flask. The crude product is dissolved in water (5 × 100 mL). The resulting clear violet solution is collected in a 1-L beaker. Acetone (200 mL) is added, and then a water solution (200 mL) containing a stoichiometric amount of $AsPh_4Cl$ (3.3 g as $AsPh_4Cl \cdot H_2O$, 7.6 mmol) is added slowly with stirring. The solution immediately becomes turbid. The stirring is continued until a microcrystalline golden precipitate forms. The mixture is allowed to stand overnight. The product is isolated by filtration through filter paper, washed with water until the filtrate is essentially colorless, and then dried in the oven at 40°C. Yield 4.2 g, 91% based on Na_2TCNE.

Purification by pressure column chromatography on silicagel is conducted as follows. A glass column (inner diameter 5 cm, length 51 cm) is filled with a slurry of silica (450 g of silicagel, Acros, 0.035–0.070 mm, pore diameter ~6 nm) in 95 : 5 (v : v) CH_2Cl_2/acetone. A 3-g portion of crude [$AsPh_4$]L′ is dissolved in a minimum amount of solvent and adsorbed on the top of the column. The column is connected with a membrane pump (output pressure 1.6 bar) and then eluted at a flow rate of about 180 mL/min. The violet fraction is collected and evaporated on a rotary evaporator. This chromatographic separation on the whole lasts 8 h.

*Although Na_2TCNE and $SiMe_3Cl$ react in the molar ratio 1 : 1, an excess amount of $SiMe_3Cl$ is necessary for the reaction to go to completion, because of the poor solubility of Na_2TCNE in the reaction solvent. The increased amount of $SiMe_3Cl$ improves the yield of NaL′, showing a limiting value when the molar ratio of $SiMe_3Cl$ to Na_2TCNE is ~3.0.

The residue is recrystallized by dissolving in acetone (100 mL) and adding water (60 mL). Slow evaporation in air, until the color of the solution turns from deep violet to orange, leaves large, golden platelike crystals of $[AsPh_4]L'$, which are collected by filtration through filter paper, washed with water, and dried in the oven at 40°C. The product weighs 2.1 g (70%). Purple needle-like crystals can be obtained by rapidly adding water ($\sim$13 mL) to a stirred dilute acetone solution of $AsPh_4L'$ (0.10 g in 10 mL) until a microcrystalline precipitate is formed, and then slowly evaporating the solvent in air.

Anal. Calcd. for $C_{35}H_{22}AsN_7$: C, 68.29; H, 3.60; N, 15.93. Found: C, 68.30; H, 3.90; N, 15.73.

Properties

$[AsPh_4]L'$ is a crystalline, air-stable solid, mp 184°C. It is very soluble in CH_2Cl_2 and $CHCl_3$, soluble in CH_3CN and oxygen containing organic solvents, and sparingly soluble in toluene, insoluble in CCl_4, petroleum ether, and water. UV–vis (acetone): Λ_{max} 550 nm, ($\varepsilon_{550} = 38 \times 10^3\ M^{-1}\ cm^{-1}$). It is best crystallized from a mixture of two solvents by slow evaporation in air. Golden, platelike, monoclinic crystals (form *b*) are readily formed either in anhydrous solvents or in the presence of water, starting from a clear solution by slow evaporation in air. On slightly changing the crystallization conditions in the acetone/water mixture, purple, needle-like, orthorombic crystals (form *a*) are formed. X-ray crystal and molecular structures of both forms *a* and *b* have been solved, and they are polymorphic modifications of the same species $[AsPh_4]L'$.[2] Moreover, forms *a* and *b* are different in other solid-state properties as reported previously: IR (form *b*, KBr): 3387 (s), 2232 (w), 2210 (sh), 2202 (vs), 1616 (s), 1521 (sh), 1510 (vs), 1473 (s), 1436 (s), 1424 (s), 1337 (vs), 1225 (m), 1184 (m), 1165 (w), 1129 (s), 1082 (s), 1022 (w), 997 (m), 788 (w), 747 (sh), 741 (s), 689 (s), 629 (w), 613 (w), 550 (w), 515 (m), 481 (m), 467 (s), 412 (w), 365 (m), 353 (m), 345 (m) cm^{-1}. ^1H NMR ($CDCl_3$) δ 7.57–8.00 (20H, m, Ph) and 4.61 ppm (2H, br, s, NH_2).

B. SODIUM 2-(5-AMINO-3,4-DICYANO-2*H*-PYRROL-2-YLIDENE)-1,1,2-TRICYANOETHANIDE, NaL′

$$[AsPh_4]L' + NaBPh_4 \xrightarrow{\text{acetone}} NaL' + [AsPh_4][BPh_4]$$

Procedure

$[AsPh_4]L'$ (3.0 g, 4.9 mmol) is placed in a 300 mL crystallizing dish and dissolved in acetone (120 mL). Then a solution of $Na[BPh_4]$ (1.66 g, 4.8 mmol) in acetone (80 mL) is added. White crystals of $[AsPh_4][BPh_4]$ immediately form. Acetone is then removed by evaporation under the hood to dryness over a period of 3 h. The residue is dissolved in water (3 × 100 mL) and filtered through a fritted-glass filter. The filtrate is evaporated to dryness in a vacuum freeze dryer, and the residue is placed in a vacuum dessicator (25°C, 24 h) over anhydrous $CaCl_2$. $NaL' \cdot 2H_2O$ is obtained as microcrystalline deep green powder (1.3 g, 92%).

Properties

NaL′ is very soluble in water and stable in the pH range 6–12. It is also soluble in oxygen containing organic solvents and CH_3CN but insoluble in CH_2Cl_2 and toluene. It may contain a variable amount of water depending on the specific preparation. Its water content is best determined by measuring the maximum absorbance at 550 nm. UV–vis (water): Λ_{max} 550 nm, (ε_{550} 34×10^3 M^{-1} cm^{-1}). IR (KBr): 3343 (m), 2221 (vs), 1643 (s), 1537 (vs), 1491 (m), 1422 (m), 1350 (vs), 1262 (w), 1234 (w), 1139 (w), 1078 (w), 798 (w), 629 (w), 517 (w), 487 (w) cm^{-1}. ^{13}C NMR (C_2D_5OD): δ 67.76, 94.62, 108.91, 114.43, 115.02, 115.37, 117.68, 117.84, 121.79, 137.34, and 161.10 ppm.

C. 5-AMINO-3-IMINO-1,2,6,7-TETRACYANO-3*H*-PYRROLIZINE, HL

Procedure

$NaL' \cdot 2\ H_2O$ (0.5 g, 1.7 mmol) is dissolved in water (100 mL). Concentrated aqueous HCl (11 M, 3 mL) is added dropwise with stirring. Immediately, a deep green crystalline solid separates, and the color of the solution changes from violet (Λ_{max} 550 nm) to red-violet (Λ_{max} 523 nm) on the formation in solution of HL′, the 2-tricyanovinyl-3,4-dicyano-5-amino-1*H*-pyrrole.[7] The solid (HL′ + HL) is then collected on filter paper, washed with water (3 × 10 mL), dried in air, and then heated in an oven at 70°C for 24 h. During this time the complete HL′ → HL transformation occurs in the solid state, and pure HL is obtained (0.38 g, 95%).

Anal. Calcd. for $C_{11}H_3N_7 \cdot 0.7H_2O$: C, 53.75; H, 1.80; N, 39.89. Found: C, 54.20; H, 1.73; N, 39.23.

Properties

HL is air-stable in the solid state (mp 270°C dec.) and in solution in the absence of water. It is very soluble in oxygen containing organic solvents and CH_3CN, sparingly soluble in CH_2Cl_2 and toluene, and minimally soluble in water, where it undergoes the dissociation HL $\rightarrow$ L$'$ + H^+. UV–vis (tetrahydrofuran): Λ_{max} 580 nm ($\varepsilon_{580} = 20 \times 10^3\ M^{-1}\ cm^{-1}$), sh 545 nm. IR (KBr): 3374 (m), 3247 (s), 3211 (m), 3145 (m), 2233 (vs), 1649 (s), 1584 (s), 1547 (m), 1511 (s), 1477 (m), 1410 (m), 1360 (s), 1255 (s), 1167 (m), 1022 (m), 924 (w), 751 (w), 720 (w), 677 (w), 596 (w), 574 (w), 540 (w), 507 (w), 473 (w), 436 (w), 373 (w) cm^{-1}. ^{13}C NMR ($D_3COCD_2CD_2OCD_3$): δ 110.8, 111.8, 112.35, 112.7, 113.1, 114.5, 114.8, 123.2, 126.6, 152.15, 154.9 ppm. HL sublimes under vacuum (200°C, 10^{-6} mmHg). The X-ray crystal and molecular structure of its 2 : 1 1-chloronaphthalene adduct has been solved.[3]

D. BIS(1,2,6,7-TETRACYANO-3,5-DIIMINO-3,5-DIHYDROPYRROLIZINIDO)NICKEL(II), NiL_2

$$2L' + Ni^{+2} \rightleftharpoons [LNi]^+ + L' \rightleftharpoons NiL_2$$

Procedure

In a 100-mL beaker, a solution of $NaL' \cdot 2H_2O$ (0.2 g, 0.7 mmol) in acetone (20 mL) is added dropwise with stirring to a solution of $NiCl_2 \cdot 6H_2O$ (2.0 g, 8.4 mmol) in water (50 mL). The color of the solution changes slowly from violet to blue on formation of the pyrrolizinato nickel(II) complexes, namely, the water-soluble chloride salt of $[NiL]^+$ as well as NiL_2. The stirring is continued until a microcrystalline black solid (NiL_2) forms. The beaker is left open in air until the volume of the solution is reduced to 50 mL. The precipitate is collected on filter paper, washed with water, and dried in air. The collected solid ($NiL_2 \cdot xH_2O$, 0.17 g) is stirred with 1,4-dioxane/water (9.9/0.1, 10 mL). The color of the solid immediately turns from black to purple on the formation of new crystals of the NiL_2 complex containing coordinated dioxane (diox) and water, of composition $NiL_2 \cdot 2H_2O \cdot 3$diox. The crystals are filtered, washed with dioxane, and dried in air at room temperature (0.16 g, 57%).*

Anal. Calcd. for $C_{32.4}H_{29.6}N_{14}NiO_{7.6}$: C, 48.93; H, 3.75; N, 24.65; Ni, 7.38. Found: C, 48.84; H, 3.92; N, 24.55; Ni, 7.13.

*The composition of this complex in bulk could be different from the composition resulting from X-ray studies on single crystals ($NiL_2 \cdot 2H_2O \cdot 3$diox), depending on the relative amount of dioxane/water. In the present case the elemental analysis values fit in with the formula $NiL_2 \cdot 2.4\ H_2O \cdot 2.6$ diox.

Properties

The X-ray crystal and molecular structure of $NiL_2 \cdot 2H_2O \cdot 3diox$ has been solved.[11] The nickel atom is six-coordinated in an octahedral NiN_4O_2 geometry, where the oxygen atoms derive from the two water molecules in the apical positions; the dioxane molecules are connected via hydrogen bonds to water and to the imino groups of L. This complex is stable in the solid state at room temperature. On heating, it loses the solvent almost completely (60–290°C) as shown by TGA measurements. It is insoluble in water and soluble in organic polar solvents: MeOH = EtOH > Me_2CO > MeCN > THF. It reacts with Ni(II) ion in organic solutions to afford the corresponding monopyrrolizinato Ni(II) complex.[6] UV–vis (MeOH, Λ_{max} in nm, ε in 10^{-3} M^{-1} cm^{-1}): 654 (81), 602 (55), 567 (14.5), 409 (23), 276 (25.7), 595 (sh), 547 (sh), 387(sh), 300 (sh). $\mu_{eff} = 3.26$ μ_B (295 K). IR (KBr): 3405 (m, br), 3289 (m), 3239 (m), 2973 (w), 2925 (w), 2855 (w), 2232 (sh), 2197 (vs), 1640 (s), 1574 (s), 1457 (s), 1386 (s), 1365 (sh), 1288 (w), 1254 (s), 1247 (s), 1222 (sh), 1198 (sh), 1119 (s), 1084 (m), 898 (w), 871 (s), 831 (w), 757 (m), 714 (s), 688 (w), 622 (w), 568 (m), 504 (w), 468 (m, br) cm^{-1}.

E. BIS(1,2,6,7-TETRACYANO-3,5-DIIMINO-3,5-DIHYDROPYRROLIZINIDO)COPPER(II), CuL_2

$$2[AsPh_4]L' + CuCl_2 \rightarrow 2[AsPh_4]Cl + CuL_2$$

Procedure

$[AsPh_4]L'$ (1.0 g, 1.6 mmol) is placed in a 200-mL crystallizing dish and dissolved in acetone (100 mL). Then a solution of $CuCl_2 \cdot 2H_2O$ (1.50 g, 8.8 mmol) in water (10 mL) is added with stirring. The color of the solution changes immediately from violet to blue, and a microcrystalline purple solid (CuL_2) starts to precipitate. The solution is evaporated to dryness under the hood. The residue is dissolved in water, the solution filtered through a fritted-glass filter, and the collected solid washed with water and then dried in air ($CuL_2 \cdot xH_2O$, 0.7 g). The crude product is recrystallized by dissolution in boiling THF (250 mL), followed by filtration, addition of toluene (70 mL), and slow evaporation in air until the final volume of solution is 90 mL. The purple crystals formed ($CuL_2 \cdot 2THF$) are collected by suction filtration and dried in air (0.42 g, 77%).

Anal. Calcd. for $C_{30}H_{20}N_{14}CuO_2$: C, 53.62; H, 3.00; N, 29.17; Cu, 9.45; O, 4.76. Found: C, 53.46; H, 3.03; N, 29.36; Cu, 9.60; O, 4.82.

Properties

The X-ray crystal and molecular structure of $CuL_2 \cdot 2THF$ has been solved.[4] The copper atom is surrounded by tro chelating imino groups in a square planar arrangement and completes its tetragonally elongated octahedral geometry with two axial cyano groups from two equivalent complex units. The THF molecules are connected via hydrogen bonds to the imino groups of L in the molecular plane. Solid-state and solution properties are comparable to those of the NiL_2 complex. UV–vis (THF, Λ_{max} in nm ε in 10^{-3} M^{-1} cm^{-1}): 653 (44.7), 620 (42.6), 394 (21.4), 272 (30.2), 600 (sh), 570 (sh), 375(sh), 292 (sh). μ_{eff} = 1.85 μ_B (290 K). IR (KBr): 3296 (m), 3242 (m), 2973 (w), 2220 (vs), 1641 (s), 1566 (s), 1476 (m), 1399 (m), 1374 (m), 1302 (w), 1247 (m), 1024 (m), 780 (w), 752 (w), 722 (m), 692 (w), 583 (m), 494 (w), 467 (m) cm^{-1}.

F. (1,2,6,7-TETRACYANO-3,5-DIIMINO-3,5-DIHYDROPYRROLIZINIDO)(DIPIVALOYLMETHANIDO) NICKEL(II), NiL(DPM)

$$Ni(DPM)_2 + HL \rightarrow DPMH + NiL(DPM)$$

Procedure

Bis(dipivaloylmethanido)Ni(II) (supplied by Strem or synthesized according to the procedure reported by Hammond[12]), $Ni(DPM)_2$ (0.420 g, 1.0 mmol), is dissolved in methanol (60 mL) and HL (0.174 g, 0.75 mmol) added. After 1 h stirring, a gray microcrystalline solid precipitates, NiL(DPM). After filtration, the solid is washed with methanol and dried in air. Yield 0.313 g (88%).

Anal. Calcd. for $C_{22}H_{21}N_7NiO_2$: C, 55.73; H, 4.46; N, 20.68. Found: C, 55.40; H, 4.42; N, 20.52.

Properties

The X-ray crystal and molecular structure of NiL(DPM) has been solved.[6] The nickel atom is four-coordinated as NiN_2O_2 in a square planar geometry with two imino groups from the pyrrolizinido ligand (L) and two oxygen atoms from DPM. This complex is stable in the solid state and also in solution even toward the ligand metathesis so that it can be recrystallized from hot saturated solutions. It shows exceptional solvatochromism; its color turns from pale green in an uncoordinating solvent (CH_2Cl_2 or acetone) to intense blue in a coordinating one

(THF) or on adding a coordinating species to the uncoordinating solvents, such as pyridine to CH_2Cl_2 or water to acetone.[6] The solid compound is diamagnetic. ^1H NMR (CD_2Cl_2) δ 5.75 (s, CH) and 1.16 ppm (s, CH_3). UV–vis (Λ_{max} in nm, ε in 10^{-3} M^{-1} cm^{-1}, CH_2Cl_2): 758 (11.5), 684 (6.0), 445 (10.0), 256 (28.0), 622 (sh), 485 (sh), 405 (sh), 342 (sh), 312 (sh), 282 (sh); (THF): 671 (44.7), 612 (12.9), 409 (8.7), 306 (16.2), 651 (sh), 587 (sh), 564 (sh). IR (KBr): 3238 (s), 2978 (m), 2969 (m), 2222 (vs), 2172 (w, br), 1627 (s), 1577 (m), 1550 (vs), 1504 (m), 1472 (s), 1406 (vs), 1376 (m), 1366 (m), 1270 (s), 1249 (s), 1230 (sh), 1197 (w), 1182 (w), 1161 (m), 1086 (w), 1023 (s), 996 (m), 975 (w), 947 (w), 935 (w), 878 (m), 852 (m), 844 (m), 808 (m), 758 (w), 734 (s), 720 (m), 657 (w), 585 (m), 525 (w), 497 (m), 477 (m) cm^{-1}.

References

1. A. Flamini and N. Poli, U.S. Patent 5,151,527 (1992).
2. V. Fares, A. Flamini, and N. Poli, *J. Chem. Res.* (S) 228 (1995).
3. V. Fares, A. Flamini, and N. Poli, *J. Am. Chem. Soc.* **117**, 11580 (1995).
4. M. Bonamico, V. Fares, A. Flamini, P. Imperatori, and N. Poli, *Angew. Chem., Int. Ed. Engl.* **28**, 1049 (1989).
5. C. B. Hoffman, V. Fares, A. Flamini, and R. L. Musselman, *Inorg. Chem.* **38**, 5742 (1999).
6. A. Flamini, V. Fares, and A. Pifferi, *Eur. J. Inorg. Chem.* 537 (2000).
7. E. Collange, A. Flamini, and R. Poli, *J. Phys. Chem. A* **106**, 200 (2002).
8. V. Fares, A. Flamini, and P. Pasetto, *J. Chem. Soc., Perkin I* 4520 (2000).
9. A. Flamini, V. Fares, A. Capobianchi, and V. Valentini, *J. Chem. Soc., Perkin I* 3069 (2001).
10. O. E. Webster, W. Mahler, and R. E. Benson, *J. Am. Chem. Soc.* **94**, 3678 (1962).
11. M. Bonamico, V. Fares, A. Flamini, and N. Poli, *Inorg. Chem.* **30**, 3081 (1991).
12. G. S. Hammond, D. C. Nonhebel, and C. S. Wu, *Inorg. Chem.* **2**, 73 (1963).

16. CATIONIC Pt(II) COMPLEXES OF TRIDENTATE AMINE LIGANDS

Submitted by GIULIANO ANNIBALE[*] and BRUNO PITTERI[*]
Checked by MICHAEL H. WILSON[†] and DAVID MCMILLIN[†]

Platinum(II) complexes containing the tridentate ligands diethylenetriamine (dien) and 2,2′:6′,2″-terpyridine (terpy) have provided useful substrates for mechanistic studies in substitution reactions at planar four-coordinated complexes. Moreover, there has been considerable interest in terpy complexes

[*]Departimento di Chimica Universitá di "CáFoscari" di Venezia, Calle Larga S. Marta 2137, 30123, Venezia, Italy.
[†]Department of Chemistry, Purdue University, West Lafayette, IN 47907.

because of their properties as intercalating metalloreagents to nucleic acids[1] as well as their intriguing spectroscopic and photophysical behavior.[2] For Pt(II) complexes containing dien and its methyl derivatives, such as *N,N,N′,N″,N″*-pentamethyldiethylenetriamine (Me_5dien), the interest is focused mainly on their interactions with nucleotide bases as analogs of the anticancer drug *cis*-$[PtCl_2(NH_3)_2]$.[3] The reported procedures for the preparation of the platinum(II) chloro derivatives [Pt(terpy)Cl]Cl·nH_2O ($n = 2,3$) (**1**),[4-7] [Pt(dien)Cl]Cl (**2**),[8] and [Pt (Me_5dien)Cl]Cl (**3**)[3] are in general rather laborious, lengthy, and with variable yields. Herein we present simpler methods that rapidly give pure **1–3** in high yield. The starting precursors, *cis*-$[Pt(dmso)_2Cl_2]$ (dmso = dimethylsulfoxide),[9] *cis/trans*-$[PtCl_2(Me_2S)_2]$,[10] and $[Pt(cod)Cl_2]$[11] (cod = 1,5-cyclooctadiene), are easily prepared in nearly quantitative yield by well established procedures.

■ **Caution.** *The following procedures release dimethyl sulfide or 1,5-cycloctadiene, which are malodorous, toxic, and flammable. All operations should be carried out in a well-ventilated fume hood.*

A. CHLORO(2,2′:6′,2″-TERPYRIDINE)PLATINUM(II) CHLORIDE DIHYDRATE, [Pt(terpy)Cl]Cl·2H₂O

Method A

$$cis/trans\text{-}[PtCl_2(SMe_2)_2] + \text{terpy} \rightarrow [Pt(\text{terpy})Cl]Cl \cdot 2H_2O + 2SMe_2$$

The synthesis of *cis/trans*-$PtCl_2(SMe_2)_2$ has been reported by Hill et al.[10] and, apart from the time employed to dry the CH_2Cl_2 solution with anhydrous $MgSO_4$, it takes about 1 h.

Procedure

$PtCl_2(SMe_2)_2$ (0.390 g, 1 mmol) is dissolved with stirring in a warm (55–60°C) methanolic–water (90/10, v/v) solution (20 mL) in a 100 mL, round-bottomed, two-necked glass flask equipped with a water-cooled reflux condenser. Terpy (Sigma–Aldrich; 0.245 g, 1.05 mmol) in methanol (2 mL) is added, and the reaction mixture is heated to 80°C using a thermostated bath and allowed to react for ~35–40 min. During this time, the color of the solution changes from bright yellow to orange. The solvent is then rapidly removed from the reaction mixture under vacuum at 70°C, and the red-orange residue is partially redissolved in boiling methanol (30 mL). The treatment with methanol and diethyl ether serves to remove the slight excess of terpy used to ensure the completeness of the reaction. After cooling the solution to room temperature and addition of diethyl ether (50 mL), the resulting yellow crystalline precipitate is washed twice with aliquots (10 mL) of diethyl ether and dried in vacuo. The dry product turns red-orange. Yield: 0.508 g (95%).

Method B[12]

$$[Pt(cod)Cl_2] + terpy + 2H_2O \rightarrow [Pt(terpy)Cl]Cl \cdot 2H_2O + cod$$

Procedure

To a suspension of $Pt(cod)Cl_2$ (0.15 g, 0.40 mmol) in water (10 mL), terpy (0.098 g, 0.42 mmol) is added with stirring, and the mixture is warmed at 60–70°C. Over 15 min all the solid dissolves, and a red-orange clear solution is obtained and is then cooled to room temperature. Water is removed under reduced pressure, leaving a red-orange solid that is collected, washed thoroughly with diethyl ether, and air-dried. Yield: 0.203 g, 95%.

Anal. Calcd. for $[Pt(terpy)Cl]Cl \cdot 2H_2O$: C, 33.7; H, 2.82, N, 7.85; Cl, 13.2. Found: C, 33.7; H, 2.62; N, 7.57; Cl, 13.5.

Properties

The complex $[Pt(terpy)Cl]Cl \cdot 2H_2O$ is obtained as a red-orange, needle-shaped microcrystalline powder. It is soluble in water, nitromethane, and dimethysulfoxide and moderately soluble in alcohols. The electronic spectrum in water has absorption maxima at 248 ($\varepsilon = 31{,}680\ dm^{-3}\ mol^{-1}\ cm^{-1}$), 278 (27,360) and 327 (14,000) nm. The Pt—Cl stretching vibration occurs at 344 cm^{-1} (polyethylene pellet). The ^1H NMR (D_2O) spectrum exhibits peaks δ 7.84–8.00 (m, 3H), 7.75–7.56 (m, 6H), and 7.18 (ddd, 2H, $J = 7.7, 5.8, 1.6$ Hz).

B. CHLORO(DIETHYLENETRIAMINE)PLATINUM(II) CHLORIDE, [Pt(dien)Cl]Cl

Method A

$$cis/trans\text{-}PtCl_2(SMe_2)_2 + dien \rightarrow [Pt(dien)Cl]Cl + 2SMe_2$$

Procedure

cis/trans-[$PtCl_2(SMe_2)_2$] (0.390 g, 1 mmol) is dissolved with stirring in a warm (55–60°C) methanolic solution (20 mL) contained in a 100-mL, round-bottomed, two-necked glass flask equipped with a water-cooled reflux condenser. Diethylenetriamine (Sigma–Aldrich; 0.112 g, 1.05 mmol) in methanol (2 mL) is added and the reaction mixture is refluxed for 2 min. The resulting colorless solution is concentrated in a rotary evaporator to ~2 mL. Then CH_2Cl_2 (120 mL) is added with stirring to precipitate the product as a gelatinous white solid, which becomes powdery on cooling the flask in an ice bath for 10 min. The solid is collected by filtration, washed with CH_2Cl_2, and dried in vacuo over P_2O_5. Yield: 0.347 g (93%).

Method B[12]

$$cis\text{-Pt(dmso)}_2\text{Cl}_2 + \text{dien} \rightarrow [\text{Pt(dien)Cl}]\text{Cl} + 2\text{dmso}$$

Procedure

A solution of diethylenetriamine (0.077 g, 0.74 mmol) in methanol (10 mL) is added with stirring to a suspension of *cis*-$Pt(dmso)_2Cl_2$ (0.30 g, 0.71 mmol) in methanol (50 mL), and the mixture is heated under reflux for ~1 h. The resulting clear solution is concentrated in a rotary evaporator to ~5 mL and then is cooled to room temperature. Chloroform (60 mL) is added with stirring to precipitate the product as a gelatinous white solid that becomes powdery on cooling the flask in an ice bath for 10 min. The solid is filtered off, washed first with chloroform and then with diethyl ether, and air-dried. Yield: 0.254 g, 96%. The product is recrystallized from an aqueous solution (2 mL) by careful addition of acetone. The precipitate is collected, washed first with acetone and then with diethyl ether, and finally dried in vacuo over P_2O_5.

Anal. Calcd. for [Pt(dien)Cl]Cl: C, 13.0; H, 3.55, N, 11.4; Cl,19.2. Found: C, 13.2; H, 3.22; N, 14.1; Cl, 19.0.

Properties

The complex [Pt(dien)Cl]Cl is obtained as a creamy white powder. The product is slightly hygroscopic and should be stored in a dessicator. It is very soluble in most common solvents. The ^1H NMR spectrum (CD_3OD) exhibits a singlet at 2.61 (s, 4H, *J*(Pt–H) = 44 Hz), and broad peaks at δ 2.85, 3.04, and 3.24 (8H). The Pt–Cl IR stretching band is at 336 cm^{-1} (polyethylene pellet).

C. CHLORO(*N,N,N′,N″,N″*-PENTAMETHYLDIETHYLENE-TRIAMINE)PLATINUM(II) CHLORIDE, $[Pt(Me_5dien)Cl]Cl$

$$cis/trans\text{-}PtCl_2(SMe_2)_2 + Me_5dien \rightarrow [Pt(Me_5dien)Cl]Cl + 2SMe_2$$

cis/trans-$PtCl_2(SMe_2)_2$ (0.390 g, 1 mmol) is dissolved with stirring in warm (55–60°C) methanol (20 mL) in a 100-mL, round-bottomed, two-necked glass flask equipped with a water-cooled reflux condenser. Me_5dien (Sigma–Aldrich; 0.182 g, 1.05 mmol) in methanol (2 mL) is added, and the reaction mixture is heated under reflux for 10 min. The resulting deep yellow solution is concentrated in a rotary evaporator to ~2 mL. Then diethyl ether (60 mL) is added to precipitate the product as a pasty cream solid, which becomes powdery on further stirring. The precipitate is collected by filtration, washed with diethyl ether, and then dried in vacuo over P_2O_5. Yield: 0.395 g (90%).

Anal. Calcd. for $[Pt(Me_5dien)Cl]Cl$: C, 24.6; H, 5.27, N, 9.56; Cl,16.1. Found: C, 24.6; H, 5.13; N, 10.1; Cl, 16.0.

Properties

The complex $[Pt(Me_5dien)Cl]Cl$ is obtained as a white-cream, very hygroscopic, powder. It is very soluble in the most common solvents. The ^1H NMR (CD_3NO_2) spectrum shows peaks at δ 2.84 (6H, s, terminal syn N—CH_3), and 3.03 (6H, s, terminal anti N—CH_3), and 3.16 (3H, s, central CH_3). The multiplets due to the methylene protons (8H) fall in the range δ 2.99–3.81. The Pt—Cl stretching vibration occurs at 342 cm^{-1} in a polyethylene pellet.

[*Note*: The same procedure proved to be successful for the synthesis of $[Pt(Me_3dien)Cl]Cl$ when isomerically pure N,N',N''-trimethyldiethylenetriamine is used as the ligand. This is not commercially available at this time but can be prepared by the procedure reported in Ref. 13.].

References

1. E. C. Constable, *Adv. Inorg.Chem. Radiochem.* **30**, 69 (1986).
2. H-K. Yip, L-K. Cheng, K-K. Cheung, and C-M. Che, *J. Chem. Soc., Dalton Trans.* 2933 (1993) and refs. therein.
3. M. Carlone, F. P. Fanizzi, F. P. Intini, N. Margiotta, L. G. Marzilli, and G. Natile, *Inorg. Chem.* **39**, 634 (2000).
4. G. T. Morgan and F. H. Burstall, *J. Chem. Soc.* 1499 (1934).
5. R. J. Mureinik and M. Bidani, *Inorg. Nucl. Chem. Lett.* **13**, 625 (1977).
6. M. H. Grant and S. J. Lippard, *Inorg. Synth.* **20**, 101 (1980).
7. D. R. Baghurst, S.R. Cooper, D. L. Greene, D. M. P. Mingos, and S. M. Reynolds, *Polyhedron* **9**, 893 (1990).

8. G. W. Watt and W. A. Cude, *Inorg. Chem.* **7**, 335 (1968).
9. (a) J. H. Price, A. S. Williamson, R. F. Schramm, and B. B. Wayland, *Inorg. Chem.* **11**, 1280 (1972); (b) R. Romeo and L. Monsú Scolaro, *Inorg. Synth.* **32**, 154 (1998).
10. G. S. Hill, M. J. Irwin, C. J. Levy, and L. M. Rendina, *Inorg. Synth.* **32**, 149 (1998).
11. (a) J. X. Mc Dermott, J. F. White, and G. M. Whitesides, *J. Am. Chem. Soc.* **98**, 6521 (1976); (b) D. Drew and J. R. Doyle, *Inorg. Synth.* **28**, 346 (1990).
12. G. Annibale, M. Brandolisio, and B. Pitteri, *Polyhedron* **14**, 451 (1995).
13. A. Bencini, A. Bianchi, E. Garcia-Espana, V. Fusi, M. Micheloni, P. Paoletti, J. A. Ramirez, A. Rodriguez, and B. Valtancoli, *J. Chem. Soc., Perkin Trans. 2* 1059 (1992).

17. PLATINUM(II) COMPLEXES OF PROPANONE OXIME

Submitted by VADIM Yu. KUKUSHKIN,[*] YOULIA A. IZOTOVA,[†] and DAVID TUDELA[‡]
Checked by MARIA F. C. GUEDES da SILVA[§] and ARMANDO J. L. POMBEIRO[§]

Complexes involving oxime ligands display a variety of reactivity modes that lead to unusual types of chemical compounds.[1] As far as the oxime chemistry of platinum is concerned, these complexes are involved in facile deprotonation of the OH group with formation of oximato complexes,[1,2] reduction of Pt(IV) species,[3] Pt(II)-assisted reactions with coordinated allene,[4] alkylation by ketones,[5] oxime-ligand-supported stabilization of Pt(III)—Pt(III) compounds,[6] oxidative conversion into rare nitrosoalkane platinum(II) species,[7,8] and coupling with organocyanamides.[9]

In view of the synthetic utility of propanone oxime complexes of platinum(II),[4,5,7–9] detailed syntheses of these starting materials should be available. The complexes *cis*- and *trans*-$PtCl_2\{(CH_3)_2C{=}NOH\}_2$ were prepared by Babaeva and Mosyagina,[10] but synthetic methods were reported schematically, and only a few physicochemical properties were given. Herein we describe procedures for the syntheses of three Pt(II)propanone oxime compounds. The reaction of $K_2[PtCl_4]$ with two equivalents of propanone oxime results in precipitation of *cis*-$PtCl_2\{(CH_3)_2C{=}NOH\}_2$ and concomitant generation in solution of $[PtCl_3\{(CH_3)_2C{=}NOH\}]^-$.[11] The latter can be precipitated as the

*Department of Chemistry, St. Petersburg State University, 198904 Stary Petergof, Russian Federation.
†St. Petersburg State Technological Institute, Zagorodny Pr. 49, 198013, St. Petersburg, Russian Federation.
‡Departamento de Química Inorgánica, Universidad Autónoma de Madrid, 28049, Madrid, Spain.
§Centro de Química Estrutural, Complexo I, Instituto Superior Técnico, Av. Rovisco Pais, 1049-001, Lisbon, Portugal.

tetra(*n*-butyl)ammonium salt. The *cis*-complex quantitatively isomerizes to *trans*-$PtCl_2\{(CH_3)_2C{=}NOH\}_2$ on heating in the solid state. An alternative method for the preparation of *trans*-$PtCl_2\{(CH_3)_2C{=}NOH\}_2$ involves heating $[PtCl\{(CH_3)_2C{=}NOH\}_3]Cl$[12] in aqueous HCl [10,13,14] or thermal extrusion of the propanone oxime ligand from solid $[PtCl\{(CH_3)_2C{=}NOH\}_3]Cl$.[12]

■ **Caution.** *Potassium tetrachloroplatinate(II) is a known sensitizing agent, and tetra(n-butyl)ammonium hydrogen sulfate is an irritant. All organic solvents used are toxic, and most of them flammable. Inhalation or contact with them should be avoided. Appropriate precautions must be taken, and an efficient hood must be used.*

A. *cis*-DICHLOROBIS(PROPANONEOXIME)PLATINUM(II) AND TETRA(*n*-BUTYL)AMMONIUM TRICHLORO (PROPANONEOXIME) PLATINATE(II), *cis*-$PtCl_2\{(CH_3)_2C{=}NOH\}_2$ AND $[(n\text{-}C_4H_9)_4N][PtCl_3\{(CH_3)_2C{=}NOH\}]$

$$K_2[PtCl_4] + (CH_3)_2C{=}NOH \rightarrow K[PtCl_3\{(CH_3)_2C{=}NOH\}] + KCl$$

$$K[PtCl_3\{(CH_3)_2C{=}NOH\}] + (CH_3)_2C{=}NOH \rightarrow$$
$$cis\text{-}PtCl_2\{(CH_3)_2C{=}NOH\}_2 + KCl$$

$$K[PtCl_3\{(CH_3)_2C{=}NOH\}] + [(n\text{-}C_4H_9)_4N][HSO_4] \rightarrow$$
$$[(n\text{-}C_4H_9)_4N][PtCl_3\{(CH_3)_2C{=}NOH\}] + KHSO_4$$

In a 50-mL beaker, propanone oxime, $(CH_3)_2C{=}NOH$, (1.07 g, 14.6 mmol) is added to a warm (40–45°C) solution of 3.02 g (7.3 mmol) of $K_2[PtCl_4]$ in 15 mL of water. The mixture is stirred at that temperature for 15 min, and then the orange-red solution is cooled to 20–25°C and left to stand for 24 h at room temperature. The lemon-yellow crystalline precipitate of *cis*-$PtCl_2\{(CH_3)_2C{=}NOH\}_2$ is collected on filter paper, washed with three 1-mL portions of water, and dried in air at room temperature. Yield of *cis*-$PtCl_2\{(CH_3)_2C{=}NOH\}_2$ is 2.02 g, 67%, based on the starting $K_2[PtCl_4]$. The filtrate and the washwater are combined and 1.15 g (3.4 mmol) of solid $[(n\text{-}C_4H_9)_4N][HSO_4]$ is added to this solution. The oily precipitate that appears immediately after the addition crystallizes on stirring with a glass rod. The precipitate of $[(n\text{-}C_4H_9)_4N][PtCl_3\{(CH_3)_2C{=}NOH\}]$ is collected on filter paper, washed with two 1-mL portions of water, and dried in air at room temperature. The yield of $[(n\text{-}C_4H_9)_4N]$

$[PtCl_3\{(CH_3)_2C{=}NOH\}]$ is 0.81 g, 18%, based on the starting $K_2[PtCl_4]$. This compound can be recrystallized from hot (80–90°C) water (1 g in 18–20 mL).

Anal. Calcd. for $PtCl_2\{(CH_3)_2C{=}NOH\}_2$: C 17.48; H 3.42; N 6.80. Found: C 17.42; H 3.38; N 6.82. Calcd. for $[(n\text{-}C_4H_9)_4N][PtCl_3\{(CH_3)_2C{=}NOH\}]$: C 36.99; H 7.02; N 4.54. Found: C 37.00; H 7.19; N 4.42.

Properties

cis-$PtCl_2\{(CH_3)_2C{=}NOH\}_2$ forms rodlike lemon-yellow crystals from water, mp 175–180°C (dec.) (lit. mp 170–175°C[10]). The DTA/TG at a heating rate of 2°C/min displays a sharp exothermic peak at 180–185°C (mass loss ~6%) and another exothermic peak at 220°C (mass loss ~9%).[8] The complex is soluble in acetone, nitromethane, and dimethylformamide; poorly soluble in water (solubility in H_2O at 25°C is 1.16 mass%);[14] and insoluble in diethyl ether and toluene. *cis*-$PtCl_2\{(CH_3)_2C{=}NOH\}_2$ isomerizes in a few days into the trans isomer in an acetone or nitromethane solution at room temperature. IR data (Nujol, cm^{-1}: 1662 (m) $\nu_{C=N}$; 340 (m-s) and 326 (m-s) ν_{Pt-Cl}. 300 MHz ^1H NMR in acetone-d_6: δ 2.20 (3H), 2.66 (3H), and 9.9 broad (OH), J_{PtH} coupling constants are not well resolved [for free $Me_2C{=}NOH$, 300 MHz ^1H NMR in acetone-d_6, δ 1.78, 1.79 (CH_3), and 9.34 (OH)]. $^{13}C\{^1H\}$ NMR in DMF-d_7: δ 18.55 ($^3J_{Pt-C} = 22$ Hz, CH_3), 24.68 ($^3J_{Pt-C} = 25$ Hz, CH_3) and 166.94 ($^2J_{Pt-C} = 52$ Hz, C=N).[8] $^{195}Pt\{^1H\}$ NMR in DMF-d_7: –2100 ppm.[8] EI (MS), *m/z* 412 $(M)^+$. Kurnakov's test of *cis*-$PtCl_2\{(CH_3)_2C{=}NOH\}_2$ with thiourea in aqueous media gives $[Pt\{(NH_2)_2CS)\}_4]Cl_2$, additionally proving the cis configuration of the starting material.[14]

$[(n\text{-}C_4H_9)_4N][PtCl_3\{(CH_3)_2C{=}NOH\}]$ crystallizes from water as orange needle-like crystals in the orthorhombic space group Pbcn with $a = 13.773(3)$, $b = 21.794(5)$, $c = 8.7460(10)$ Å, $V = 2625.3(9)$ Å^3, $Z = 4$, $\rho_{calcd.} = 1.561$ Mg·m^{-3}; mp = 133°C. 300 MHz ^1H NMR in D_2O: δ 2.26 (3H) and 2.75 (3H). IR (Nujol, cm^{-1}): 1662 (vw), $\nu_{C=N}$; 350 (m), 335 (m-s), and 325 (s), ν_{Pt-Cl}. FAB(–)-MS, *m/z* 374, $[PtCl_3(Me_2C{=}NOH)]^-$).

B. *trans*-DICHLOROBIS(PROPANONEOXIME)PLATINUM(II), $PtCl_2\{(CH_3)_2C{=}NOH\}_2$

$$cis\text{-}PtCl_2\{(CH_3)_2C{=}NOH\}_2 \xrightarrow[\text{solid}]{\Delta} trans\text{-}PtCl_2\{(CH_3)_2C{=}NOH\}_2$$

Finely powdered *cis*-$PtCl_2\{(CH_3)_2C{=}NOH\}_2$ (1.00 g, 2.4 mmol) is spread as a uniform thin layer on a petri dish and placed into an air thermostat or oven at

140°C for up to 4 h to complete conversion into *trans*-$PtCl_2\{(CH_3)_2C{=}NOH\}_2$. The process of cis–trans isomerization can be monitored by TLC. The yield of *trans*-$PtCl_2\{(CH_3)_2C{=}NOH\}_2$ is almost quantitative. The beige product can be recrystallized from a boiling acetone/water mixture (3 : 4 in v/v, ~1 g in 35 mL) to give pale yellow crystals.

Anal. Calcd. for $PtCl_2\{(CH_3)_2C{=}NOH\}_2$: C 17.48; H 3.42; N 6.80. Found: C 17.55; H 3.48; N 6.91.

Properties

trans-$PtCl_2\{(CH_3)_2C{=}NOH\}_2$ forms pale yellow, needle-like crystals from acetone/water. It melts with decomposition at 230°C, and the principal feature of a DTA/TG at a heating rate of 2°C/min is an exothermic peak with ~15% mass loss at 225°C.[8] The complex is soluble in acetone, nitromethane, and dimethylformamide; poorly soluble in cold water (solubility in H_2O at 25°C is 0.11 mass%[14]); and insoluble in diethyl ether and toluene. TLC on Riedel–deHaën 60 F 254 plates (silicagel on aluminum plates with layer thickness 200 μm, particle size 2–25 μm, and pore size 60 Å), $CHCl_3$: $(CH_3)_2CO$ = 1 : 1, $R_f = 0.71$ [cf. R_f (cis) = 0.38]. IR data (Nujol, cm^{-1}): 1662 (m), $\nu_{C=N}$; 330 (m-s), ν_{Pt-Cl}. 300 MHz ^1H NMR in acetone-d_6: δ 2.18 (3H), 2.64 (3H), and 10.01 (broad, OH). $^{13}C(^1H)$ NMR in DMF-d_7: δ 18.17 and 24.20 (CH_3), 164.55 (C=N). $^{195}Pt(^1H)$ NMR in DMF-d_7, −2104 ppm.[8] EI (MS), *m/z* 412 $(M)^+$. Kurnakov's test of *trans*-$PtCl_2\{(CH_3)_2C{=}NOH\}_2$ with thiourea in acidic aqueous media gives $Pt\{(NH_2)_2CS\}_2\{(CH_3)_2C{=}NOH\}_2$, additionally proving the trans configuration of the starting material.[14]

Acknowledgments

The authors thank the Russian RFBR and the Spanish DGI for grants 03-03-32362 and MAT2001-2112-C02-01.

References

1. (a) V. Yu. Kukushkin, D. Tudela, and A. J. L. Pombeiro, *Coord. Chem. Rev.* **156**, 333 (1996); (b) V. Yu. Kukushkin and A. J. L. Pombeiro, *Coord. Chem. Rev.* **181**, 147 (1999).
2. A. I. Stetsenko and B. S. Lipner, *Zh. Obstch. Khim.* **44**, 2289 (1974) and refs. therein.
3. P. W. Bléaupré and W. J. Holland, *Mikrochim. Acta* **3**, 341 (1983).
4. A. T. Hutton, D. M. McEwan, B. L. Shaw, and S. W. Wilkinson, *J. Chem. Soc., Dalton Trans.* 2011 (1983).
5. V. Yu. Kukushkin, V. K. Belsky, and D. Tudela, *Inorg. Chem.* **35**, 510 (1996).
6. L. A. M. Baxter, G. A. Heath, R. G. Raptis, and A. C. Willis, *J. Am. Chem. Soc.* **114**, 6944 (1992).
7. V. Yu. Kukushkin, D. Tudela, Yu. A. Izotova, V. K. Belsky, and A. I. Stash, *Inorg. Chem.* **35**, 4926 (1996).
8. V. Yu. Kukushkin, V. K. Belsky, E. A. Aleksandrova, V. E. Konovalov, and G. A. Kirakosyan, *Inorg. Chem.* **31**, 3836 (1992).
9. C. M. P. Ferreira, M. F. C. Guedes da Silva, J. J. R. Fraústo da Silva, A. J. L. Pombeiro, V. Yu. Kukushkin, and R. A. Michelin, *Inorg. Chem.* **40**, 1134 (2001).

10. A. V. Babaeva and M. A. Mosyagina, *Dokl. Akad. Nauk SSSR* **89**, 293 (1953).
11. M. F. C. Guedes da Silva, Y. A. Izotova, A. J. L. Pombeiro, and V. Yu. Kukushkin, *Inorg. Chim. Acta* **277**, 83 (1998).
12. V. Yu. Kukushkin, T. Nishioka, D. Tudela, K. Isobe, and I. Kinoshita, *Inorg. Chem.* **36**, 6157 (1997).
13. A. V. Babaeva and M. A. Mosyagina, *Izv. Sektora Platini AN SSSR* **28**, 203 (1954).
14. I. I. Chernyaev, ed., Sintez Kompleksnikh Soedinenii Metallov Platinovoii Gruppi (*Synthesis of Complex Compounds of the Platinum Group Metals*), Nauka, Moscow, 1964.

18. TRIFLUOROMETHANESULFONATOSILVER(I) DERIVATIVES

Submitted by ARÁNZAZU MENDÍA,[*] ELENA CERRADA,[†] and MARIANO LAGUNA[†]
Checked by DEAN H. JOHNSTON[‡] and ROBERT B. LETTAN II[‡]

Silver salts have widespread applications in synthetic procedures as halogen abstractors,[1] as one-electron oxidants,[2] or as building blocks in the preparation of heteropolynuclear complexes.[3–6] For all these purposes $Ag(OClO_3)$ has been widely used, and some derivatives such as $Ag(OClO_3)L$ (L = PPh_3, PPh_2Me or tetrahydrothiophene)[7] have proved to be appropriate synthons for heteronuclear silver complexes. Since metal perchlorates are potential explosives, their use should be accordingly restricted,[8] and we need to seek similar complexes with another poorly coordinating counterion. The compound $Ag(O_3SCF_3)$ has been used by many authors[9] as a safer source of silver ions, and different $Ag(O_3SCF_3)L$ derivatives have been used as silver synthons for heteronuclear complexes.[10,11]

More recently, studies on the synthesis and spectroscopic properties of $Ag(O_3SCF_3)L$ derivatives (L = PPh_3,[11–13] PPh_2Me,[12,13] tht[11,13]), as well as the determination of their solid-state structures, have been carried out. The addition of the corresponding neutral ligand to a diethyl ether solution of silver triflate, in a 1 : 1 ratio, is an easier way to synthesize these trifluoromethanesulfonate (triflate) derivatives than that previously reported.[11] The improved procedure described below is carried out in contact with air and at room temperature and results in the rapid formation of $Ag(O_3SCF_3)L$. The starting material, trifluoromethanesulfonatosilver(I), which is also commercially available (Aldrich), can easily be obtained from $AgNO_3$ through the formation of silver carbonate and

*Departamento de Química área de Química Inorgánica, Universidad de Burgos, E-09001 Burgos, Spain.
†Departamento de Química Inorgánica, Instituto de Ciencia de Materiales de Aragón, Universidad de Zaragoza CSIC, E50009 Zaragoza, Spain.
‡Department of Chemistry, Otterbein College, Westerville, OH 43081.

the addition of trifluoromethanesulfonic acid. The ligands may be commercial materials (Aldrich) used as received.

■ **Caution.** *Trifluoromethanesulfonic acid is harmful by inhalation, in contact with skin, and if swallowed. It causes burns. Precautions should be taken to minimize inhalation of the corrosive vapors given off from the acid. Trifluoromethanesulfonatosilver(I) is an eye, respiratory system, and skin irritant. Methyldiphenylphosphine is harmful by inhalation, in contact with skin, and if swallowed. Tetrahydrothiophene is an eye, respiratory system, and skin irritant. It is necessary to wear suitable protective clothing, gloves, and eye/face protection. All the reactions must be conducted in a well-ventilated fume hood.*

(*Note*: The reactions must be carried out by avoiding light exposure, because of the light-sensitive nature of the silver reagents. This is achieved by simply wrapping the glassware in aluminum foil.)

A. TRIFLUOROMETHANESULFONATOSILVER(I), $Ag(O_3SCF_3)$

$$2AgNO_3 + Na_2CO_3 \longrightarrow Ag_2CO_3 + 2NaNO_3$$

$$Ag_2CO_3 + 2CF_3SO_3H \longrightarrow 2Ag(O_3SCF_3) + CO_2 + H_2O$$

Procedure

A 250-mL beaker equipped with a magnetic stirring bar is charged with silver nitrate (3.40 g, 20 mmol) and 50 mL of water. To the resulting colorless solution is added a water solution (50 mL) of Na_2CO_3 (1.06 g, 10 mmol). A yellow solid promptly precipitates, and the mixture is allowed to stir for 5–10 min at room temperature. This solid is separated by filtration through a glass frit, washed with water (2 × 10 mL), and dried by suction.* The Ag_2CO_3 obtained is placed in an porcelain vessel equipped with a magnetic stirring bar and CF_3SO_3H (2.85 g, 1.67 mL, 19 mmol) is slowly added. (*Note*: Vigorous gas evolution). The resulting off-white paste is heated until dry. After 24 h in a heater at 110°C the crude product can be used without further purification. Yield: 4.06 g, 79% based on the starting silver nitrate. This synthesis can be scaled up to 40 mmol without lost of yield.

Anal. Calcd. for $CAgF_3O_3S$: C, 4.67; S 12.48. Found: C, 4.91; S, 11.93.

Properties

The compound $Ag(O_3SCF_3)$ is a white, hygroscopic, and light-sensitive solid. Its acetone solution displays the conductivity of a 1 : 1 electrolyte

*The checkers dried the Ag_2CO_3 at 60°C for 6 h in a vacuum oven and obtained 83% yield.

($\Lambda_M = 132\ \Omega^{-1}cm^2\ mol^{-1}$) because of the replacement of trifluoromethanesulfonate by solvent molecules.[10,11,14] It can be stored without decomposition for several months at room temperature in a vacuum desiccator when protected from light. It is soluble in water, acetone, and diethyl ether and insoluble in dichloromethane, chloroform, and *n*-hexane. The infrared spectrum shows strong and very broad absorption at 1310–1040 cm^{-1} that can be assigned to a covalent coordination of the trifluoromethanesulfonate group.[15] The liquid secondary ion mass spectrum (LSIMS+) shows the silver ion *m/z* 107(100%) as the base peak.

B. TRIFLUOROMETHANESULFONATO (TRIPHENYLPHOSPHINE)SILVER(I), $Ag(O_3SCF_3)(PPh_3)$

$$Ag(O_3SCF_3) + PPh_3 \rightarrow Ag(O_3SCF_3)(PPh_3)$$

Procedure

A 250-mL, one-necked, round-bottomed flask equipped with a magnetic stirring bar is charged with $Ag(O_3SCF_3)$ (2.569 g, 10 mmol) and 35 mL of diethyl ether. To the resulting colorless solution is added PPh_3 (2.623 g, 10 mmol). A white solid begins to precipitate, and the mixture is allowed to stir for 1 h at room temperature. After the solution volume is reduced under vacuum to 10 mL, the solid precipitate is separated by filtration with a glass frit. It is washed with diethyl ether (2 × 3 mL) and then hexane (2 × 5 mL) and finally dried by suction. The crude product can be used without further purification. Yield: 4.41 g, 85%. The complex can be synthesized in smaller quantities (down to 5 mmol) without significant decrease in yield.

Anal. Calcd. for $C_{57}H_{45}Ag_3F_9O_9PS_3$: C, 43.95; H, 2.91; S, 6.18. Found: C, 44.14; H, 3.02; S, 6.57.

Properties

The compound $Ag(O_3SCF_3)(PPh_3)$ is a white solid that is air- and light-stable in both solid state and solution. It behaves as a nonconductor ($\Lambda_M = 3.8\ \Omega^{-1}cm^2\ mol^{-1}$) in dichloromethane but in acetone displays the conductivity typical of a 1:1 electrolyte ($\Lambda_M = 122\ \Omega^{-1}cm^2\ mol^{-1}$).[11,12,14] It can be stored without decomposition for several months at room temperature in a vacuum desiccator. It is soluble in dichloromethane, acetone, and chloroform and insoluble in diethyl ether and *n*-hexane. The infrared spectrum shows absorptions at 1297 (vs), 1275 (vs), 1250 (vs, br), 1224 (vs), 1209 (vs) and 1170 (vs, br) cm^{-1} that can be assigned to a covalent coordination of the triflate group.[15] The ^{1}H NMR spectrum in CD_2Cl_2 exhibits a multiplet for the phosphine ligand at δ 7.6–7.3 (m, 15H, Ph). The $^{31}P\{^1H\}$ NMR spectrum in the same solvent shows a broad singlet at 14.5 (s) ppm at room temperature, which splits into two doublets at 14.7 (dd, $^2J_{107Ag-P} = 740.6$ Hz, $^2J_{109Ag-P} = 855.0$ Hz) ppm at -80°C because of the presence of two silver isotopomers (^{107}Ag, 51.82%, γ -1.0828; ^{109}Ag, 48.18%, γ -1.2448). The ^{19}F NMR spectrum shows a singlet at -78.0 (s) ppm at room temperature. The liquid secondary-ion mass spectrum (LSIMS+,VG autospec, nitrobenzyl alcohol as matrix), shows the trinuclear ion less one trifluoromethanesulfonate anion at *m/z* (%) 1409(2) $[Ag_3(O_3SCF_3)_2\ (PPh_3)_3]^+$, as well as the corresponding di- and mononuclear species 889(15) $[Ag_2(O_3SCF_3)\ (PPh_3)_2]^+$ and 369(100) $[Ag(PPh_3)]^+$. The X-ray crystal structure shows a trinuclear silver complex with the three trifluoromethanesulfonates in three different coordination modes, and the presence of two diastereoisomers in the same unit cell.[12,13]

C. TRIFLUOROMETHANESULFONATO (METHYLDIPHENYLPHOSPHINE)SILVER(I), $Ag(O_3SCF_3)(PPh_2Me)$

$$Ag(O_3SCF_3) + PPh_2Me \rightarrow Ag(O_3SCF_3)(PPh_2Me)$$

Procedure

A 250-mL, one-necked, round-bottomed flask equipped with a magnetic stirring bar is charged with $Ag(O_3SCF_3)$ (2.569 g, 10 mmol) and 35 mL of diethyl ether. To the resulting colorless solution is added PPh_2Me (2.002 g, 1.88 mL, 10 mmol). A white solid begins to precipitate, and the mixture is allowed to stir for 1 h at room temperature. After partial solvent removal under vacuum to 10 mL, the solid is collected by filtration with a glass frit, washed with diethyl ether (2×3 mL), and dried by suction. The crude product can be used without further purification. Yield: 3.75 g, 82%. The complex can be synthesized in smaller quantities (down to 5 mmol) without significant decrease in the yield.

Anal. Calcd. for $C_{56}H_{52}Ag_4F_{12}O_{12}P_4S_4$: C, 36.79; H, 2.87; S, 7.01. Found: C, 36.53; H, 2.80; S, 7.37.

Properties

The compound $Ag(O_3SCF_3)(PPh_2Me)$ is air- and light-stable in both solid state and solution. It behaves as a nonconductor ($\Lambda_M = 2.4\ \Omega^{-1}\ cm^2\ mol^{-1}$) in dichloromethane but in acetone displays the conductivity typical of a 1 : 1 electrolyte ($\Lambda_M = 124\ \Omega^{-1}\ cm^2\ mol^{-1}$).[10,11,14] It can be stored without decomposition for several months at room temperature in a vacuum desiccator. It is soluble in dichloromethane, acetone, and chloroform and insoluble in diethyl ether and *n*-hexane. The infrared spectrum shows absorptions at 1316 (vs), 1296 (vs), 1281 (vs), 1223 (vs), and 1184 (vs, br) cm^{-1} that can be assigned to a covalent coordination of the trifluoromethanesulfonate group.[15] The ^1H NMR spectrum in CD_2Cl_2 exhibits a multiplet for the phenyl protons of the phosphine ligand at δ 7.55–7.35 and a broad singlet assigned to the protons of the methyl group at δ 1.95. The $^{31}P\{^1H\}$ NMR spectrum in the same solvent, shows a broad singlet at −1.7 (s) ppm at room temperature which splits into two doublets at −1.8 ppm (dd, $^2J_{^{107}Ag-P} = 768.0$ Hz, $^2J_{^{109}Ag-P} = 875.9$ Hz) at −80°C. The ^{19}F NMR spectrum shows a singlet at −77.7 ppm at room temperature. The liquid secondary-ion mass spectrum (LSIMS+), *m/z* (%), shows the trinuclear ion less one trifluoromethanesulfonate anion 1223 (1) $[Ag_3\ (O_3SCF_3)_2\ (PPh_2Me)_3]^+$, as well as higher nuclearities as 1557 (3) $[Ag_4\ (O_3SCF_3)_2\ (S)(PPh_2Me)_4]^+$ and other lower peaks as the corresponding di- and mononuclear species 765 (9) $[Ag_2(O_3SCF_3)(PPh_2Me)_2]^+$ and 307 (100) $[Ag(PPh_2Me)]^+$. The X-ray crystal structure shows a tetranuclear silver complex, with two dinuclear units connected over an inversion center, forming tetranuclear assemblies at the secondary level. The unexpected tertiary level involves contacts of the form Ag . . . phenyl, forming a polymeric chain.[13]

D. TRIFLUOROMETHANESULFONATO (TETRAHYDROTHIOPHENE)SILVER(I), $Ag(O_3SCF_3)(tht)$

$$Ag(O_3SCF_3) + C_4H_8S(tht) \rightarrow Ag(O_3SCF_3)(tht)$$

Procedure

A 250-mL, one-necked, round-bottomed flask equipped with a magnetic stirring bar is charged with $Ag(O_3SCF_3)$ (2.569 g, 10 mmol) and 100 mL of diethyl ether. To the resulting colorless solution is added tetrahydrothiophene (tht) (0.88 g, 0.88 mL, 10 mmol). A white solid begins to precipitate, and the mixture is allowed to stir for 1 h at room temperature. This solid is collected by filtration with a glass frit, washed with diethyl ether (2 × 15 mL), and dried by suction. The crude product can be used without further purification. Yield: 3.19 g, 92%. The complex can be synthesized in smaller quantities (down to 5 mmol) without significant decrease in the yield.

Anal. Calcd. for $C_5H_8AgF_3O_3S_2$: C, 17.40; H, 2.34; S, 18.58. Found: C, 17.56; H, 2.48; S, 18.21.

Properties

The compound $Ag(O_3SCF_3)(tht)$ is a white sold that is air- and light-stable in both solid state and solution. Its dichloromethane solution behaves as a nonconductor ($\Lambda_M = 0.4\ \Omega^{-1}\ cm^2\ mol^{-1}$), but its acetone solution has the conductivity typical of a 1 : 1 electrolyte ($\Lambda_M = 127\ \Omega^{-1}\ cm^2\ mol^{-1}$).[10,11,14] It can be stored without decomposition for several months at room temperature in a vacuum desiccator. It is soluble in dichloromethane and acetone, slightly soluble in chloroform, and insoluble in diethyl ether and *n*-hexane. The infrared spectrum shows absorptions at 1285 (vs, br) 1240 (vs, br), 1220 (sh), and 1180 (vs, br) cm^{-1} that can be assigned to a covalent coordination of the trifluoromethanesulfonate group.[15] The ^1H NMR spectrum in deuterated acetone exhibits two multiplets for the tetrahydrothiophene ligand at δ 3.22 (4H) and 2.11 (4H); the former consist of the methylene groups nearer the sulfur. The ^{19}F NMR spectrum shows a singlet at -78.7 at room temperature. The liquid secondary-ion mass spectrum (LSIMS+), *m/z* %) shows the dinuclear ion 453(25) $[Ag_2(O_3SCF_3)(tht)]^+$ as well as other peaks with higher nuclearities, 1287 (9) $[Ag_4(O_3SCF_3)_4(tht)_3]^+$ and 1395 (5) $[Ag_5(O_3SCF_3)_4(tht)_3]^+$. The X-ray crystal structure consists of a polymeric silver complex with the monomeric units linked to form a ribbon structure, where the coordination geometry at silver is close to square pyramidal.[13]

References

1. (a) J. M. Vila, M. Gayoso, M. T. Pereira, M. L. Torres, J. J. Fernández, A. Fernández, and J. M. Ortigueira, *J. Organomet. Chem.* **532**, 171 (1997); (b) R. Contreras, M. Valderrama, A. Nettle, and D. Boys, *J. Organomet. Chem.* **527**, 125 (1997).
2. (a) S. K. Chandra and E. S. Gould, *Inorg. Chem.* **35**, 3881 (1996); (b) J. P. Pérez, P. S. Barrera, M. Sabat, and W. D. Harman, *Inorg. Chem.* **33**, 3026 (1994).
3. (a) M. Contel, J. Garrido, M. C. Gimeno, P. G. Jones, A. Laguna, and M. Laguna, *Inorg. Chim. Acta* **200**, 167 (1992); (b) J. A. Whiteford, C. V. Lu, and P. J. Stang, *J. Am. Chem. Soc.* **119**, 2524 (1997); (c) M. E. Olmos, A. Schier, H. Schmidbaur, A. L. Spek, D. M. Grove, H. Lang, and G. van Koten, *Inorg. Chem.* **35**, 2476 (1996); (e) A. A. Donapoulus, G. Wilkinson, T. K. Seweet, and M. B. Hursthouse, *J. Chem. Soc., Dalton Trans.* 2995 (1996); (e) S. Allshouse, R. C. Haltiwanger, V. Allured, and M. R. Dubois, *Inorg. Chem.* **33**, 2505 (1994).
4. J. Forniés and E. Lalinde, *J. Chem. Soc., Dalton Trans.* 2587 (1996).
5. R. Usón, and J. Forniés, *Inorg. Chim. Acta* **254**, 157 (1997).
6. R. Usón, A. Laguna, M. Laguna, M. C. Gimeno, P. G. Jones, C. Fittschen, and G. M. Sheldrick, *J. Chem. Soc., Chem. Commun.* 509 (1986).
7. (a) M. Bardají, P. G. Jones, A. Laguna, and M. Laguna, *Organometallics* **14**, 1310 (1995); (b) M. Bardají, A. Laguna, and M. Laguna, *J. Organomet. Chem.*, **496**, 245 (1995).

8. (a) Safety notes, *Inorg. Chem.* **37**, 11A (1998) and refes. therein; (b) J. L. Ennis and E. S. Shanley, *J. Chem. Educ.* **68**, A6 (1991); (c) G. J. Grant, P. L. Mauldin, and W. N. Setzer, *J. Chem. Educ.* **68**, 605 (1991).
9. (a) Y. Kayaki, I. Shmizu, and A. Yamamoto, *Bull. Chem. Soc. Jpn.* **70**, 917 (1997); (b) T. G. Back and B. P. Dyck, *J. Am. Chem. Soc.* **119**, 2079 (1997); (c) S. Morrone, D. Guillon, and D. W. Bruce, *Inorg. Chem.* **35**, 7041 (1996); (d) D. Veghini and H. Berke, *Inorg. Chem.* **35**, 4770 (1996).
10. M. Contel, J. Garrido, M. Laguna, and M. D. Villacampa, *Inorg. Chem.* **37**, 133 (1998).
11. M. Contel, J. Jiménez, P. G. Jones, A. Laguna, and M. Laguna, *J. Chem. Soc., Dalton Trans.* 2515, (1994).
12. R. Terroba, M. B. Hursthouse, M. Laguna, and A. Mendía, *Polyhedron* **18**, 807 (1999).
13. M. Bardají, O. Crespo, A. Laguna, and A. K. Fischer, *Inorg. Chim. Acta* **304**, 7 (2000).
14. W. J. Geary, *Coord. Chem. Rev.* **7**, 81 (1971).
15. (a) D. H. Johnston and D. F. Shriver, *Inorg. Chem.* **32**, 1045 (1993); (b) G. A. Lawrance, *Chem. Rev.* **86**, 17 (1986).

19. PENTAKIS(TETRAHYDROFURAN)DICADMIUM TETRAKIS(TETRAFLUOROBORATE), $[Cd_2(THF)_5][BF_4]_4$

Submitted by DANIEL L. REGER[*] and JAMES E. COLLINS[*]
Checked by GERALD PARKIN[†] and THOMAS G. CARREL[†]

$$Cd(acac)_2 + 4HBF_4 \cdot Et_2O \xrightarrow{THF} [Cd_2(THF)_5](BF_4)_4 + Hacac$$

Considerable attention has been directed toward the synthesis of monomeric cadmium(II) complexes that model biologically relevant zinc enzymes such as carbonic anhydrase.[1] Whereas complexes prepared from zinc(II) are generally spectroscopically silent,[2] those of cadmium(II) offer the possibility of characterization through ^{113}Cd NMR spectroscopy.[3] A synthetic challenge in preparing cadmium(II) complexes has been to overcome the low solubility of standard cadmium(II) precursors, such as $CdCl_2$ or $Cd(NO_3)_2$, in organic solvents. Also, in many cases it is desirable to use a cadmium precursor that does not contain coordinating anions.[4] Reported here is the preparation of $[Cd_2(THF)_5][BF_4]_4$,[4a] a useful starting material for the synthesis of cadmium(II) complexes in nonaqueous solutions, starting from commercially available $Cd(acac)_2$ (acac = acetylacetonate) and $HBF_4 \cdot Et_2O$. The complex has been used in the syntheses of $[\{HC(3,5\text{-}Me_2pz)_3\}_2Cd][BF_4]_2$,[4a] $[\{H(3\text{-}Ph)_2pz)_3\}_2Cd][BF_4]_2$,[4a] $[HB(3\text{-}{}^tBupz)_3\}Cd(NCS)$,[4c] and $[(2,6\text{-pyridyl-diimine})Cd(\mu\text{-}NCS]_2[BF_4]_2$.[4c] These preparations were not successful when starting with $CdCl_2$ or $Cd(NO_3)_2 \cdot 4H_2O$ in THF, indicating the usefulness of this THF adduct as a starting material. We

[*]Department of Chemistry and Biochemistry, University of South Carolina, Columbia, SC 29208.
[†]Department of Chemistry, Columbia University, New York, NY 10027.

have also reported the synthesis and use of $[Cd(THF)_3][\{3,5\text{-}(CF_3)_2\ C_6H_3\}_4B]_2$ in the preparation of more soluble cadmium(II) compounds.[5]

■ **Caution.** *Cadmium(II) compounds are extremely toxic, and care should be used when handling them. The reactions should be carried out in a well-ventilated fume hood, and gloves should be worn at all times.*

Procedure

A THF (8 mL) suspension of $Cd(acac)_2$ (Strem, 0.50 g, 1.6 mmol) is treated with 85% $HBF_4 \cdot Et_2O$ (Aldrich, 0.51 mL, 3.2 mmol). The reaction mixture is stirred for 2 h, and then the solvent is removed by cannula filtration. The remaining white solid is washed with hexanes (5 mL) and dried overnight under vacuum (0.66 g, 0.71 mmol, 88%); mp = 183–187°C.

Anal. Calcd. for $C_{20}H_{40}B_4Cd_2F_{16}$: C, 25.76; H, 4.32. Found: C, 25.92; H, 4.51.

Properties

The compound is soluble in CH_2Cl_2 and acetone, but not in ethers or hydrocarbon solvents. It is moderately air-stable as a powder. ^{1}H NMR (acetone-d_6): δ 3.64 (m, 4H, d-H), 1.79 (m, 4H, B-H). Crystals of $[Cd(THF)_4](BF_4)_2$ (see Ref. 4a for an X-ray structural analysis of this compound) were obtained by layering a saturated CH_2Cl_2 solution of $[Cd_2\ (THF)_5][BF_4]_4$ with THF.

References

1. (a) D. L. Reger, S. S. Mason, A. L. Rheingold, and R. L. Ostrander, *Inorg. Chem.* **32**, 5216 (1993); (b) A. S. Lipton, S. S. Mason, D. L. Reger, and P. D. Ellis, *J. Am. Chem. Soc.* **116**, 10182 (1994); (c) D. L. Reger and S. S. Mason, *Polyhedron* **13**, 3059 (1994); (d) D. L. Reger, S. M. Myers, S. S. Mason, A. L. Rheingold, B. S. Haggerty, and P. D. Ellis, *Inorg. Chem.*, **34**, 4996 (1995); (e) A. S Lipton, S. S. Mason, S. M. Myers, D. L. Reger, and P. D. Ellis, *Inorg. Chem.*, **35**, 7111 (1996); (f) D. L. Reger, S. M. Myers, S. S. Mason, D. J. Darensbourg, M. W. Holtcamp, J. H. Reibenspeis, A. S. Lipton, and P. D. Ellis, *J. Am. Chem. Soc.*, **117**, 10998 (1995); (g) C. Kimblin and G. Parkin, *Inorg. Chem.* **35**, 6912 (1996); (h) A. Looney, A. Saleh, Y. Zhang, and G. Parkin, *Inorg. Chem.* **33**, 1158 (1994); (i) R. A. Santos, E. S. Gruff, S. A. Koch, and G. S. Harbison, *J. Am. Chem. Soc.* **112**, 9257 (1990); (j) R. A. Santos, E. S. Gruff, S. A. Koch, and G. S. Harbison, *J. Am. Chem. Soc.* **113**, 469 (1991); (k) H. Vahrenkamp, *Acc. Chem. Res.* **32**, 589 (1999).
2. For a report on the use of ^{67}Zn NMR, see A. S. Lipton, T. A. Wright, M. K. Bowman, D. L. Reger, and P. D. Ellis, *J. Am. Chem. Soc.* **124**, 5850 (2002).
3. (a) J. E. Coleman, in *Metallobiochemistry,* Part D, J. F. Riordan and B. L. Vallee, eds., *Meth. Enzymol.* Vol. 227; Academic Press, San Diego, CA, 1993, pp. 16–43; (b) E. Rivera, M. A. Kennedy, and P. D. Ellis, *Adv. Magn. Reson.* **13**, 257 (1989); (c) M. F. Summers, *Coord. Chem. Rev.* 86, 43 (1988).
4. (a) D. L. Reger, J. E. Collins, S. M. Myers, A. L. Rheingold, and L. M. Liable-Sands, *Inorg. Chem.* **35**, 4904 (1996); (b) D. L. Reger, Comments *Inorg. Chem.* **21**, 1, (1999); (c) D. L. Reger, T. D. Wright,

M. D. Smith, A. L. Rheingold, S. Kassel, T. Concolino, and B. Rhagitan, *Polyhedron* **21**, 1795, (2002).
5. D. L. Reger, J. E. Collins, A. L. Rheingold, L. M. Liable-Sands, *Inorg. Chem.* **38**, 3235 (1999).

20. DI(μ-CHLORO)BIS{CHLORODIOXOBIS (TETRAHYDROFURAN)URANIUM(VI)}, $\{UO_2Cl_2(THF)_2\}_2$

Submitted by MARIANNE P. WILKERSON,[*] CAROL J. BURNS,[*] and ROBERT T. PAINE[†]
Checked by LAURA L. BLOSCH[‡] and RICHARD A. ANDERSEN[‡]

$$UO_3 \xrightarrow{HCl} UO_2Cl_2 \cdot x\,H_2O \; (x = 1\text{–}3)$$

$$UO_2Cl_2 \cdot x\,H_2O + 6\,ClSiMe_3 \xrightarrow{THF} UO_2Cl_2(THF)_3 + x\,O(SiMe_3)_2 + 2x\,HCl$$

$$2\,UO_2Cl_2(THF)_3 \xrightarrow{vac} \{UO_2Cl_2(THF)_2\}_2 + 2\,THF$$

The development of nonaqueous uranyl chemistry has demonstrated the existence of novel coordination geometries in the absence of hydrogen bonding and/or inner-sphere water coordination.[1] The anhydrous compound UO_2Cl_2 is necessary as a precursor to uranyl compounds that contain coordinated ligands unstable in aqueous solution. Although various routes for the synthesis of anhydrous UO_2Cl_2 have been reported in the literature,[2] many are unsuitable because of the difficulties involved in separating the starting material from other products, the limited scale of preparation, or the lack of a commercial source for the common starting material, UCl_4.[3] The alternative, facile approach described here is based on reports of the removal of coordinated water from transition metal halides.[4] Additional advantages offered by this procedure are that the precursor, UO_3, is readily available, and the synthesis avoids the use of oxygen gas at elevated temperatures. Because we have experienced inconsistent results utilizing commercially available hydrated uranyl chloride for this procedure, we also describe here the preparation of $UO_2Cl_2 \cdot x\,H_2O$ ($x = 1$–3) from UO_3.[5] The subsequent transformation to $UO_2Cl_2(THF)_3$ and $\{UO_2Cl_2(THF)_2\}_2$ uses standard Schlenk line techniques.[6,7] Metathesis reactions between uranyl chloride and alkali metal alkoxides or amides have been described elsewhere.[1]

[*]Chemistry Division, Los Alamos National Laboratory, Los Alamos, NM 87545.
[†]Department of Chemistry, University of New Mexico, Albuquerque, NM 87131.
[‡]Department of Chemistry, University of California, Berkeley, CA 94720.

■ **Caution.** *Handling radioactive materials. Depleted uranium as purchased consists primarily of the isotope ^{238}U, a weak α emitter with a half-life of 4.47×10^9 years. As such, it is principally an internal radiation hazard, and ingestion or inhalation of ^{238}U should be avoided. All reactions and manipulations must be performed in a well-ventilated fume hood with appropriate workplace monitoring for contamination control. As with all synthetic procedures involving hazardous chemicals, appropriate eye protection, gloves, and lab coats should be worn when handling this material. Practitioners should also adhere to their local institutional requirements for handling radioactive waste that is generated from this procedure.*

Starting Materials and Reagents

UO_3 is obtained from Strem Chemicals Inc., and concentrated HCl is available from Baker Chemical Co. Tetrahydrofuran and hexane are freshly distilled from benzophenone ketyl under nitrogen prior to use. Chlorotrimethylsilane is purchased from Aldrich and used without further purification.

Procedure

Approximately 10 g (35 mmol) of UO_3 is transferred to a 250-mL round-bottomed flask fitted with a ground-glass neck, a sidearm with a ground-glass stopcock, and a magnetic stirring bar. With stirring, 16 mL concentrated HCl is added at room temperature, and within a few minutes, a clear yellow solution is formed. The reaction is stirred for ~10 min, and then the flask is placed in an oil bath and connected to a Schlenk line. Water and excess acid are removed under reduced pressure (10^{-3} torr) at a slightly elevated temperature (50°C). The solid residue is redissolved in 20 mL distilled water to remove any adsorbed HCl, and the water is removed under reduced pressure with slight heating. Thermogravimetric analysis of this bulk material suggests that the degree of hydration is approximately two water molecules per uranyl chloride. When freshly isolated from aqueous solution, uranyl chloride is reported to exist as the trihydrate, $UO_2Cl_2 \cdot 3\,H_2O$.[2] Other references suggest that partial dehydration occurs under reduced pressure at temperatures above 35°C to give rise to the monohydrate, $UO_2Cl_2 \cdot H_2O$.[2,8] An excess of drying agent, based on the uranyl chloride trihydrate formula, is used in the next step.

The preparation of $UO_2Cl_2(THF)_3$ is performed in dry tetrahydrofuran. The Schlenk flask containing the crude $UO_2Cl_2 \cdot x\,H_2O$ is evacuated, and then back-filled with nitrogen. The evacuation/refill is repeated twice. After 100 mL of tetrahydrofuran is introduced via cannula, the yellow mixture is stirred and a suspension forms. At room temperature, an excess of $ClSiMe_3$ (30 mL, 0.32 mol) is added by using a syringe with a nitrogen flow, and the subsequent mixture is allowed to stir for 1 h. The suspension is concentrated to ~50 mL by removal of volatiles under reduced pressure, 50 mL hexane is added, and the suspension is stirred vigorously for 5 min to allow for precipitation of a fine yellow powder. The liquid layer is decanted away, and the powder is washed with hexane (2 × 50 mL). The product is dried under reduced pressure to yield 13.5 g (79% based on UO_3) of analytically pure $[UO_2Cl_2(THF)_2]_2$.[9]

Anal. Calcd. for $C_8H_{16}Cl_2O_4U$: C, 19.81; H, 3.32. Found: C, 19.60; H, 3.30.

Properties

$\{UO_2Cl_2(THF)_2\}_2$ is a bright yellow, moisture-unstable powder that is soluble in tetrahydrofuran, giving rise to $UO_2Cl_2(THF)_3$.[6] Single-crystal X-ray diffraction analysis of crystals removed directly from the mother liquor to a cold goniometer reveals the structure of $UO_2Cl_2(THF)_3$, but elemental analysis of crystals allowed to dry under inert atmosphere indicates the presence of only two THF molecules per uranyl chloride. The ^{1}H NMR spectrum of $UO_2Cl_2(THF)_3$ in THF-d_8 shows only two resonances (δ 3.62 (br m, α-H) and 1.77 (br m, β-H) for THF ligands. The UV–visible spectrum of $UO_2Cl_2(THF)_3$ in THF has an ε (at 429 nm) of 60 M^{-1} cm^{-1}. Comparison of the powder X-ray diffraction data of the dried material with the calculated *d* spacings from single-crystal X-ray diffraction analysis of $\{UO_2Cl_2(THF)_2\}_2$[9] shows that the compounds are the same. The IR spectrum of the dimer, recorded as a Nujol mull between KBr plates, has absorption bands at 1347 (s), 1295 (m), 1246 (m), 1214 (w), 1172 (m), 1136 (w), 1042 (m), 1007 (s), 961 (s), 949 (s), 921 (s), 891 (w), 875 (s), 841 (s), 683 (w) cm^{-1}. Both $UO_2Cl_2(THF)_3$ and $\{UO_2Cl_2(THF)_2\}_2$ must be protected from atmospheric moisture because each compound readily exchanges coordinated tetrahydrofuran for water.

References

1. C. J. Burns and A. P. Sattelberger, *Inorg. Chem.* **27**, 3692 (1988); C. J. Burns, D. C. Smith, A. P. Sattelberger and H. B. Gray, *Inorg. Chem.* **31**, 3724 (1992); M. P. Wilkerson, C. J. Burns, H. J. Dewey, J. M. Martin, D. E. Morris, R. T. Paine, and B. L. Scott, *Inorg. Chem.* **39**, 5277 (2000). M. P. Wilkerson, C. J. Burns, D. E. Morris, R. T. Paine, and B. L. Scott, *Inorg. Chem.* **41**, 3110 (2002).
2. G. Prins, *Investigations on Uranyl Chloride, its Hydrates, and Basic Salts, Reactor Center Nederlands,* RCN-186, 1973 and refs. therein.
3. J. A. Leary and J. F. Suttle, *Inorg. Synth.* **5**, 148 (1957).
4. J.-H. So and P. Boudjouk, *Inorg. Chem.* **29**, 1592 (1990).
5. J. J. Katz, G. T. Seaborg, and L. R. Morss, *The Chemistry of the Actinide Elements*, 2nd ed., Vol. 1, Chapman & Hall, New York, 1986.
6. M. P. Wilkerson, C. J. Burns, R. T. Paine, and B. L. Scott, *Inorg. Chem.* **38**, 4156 (1999).
7. D. F. Shriver and M. A. Drezdzon, *The Manipulation of Air-Sensitive Compounds,* 2nd ed., Wiley-Interscience, New York, 1986.
8. P. C. Debets, *Acta Cryst.* **B24**, 400 (1968); J. C. Taylor and P. W. Wilson, *Acta Cryst.*, **B30**, 169 (1974).
9. P. Charpin, M. Lance, M. Nierlich, D. Vigner, and C. Baudin, *Acta Cryst.* **C43**, 1832 (1987); R. D. Rogers, L. M. Green, and M. M. Benning, *Lanthanide Actinide Research* **1**, 185 (1986).

Chapter Three

TRANSITION METAL CARBONYL COMPOUNDS

21. HEXACARBONYLVANADATE(1-) AND HEXACARBONYLVANADIUM(0)

Submitted by XIN LIU[*] and JOHN E. ELLIS[*]
Checked by TRENT D. MILLER,[†] PRANSANA GHALASI,[†] and JOEL S. MILLER[†]

Since the discovery of hexacarbonylvanadium(0) and hexacarbonylvanadate(1-) by Calderazzo and co-workers in 1959 and 1960,[1,2] these substances have been key precursors to a variety of vanadium compounds, including inorganic noncarbonyl species,[3] organovanadium complexes,[4] and other vanadium carbonyls.[5] Neutral $V(CO)_6$ is of special interest in that it is the only isolable 17-electron homoleptic metal carbonyl and exhibits fascinating chemical properties that are often reminiscent of iodine and classic pseudohalogens.[6]

Although $V(CO)_6$ has been obtained by the direct carbonylation of bis (naphthalene)vanadium,[7] the best preparations involve the oxidation of $[V(CO)_6]^-$ by anhydrous HCl[8] or 100% orthophosphoric acid.[9] The procedure presented herein is based on a prior study,[9c] but has been improved by the use of an easily constructed Schlenk tube receiver that should also be useful for the isolation of other volatile, air-sensitive, and/or thermally unstable substances. Original syntheses of $[V(CO)_6]^-$ entailed high-pressure reductive carbonylations of VCl_3, requiring special equipment and potentially dangerous conditions.[2,9a] Later Calderazzo and Pampaloni developed a far safer atmospheric-pressure

[*]University of Minnesota, Department of Chemistry, Minneapolis, MN 55455.
[†]University of Utah, Department of Chemistry, Salt Lake City, UT 84112.

Inorganic Syntheses, Volume 34, edited by John R. Shapley
ISBN 0-471-64750-0

reductive carbonylation of VCl_3, mediated by 1,3,5,7-cyclooctatetraene, COT.[10] This method provides up to 85% yields of $[V(CO)_6]^-$, but requires an optimum COT/V ratio of 0.5 to achieve the best yields. Since the COT is expensive and not easily recovered from this synthesis, an alternative atmospheric pressure route to $[V(CO)_6]^-$ has been developed that involves alkali metal anthracene or naphthalene mediated carbonylations of VCl_3.[11] However, these procedures are suitable only for the synthesis of rather small quantities (1–2 g) of product. By employing lithium 1-methylnaphthalene as the reducing agent in the carbonylation of VCl_3, the alkali metal polyarene route to $[V(CO)_6]^-$ is easily scaled up by a factor of 10, and larger-scale reactions should also be possible with commercially available glassware. Lithium 1-methylnaphthalene is used in similar syntheses of $[Nb(CO)_6]^-$ and $[Ta(CO)_6]^-$,[12] and the excellent solubility of $Li[1\text{-}MeC_{10}H_7]$ in THF and 1,2-dimethoxyethane at low temperature, compared to that of alkali metal naphthalenes or anthracenes, is likely responsible for its unusual efficacy in these reductions. Since $Li[V(CO)_6]$ and solvated versions thereof are difficult to isolate in pure form, a cation exchange with inexpensive tetraethylammonium bromide is carried out to provide $[Et_4N][V(CO)_6]$. The latter salt is easily isolated, purified, stored, and converted to $V(CO)_6$. Other tetraalkylammonium salts of $[V(CO)_6]^-$ are available by analogous metathesis reactions. Since $V(CO)_6$ and $[V(CO)_6]^-$ are no longer commercially available at this time and prior syntheses of these species have not appeared in *Inorganic Syntheses*, the inclusion of these procedures should help facilitate further development of the chemistry of these useful compounds.

General Procedures and Starting Materials

All operations are conducted under an atmosphere of 99.9% argon or 99.5% carbon monoxide, as specified, with further purification by passage through columns of activated BASF catalyst and $13\times$ molecular sieves. Also, the CO is passed through a column of Ascarite (VWR Scientific), which is a trade name for a self-indicating sodium hydroxide–coated nonfibrous silicate formulation, for the quantitative absorption of CO_2.* Similar products used by microanalysts should provide satisfactory results. Standard Schlenk techniques are employed with a double-manifold vacuum line.[13] Unless otherwise stated, all starting materials were obtained from Aldrich Chemical Co. or Strem Chemicals, Inc. Tetrahydrofuran (THF) is distilled from sodium benzophenone ketyl, and 1-methylnaphthalene is stirred over molten sodium metal at $\sim$100°C for 12 h, cooled, and then distilled in vacuo. All other solvents were deaerated by purging with nitrogen. Unless otherwise stated, all solid reagents were transferred in an inert gas-filled drybox.

*Passage of CO through Ascarite was not necessary for the checkers to obtain the yields reported.

A. TETRAETHYLAMMONIUM HEXACARBONYLVANADATE, $[Et_4N][V(CO)_6]$

$$\mathrm{Li + 1\text{-}MeC_{10}H_7 \xrightarrow[20^\circ C]{THF} Li[1\text{-}MeC_{10}H_7]}$$

$$\mathrm{VCl_3 + 3\ THF \xrightarrow[reflux]{THF} VCl_3(THF)_3}$$

$$\mathrm{Li[1\text{-}MeC_{10}H_7] + VCl_3(THF)_3 \xrightarrow[CO]{-60^\circ C} [Li(THF)_x][V(CO)_6] + 1\text{-}MeC_{10}H_7}$$

$$\mathrm{[Li(THF)_x][V(CO)_6] \xrightarrow[acetone/EtOH/H_2O]{[Et_4N]Br} [Et_4N][V(CO)_6] + LiBr + THF}$$

Procedure

■ **Caution.** *Tetrahydrofuran (THF) is extremely flammable and hygroscopic and forms explosive peroxides; only anhydrous peroxide-free solvent should be used. Lithium wire is a hazardous substance and must be handled under strictly anaerobic conditions. Further, since it slowly reacts with dinitrogen at room temperature, lithium metal is best handled under an atmosphere of dry, oxygen-free argon. Vanadium trichloride is air-sensitive and should be transferred under an inert atmosphere. Carbon monoxide is a toxic and flammable gas and must be handled in a well-ventilated fume hood.*

[*Note*: In the following procedure, the THF solutions of $Li[1\text{-}MeC_{10}H_7]$ and $VCl_3(THF)_3$ should be prepared and cooled to $-60^\circ C$ at about the same time. Subsequently, they should be mixed without delay, as described below, and carbonylated to achieve optimum results.]

A 1-L round-bottomed flask, equipped with a water-cooled condensor and a Teflon-coated magnetic stirring bar, is charged with finely ground unsolvated VCl_3 (7.50 g, 47.7 mmol) and 350 mL of anhydrous THF under an argon atmosphere.* The reaction mixture is then heated to about 70–80°C with brisk stirring in an oil bath for 8 h.† After this period, almost all of the solid should have dissolved to provide a blood-red solution of $VCl_3(THF)_3$.[14]

1-Methylnaphthalene (40.5 mL, 285 mmol) is transferred via cannula to a stirred suspension of lithium wire, 3.2 mm diameter, cut into ~2-mm lengths (1.34 g, 193 mmol) in 200 mL of anhydrous THF in a 1-L round-bottomed flask, equipped with a sidearm stopcock for attachment to the vacuum line and a

*Checkers obtained reduced yields of $[V(CO)_6]^-$ in the range of 40–45% when the initial steps were carried out under a nitrogen atmosphere.

†Checkers found that 16–20 h was necessary for all VCl_3 to dissolve, in accord with Ref. 14.

glass-covered magnetic stirring bar. Within seconds the characteristic deep green color of $Li[1\text{-}MeC_{10}H_7]$ forms, if reactants, solvent, and argon atmosphere are satisfactorily anaerobic. The reaction mixture is briskly stirred for 8 h at room temperature to ensure dissolution of all lithium metal. (Although unreacted lithium particles float on the surface of the reaction mixture, the latter becomes so dark that they are difficult to observe, especially near the end of the reaction.) Since alkali metal naphthalenes slowly attack THF at room temperature, this solution should be cooled to −60°C with stirring in preparation for the addition of $VCl_3(THF)_3$.

The flask containing $VCl_3(THF)_3$ is cooled to −60°C with stirring, and the contents are added via a 16-gauge (or wider-diameter) cannula as a solution/slurry to the cold solution of $Li[1\text{-}MeC_{10}H_7]$ with efficient stirring. Cannulas that are longer than 1 m or have inside diameters of less than 1.19 mm should not be used to minimize clogging by the finely divided $VCl_3(THF)_3$. The resulting brown mixture is vigorously stirred while slowly warming to 0°C over a period of about 12 h. Then the reaction mixture is cooled to −60°C again with stirring, the argon atmosphere is removed in vacuo, and CO gas is introduced at about ambient pressure. The reaction mixture is constantly and vigorously stirred under an atmosphere of CO gas while slowly warming to room temperature over a period of 12–15 h. After the CO gas is removed in vacuo and argon is introduced, the reaction mixture is filtered through a 2-cm pad of filter aid (i.e., diatomaceous earth or kieselguhr) in a large (70-mm-diameter) medium-porosity filtration apparatus at room temperature. The filtercake is thoroughly washed with several 30-mL portions of THF until the washings are very pale yellow or nearly colorless. All solvent is removed in vacuo from the yellow-brown filtrate, until ~50–60 mL remains (including the 1-methylnaphthalene). The crude lithium salt is then precipitated with pentane (500 mL), and the residue is thoroughly triturated with more pentane (200-mL portions) to remove 1-methylnaphthalene. Each solvent aliquot is removed by cannula filtration. Finally, the residue is dried in vacuo to give a yellow-brown tar. This material has good thermal stability, although it is quite air-sensitive, and it can be stored prior to the cation metathesis.

The tar is dissolved in acetone (300 mL, O_2-free) and added to a solution of excess $[Et_4N]Br$ (12.0 g) in ethanol (200 mL, O_2-free). After evaporation of about 400 mL of solvent in vacuo, 600 mL of water (O_2-free) is added to precipitate the product. [*Note*: A dry-ice/acetone trap (−78°C) is needed to prevent moisture from diffusing back into the inert-gas manifold of the vacuum line during this and subsequent operations involving water!] The yellow-brown solid is collected on a large (70-mm-diameter) coarse-porosity frit, thoroughly washed with 2 × 20 mL of water, and dried in vacuo for 12 h.* (*Note*: If appreciable

*The checkers found that filtration using a Büchner funnel with filter paper was effective and took much less time than filtering with a glass frit.

amounts of 1-methylnaphthalene are present as a result of ineffective washing with pentane in the previous step, the product will be oily and possibly intractable at this stage.) The product is crystallized from THF/Et_2O,* washed with 3×20 mL of pentane, and dried in vacuo at room temperature to provide a bright yellow, free-flowing solid. Yield: 12.2 g (73% based on VCl_3).

Anal. Calcd. for $C_{14}H_{20}NO_6V$: C, 48.15; H, 5.77. Found: C, 48.26; H, 5.83.

Properties

Bright yellow microcrystalline $[Et_4N][V(CO)_6]$ is stable in dry air for days at room temperature; however, the moist salt reacts slowly with air to give an impure brownish yellow-green product. It is stable indefinitely at room temperature when protected from light and stored under an inert atmosphere. The salt has good solubility and stability in anaerobic dichloromethane, tetrahydrofuran, pyridine, acetonitrile, and dimethylsulfoxide. It is poorly soluble in ethanol and insoluble in water, diethyl ether, and hydrocarbons. Although solutions are only slightly air-sensitive at room temperature, they should be handled under an inert atmosphere for extended periods. Solutions of $[V(CO)_6]^-$ should be protected from light, since they slowly degrade when exposed to fluorescent lamp radiation or sunlight and are quickly decomposed by UV radiation. IR, ν_{CO}: (CH_2Cl_2), 1860 (vs) cm^{-1}; (mineral oil mull), 1843 (vs br) cm^{-1}. In general, alkali metal salts of $[V(CO)_6]^-$ are more difficult to handle and purify than the tetraethylammonium salt. For example, unsolvated $Na[V(CO)_6]$ is pyrophoric,[10b] while $[Na(diglyme)_2][V(CO)_6]$, diglyme = bis(2-methoxyethyl)ether[9a] is appreciably hygroscopic and is readily oxidized by air. Unsolvated $Na[V(CO)_6]$ may be easily obtained from $[Et_4N][V(CO)_6]$ by metathesis with an equivalent of $Na[BPh_4]$ in methanol, followed by filtration to remove insoluble $[Et_4N][BPh_4]$, and evaporation of solvent.[15] This same procedure is described in detail for the conversion of $[Et_4N][Ta(CO)_6]$ to the pyrophoric $Na[Ta(CO)_6]$.[12]

B. HEXACARBONYLVANADIUM, $V(CO)_6$

$$[Et_4N][V(CO)_6] + H_3PO_4 \xrightarrow{20^\circ C} V(CO)_6 + [Et_4N][H_2PO_4] + \tfrac{1}{2} H_2$$

Procedure

■ **Caution.** *Highly flammable H_2 is evolved in this procedure.*

*Before the hexacarbonylvanadate(1-) was crystallized, the checkers found it necessary to dissolve the product in THF and filter through Celite to remove approximately 2–3 g of insoluble brown material.

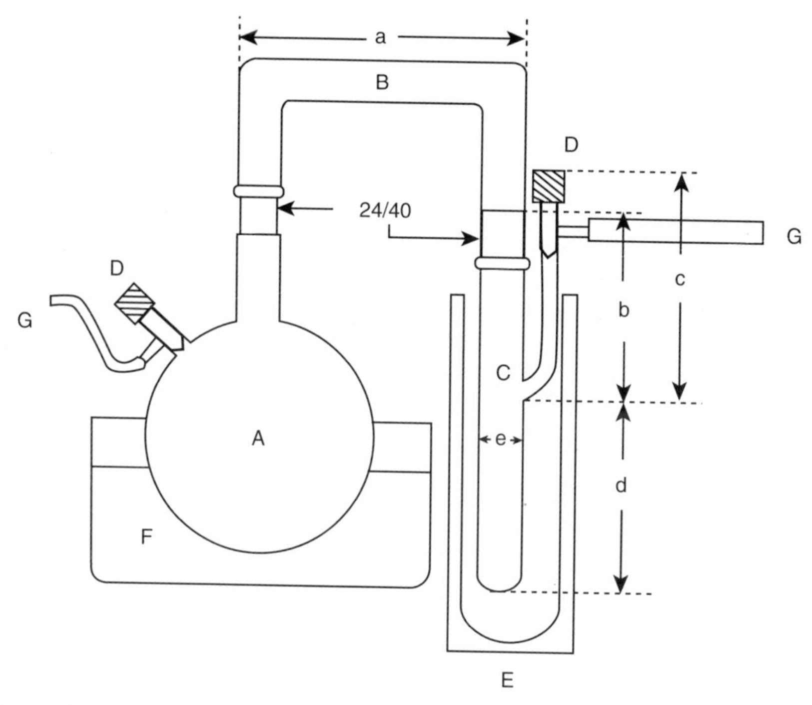

Figure 1. Diagram of the apparatus used in the synthesis and isolation of $V(CO)_6$. Components: (A) round-bottom flask, 500 mL; (B) connecting tube, (C) specially adapted Schlenk tube receiver; (D) stopcock; (E) Dewar flask; (F) water bath; (G) butyl rubber vacuum tubing connection to vacuum/argon manifold. Dimensions: a = 165 mm; b = 130 mm; c = 170 mm; d = 75 mm; e = 20 mm od.

Crystalline 100% orthophosphoric acid (40 g, 0.41 mol) and $[Et_4N][V(CO)_6]$ (6.0 g, 17.2 mmol) were intimately mixed together in an inert-atmosphere box at room temperature in a 500-mL round-bottom flask equipped with a large Teflon-coated magnetic stirring bar. The flask is attached to a connecting tube and a specially adapted Schlenk tube receiver as shown in Fig. 1.

Specifically, the glass connection between the Schlenk tube receiver (C) and stopcock (D) should extend about 130 mm below the top of the 24/40 standard taper inner joint and approximately 50 mm into Dewar flask (E) to prevent loss of the especially volatile $V(CO)_6$ to the vacuum line during the reaction and sublimation. The Schlenk tube receiver is cooled to about −70°C with dry ice/acetone or dry ice/isopropyl alcohol. The level of coolant should be near the top of the Dewar so that the glass connection between the stopcock (D) and the Schlenk tube is entirely covered. The apparatus is evacuated very slowly and cautiously to minimize entrainment of the finely divided $[Et_4N][V(CO)_6]$ by the hydrogen gas as it is evolved. The reaction mixture is slowly warmed to 40°C with a water bath, while continuously stirring the mixture and maintaining a dynamic vacuum.

■ **Caution.** *Considerable frothing of the viscous phosphoric acid melt will occur during this operation. Careful control of the vacuum is necessary to prevent contamination of $V(CO)_6$ by the reaction mixture melt. Do not heat the water bath above 40°C, and do not use a heat gun or other external heat source to facilitate the transfer, since $V(CO)_6$ will quickly decompose to a vanadium mirror.*

The sublimation is continued until all $V(CO)_6$ is transferred into the Schlenk receiver. The reaction and collection process requires approximately 1.5 h. The $V(CO)_6$ product is transferred in an inert-atmosphere box into a conventional Schlenk tube and stored in a freezer (−30°C or lower) in the dark. Yield: 3.25 g (86%).

Properties

Paramagnetic $V(CO)_6$ forms lustrous bluish green–black crystals. Magnetic data (solid state): $\chi_m^{cor}(20°C) = 1230 \times 10^{-6}$ cm^3/mol, giving $\mu_{eff} \times 1.71$ μ_B ($\chi_{diam} = -56 \times 10^{-6}$ cm^3/mol), lit. $\mu_{eff} \times 1.73$ μ_B.[16] The compound dissolves in alkanes and CH_2Cl_2 to provide yellow to orange solutions; it is also pale orange when doped into crystalline $Cr(CO)_6$.[17] It is slightly soluble (≈5 mg $V(CO)_6$/mL) in saturated hydrocarbons and appreciably more soluble in toluene and CH_2Cl_2 and shows a single sharp IR ν_{CO} absorption at about 1972 cm^{-1} in these solvents. Sample purity may be assessed by examining its solubility in pure hexane or heptane; for example, 5 mg of pure $V(CO)_6$ should quickly and completely dissolve in about 1 mL of these solvents at room temperature under an inert atmosphere. Any insoluble residue is likely to be vanadium metal, $[Et_4N][V(CO)_6]$, or an oxidation product. $V(CO)_6$ is the most unstable and reactive of all isolable neutral homoleptic carbonyls of the 3*d* metals and requires special care in handling because of its high sensitivity toward light, air, heat, and most polar organic solvents. For example, we have found that solid $V(CO)_6$ slowly decomposes even at −30°C in the dark, so for critical applications freshly sublimed $V(CO)_6$ must be used. Hexacarbonylvanadium undergoes facile nucleophilic substitution[18] and Lewis base–induced disproportionation reactions,[19] consistent with its metal carbonyl radical nature. $V(CO)_6$ also has a rich redox chemistry; for instance, it functions as both a mild one-electron oxidant, due to its tendency to form the 18-electron $[V(CO)_6]^-$, and is a strong reductant because of its formal zerovalent vanadium character.[20]

Acknowledgments

The submitters thank the National Science Foundation and the Petroleum Research Fund for their continuing support of research that led to these preparations. Ms. Christine Lundby is thanked for her expert assistance in the preparation of this manuscript.

References

1. G. Natta, R. Ercoli, F. Calderazzo, A. Alberola, P. Corrandini, and G. Allegra, *Atti. Acad. Naz. Lincei, Classe Sci. Fis. Mater. Nat.* **27**, 107 (1959).
2. R. Ercoli, F. Calderazzo, and A. Alberola, *J. Am. Chem. Soc.* **82**, 2966 (1960).
3. (a) A. Davison, N. Edelstein, R. H. Holm, and A. H. Maki, *Inorg. Chem.* **4**, 55 (1965); (b) A. Davison and E. T. Shawl, *Chem. Commun.* 670 (1967); (c) F. Calderazzo and G. Pampaloni, *J. Organomet. Chem.* **330**, 47 (1987); (d) F. Calderazzo, G. Pampaloni, G. Pelizzi, and F. Vitali, *Polyhedron* **7**, 2039 (1988); (e) J. M. Manriquez, G. T. Yee, R. S. McLean, A. J. Epstein, and J. S. Miller, *Science* **252**, 1415 (1991); (f) B. B. Kaul and G. T. Yee, *Inorg. Chim. Acta* **326**, 9 (2001) and refs. therein.
4. (a) F. Calderazzo, *Inorg. Chem.* **3**, 1207 (1964); (b) F. Calderazzo, *Inorg. Chem.* **4**, 223 (1965); (c) R. P. M. Werner and S. A. Manastyrskyj, *J. Am. Chem. Soc.* **83**, 2023 (1961); (d) F. Calderazzo, G. Pampaloni, L. Rocchi, J. Strähle, and K. Wurst, *J. Organomet. Chem.* **413**, 91 (1991); (e) F. Calderazzo, I. Ferri, G. Pampaloni, and U. Englert, *Organometallics* **18**, 2452 (1999).
5. F. Calderazzo, "Carbonyl complexes of the transition metals," in *Encyclopedia of Inorganic Chemistry*, R. B. King, ed., Wiley, Chichester, UK, 1994, Vol. 2.
6. (a) J. Ellis, *J. Organomet. Chem.* **86**, 1 (1975); (b) J. Ellis, *J. Chem. Educ.* **53**, 2 (1976).
7. M. K. Pomije, C. J. Kurth, J. E. Ellis, and M. V. Barybin, *Organometallics* **16**, 3582 (1997).
8. F. Calderazzo and G. Pampaloni, *Organomet. Synth.* **4**, 49 (1988).
9. (a) R. P. M. Werner and H. E. Podall, *Chem. Ind.* (London) 144 (1961); (b) G. Silvestri, S. Gambino, M. Guaninazzi, and R. Ercoli, *J. Chem. Soc., Dalton Trans.* 2558 (1972); (c) J. E. Ellis, R. A. Faltynek, G. L. Rochfort, R. E. Stevens, and G. A. Zank, *Inorg. Chem.* **19**, 1082 (1980).
10. (a) F. Calderazzo and G. Pampaloni, *J. Organometal. Chem.* **250**, C33 (1983); (b) F. Calderazzo and G. Pampaloni, *Organomet. Synth.* **4**, 46 (1988).
11. M. V. Barybin, M. K. Pomije, and J. E. Ellis, *Inorg. Chem. Acta* **269**, 58 (1998).
12. C. G. Dewey, J. E. Ellis, K. L. Fjare, K. M. Pfahl, and G. F. Warnock, *Organometallics* **2**, 388 (1983).
13. (a) D. F. Shriver and M. A. Drezdon, *The Manipulation of Air-Sensitive Compounds*, 2nd ed., Wiley, New York, 1986; (b) A. L. Wayda and M. Y. Darensbourg, eds., *Experimental Organometallic Chemistry*; ACS Symposium Series Vol. 357, American Chemical Society: Washington, DC, 1987.
14. L. E. Manzer, *Inorg. Synth.* **21**, 135 (1982).
15. J. E. Ellis and K. L. Fjare, unpublished research.
16. F. Calderazzo, R. Cini, P. Corradini, R. Ercoli, and G. Natta, *Chem. Ind.* (London) **79**, 500 (1960).
17. G. F. Holland, M. C. Manning, D. E. Ellis, and W. C. Trogler, *J. Am. Chem. Soc.* **105** 2308 (1983).
18. Q.-Z. Shi, T. G. Richmond, W. C. Trogler, and F. Basolo, *J. Am. Chem. Soc.* **106**, 71 (1984).
19. T. G. Richmond, Q.-Z. Shi, W. C. Trogler, and F. Basolo, *J. Am. Chem. Soc.* **106**, 76 (1984).
20. F. Calderazzo and G. Pampaloni, *J. Organomet. Chem.* **303**, 111 (1986).

22. TRICARBONYLTRIS(PYRIDINE)MOLYBDENUM(0) AND ITS USE AS A Mo(0) SOURCE FOR ALLYLMOLYBDENUM(II) COMPLEXES

Submitted by EUGEN F. MESAROS,* ANTHONY J. PEARSON,* and ELKE SCHOFFERS*

Checked by KENNETH C. STONE† and J. L. TEMPLETON†

Tricarbonyltris(pyridine)molybdenum(0) was prepared and isolated for the first time in 1935 by Hieber and Mühlbauer.[1] Typically used in the synthesis of η^6-arenemolybdenum complexes, it was considered until 1997[2] to be an inferior precursor of π-allylmolybdenum complexes,[3] in comparison with the more traditional Mo(0) sources for these complexes, tricarbonyltris(acetonitrile)molybdenum(0)[4] or tricarbonyltris(dimethylformamide)molybdenum(0).[5] Parallel reactions performed in our laboratory,[2] involving all these Mo(0) sources and a series of organic allylic substrates, have shown that use of the pyridine complex in the synthesis of π-allylmolybdenum complexes results in comparable and sometimes better yields than using the acetonitrile or dimethylformamide complexes. In addition, tricarbonyltris(pyridine)molybdenum has a longer shelf life and is much more stable in air, so that its synthesis and manipulation are considerably less complicated. Unlike the other two complexes, tricarbonyltris(pyridine)molybdenum(0) is also soluble and stable in a variety of organic solvents. A straightforward procedure for the preparation of $Mo(CO)_3(py)_3$ is presented here together with the synthesis of a representative π-allylmolybdenum(II) complex[2,6,7] that uses $Mo(CO)_3(py)_3$ as a Mo(0) source and a scorpionate ligand[8] for stability.

Starting Materials

All the starting materials are commercially available (Aldrich) and used as received, unless otherwise specified.

■ **Caution.** *Because pyridine is a carcinogenic and malodorous liquid and CO is a toxic gas, all the operations should be performed in a well-ventilated fume hood.*

A. TRICARBONYLTRIS(PYRIDINE)MOLYBDENUM(0), $Mo(CO)_3(py)_3$

$$Mo(CO)_6 + 3\ py \longrightarrow Mo(CO)_3(py)_3$$

*Department of Chemistry, Case Western Reserve University, Cleveland, OH 44106.

†Department of Chemistry, University of North Carolina, Chapel Hill, NC 27599.

Procedure

Hexacarbonylmolybdenum(0) (13.2 g, 50 mmol) is placed in a two-necked 125-mL round-bottomed flask equipped with a water condenser and a magnetic stirring bar, under argon.* All equipment is oven-dried and then assembled and cooled under a stream of argon prior to use. The argon line is connected to the sideneck of the flask, and the condenser is connected at the top to a mineral oil bubbler. After adding the solid hexacarbonylmolybdenum(0), the sideneck of the flask is closed with a rubber septum, and a stream of argon is passed through the apparatus for another 10 min (through an 18-gauge needle), then disconnected. Freshly distilled (from KOH, under argon) pyridine† (35 mL) is added via syringe (glass, oven-dried, and cooled under argon) and stirring is commenced. The flask is heated in an oil bath at 80°C for 1 h and at 130°C for an additional 2 h. The reaction mixture gradually turns from colorless to yellow and eventually dark red. It is allowed to cool to room temperature, under argon without stirring, and crystals are formed. The crystallization is completed by adding pentane (20 mL) and cooling the flask in an ice bath. The crystals are filtered and washed with ice-cold pyridine (2 × 10 mL) and then cold pentane (2 × 10 mL), and dried overnight under vacuum. The yield is 18.5 g (89%). The reaction can be scaled both up and down by a factor of 2 without affecting the yield significantly.

Properties

Tricarbonyltris(pyridine)molybdenum(0) is a yellow to orange crystalline compound that can be handled in the ambient atmosphere without noticeable decomposition. It can be stored indefinitely at 0–5°C under an inert atmosphere. It decomposes at 205–210°C (a gradual change in color from yellow to brown is already observed starting at 100–105°C). Its IR spectrum shows two carbonyl bands at 1901 and 1764 cm^{-1}.

B. DICARBONYL{HYDRIDOTRIS(1-PYRAZOLYL)BORATO}{η-(1,2,3)-(+)-(1R,2R)-1-(METHOXYCARBONYL)-2-PROPEN-1-YL}MOLYBDENUM(II), $Mo(CO)_2(\eta^3\text{-}CH_2CH{=}CHCO_2CH_3)Tp$

$$Mo(CO)_3(py)_3 + BrCH_2CH{=}CHR\ (R = CO_2Me) \longrightarrow Mo(CO)_2(py)_2(\eta^3\text{-}CH_2CHCHR)Br + py + CO$$

$$Mo(CO)_2(py)_2(\eta^3\text{-}CH_2CHCHR)Br + NaTp \longrightarrow Mo(CO)_2(\eta^3\text{-}CH_2CHCHR)Tp + NaBr + 2\ py$$

*Checkers used nitrogen atmosphere throughout instead of argon.

†Checkers used high purity pyridine (Aldrich) without further purification.

$Mo(CO)_3(py)_3$ (125 mg, 0.3 mmol) is placed in a two-necked 25-mL round-bottomed flask equipped with a water condenser and a stirring bar, under argon. All equipment is oven-dried, then assembled and cooled under a stream of argon prior to use. The argon line is connected to the sideneck of the flask and the condenser is connected at the top to a mineral oil bubbler. After adding the solid $Mo(CO)_3(py)_3$, the sideneck of the flask is closed with a rubber septum and a stream of argon is passed through the apparatus for another 10 min (through an 18-gauge needle), then disconnected. Methyl 4-bromocrotonate (43 μL, 0.36 mmol, 1.2 equiv) and freshly distilled dichloromethane (from CaH_2, under nitrogen, 5 mL) are then added via syringe (glass, oven-dried, and cooled under argon), and the reaction mixture is refluxed under a positive pressure of argon (the bubbler on top of the condenser is replaced with an argon-filled balloon) for 1 h. The solvent is then evaporated at room temperature under a stream of argon (the bubbler is reconnected to the apparatus) until a brown oily residue remains. In a separate operation, freshly distilled THF (from Na/benzophenone under nitrogen) (5 mL) is added via syringe (glass, oven-dried, and cooled under argon) to a 10-mL oven-dried flask, containing sodium hydridotris(1-pyrazolyl) borate* (NaTp, 85 mg, 0.36 mmol, 1.2 equiv) under a stream of argon, and the resulting solution is transferred to the residue in the reaction flask via cannula (oven-dried). The reaction mixture is stirred for 3 h at room temperature under a positive pressure of argon (balloon), then diluted with CH_2Cl_2 (5 mL) and filtered through a plug of alumina. Removal of solvent by rotary evaporation provides the crude product as a brown oil, which is purified by flash chromatography (silicagel, EtOAc/hexanes: 1/4).† The yield is 106 mg (76%).

(*Note*: The reaction can also be performed at multigram scale. Owing to the CO released in this reaction, a bubbler should be connected to the condenser in the scaled-up protocol rather than an argon-filled balloon.)

Properties

Dicarbonyl{hydridotris(1-pyrazolyl)borato}{η-(1,2,3)-($\pm$)-(1*R*, 2*R*)-1-(methoxycarbonyl)-2-propen-1-yl}molybdenum is a yellow solid, stable under ambient conditions for several days and indefinitely under argon at 0–5°C. It has the following physical and spectroscopic properties: mp 140–142°C; ^1H NMR ($CDCl_3$, 300 MHz), δ 7.93 (br s, 3H), 7.53 (d, $J = 2.0$ Hz, 3H), 6.20 (t, $J = 2.3$ Hz, 3H), 4.83 (dt, $J = 6.5, 9.5$ Hz, 1H), 3.65 (dd, $J = 6.5, 2.8$ Hz, 1H), 3.55 (s, 3H), 2.49 (d, $J = 9.5$ Hz, 1H), 1.42 (dd, $J = 9.5, 2.8$ Hz, 1H); ^{13}C NMR ($CDCl_3$, 75.5 MHz), δ 234.5, 224.8, 171.9, 143.7, 135.4, 105.4, 85.5, 66.8, 51.3, 47.9. (The

*Checkers used the potassium salt, KTp.

†Checkers used column chromatography on alumina and observed ~5% anti isomer eluting first.

signals of the pyrazole rings in the Tp ligand are averaged by a dynamic process, so that only three distinct pyrazole signals are observed in both ^{1}H NMR and ^{13}C NMR spectra.[6]) IR ($CHCl_3$), 2498 (w, BH), 1945 (s), 1858 (s), 1714 (s) cm^{-1}. HRMS (EI, 20 eV), m/z calcd. for $C_{16}H_{17}O_4N_6MoB$ (^{98}Mo) 466.0464, found 466.0409.

References

1. W. Hieber and F. Z. Mühlbauer, *Anorg. Allg. Chem.* **221**, 337 (1935).
2. A. J. Pearson and E. Schoffers, *Organometallics* **16**, 5365 (1997).
3. R. G. Hayter, *J. Organomet. Chem.* **13**, P1 (1968).
4. D. P. Tate, W. R. Knipple, and J. M. Augl, *Inorg. Chem.* **1**, 433 (1962); R. B. J. King, *Organomet. Chem.* **8**, 139 (1967).
5. M. Pasquali, P. Leoni, P. Sabatino, and D. Braga, *Gazz. Chim. Ital.* **122**, 275 (1992).
6. Y. D. Ward, L. A. Villanueva, G. D. Allred, S. C. Payne, M. A. Semones, and L. S. Liebeskind, *Organometallics* **14**, 4132 (1995).
7. A. J. Pearson, I. B. Neagu, A. A. Pinkerton, K. Kirschbaum, and M. J. Hardie, *Organometallics* **16**, 4346 (1997).
8. (a) S. Trofimenko, *Chem. Rev.* **93**, 943 (1993); (b) S. Trofimenko, *Inorg. Synth.* **12**, 99 (1970).

23. A TUNGSTEN(0)–FULLERENE-60 DERIVATIVE

Submitted by JANICE M. HALL,[*] JOSHUA H. HOYNE,[*] and JOHN R. SHAPLEY[*]

Checked by CHANG YEON LEE[†] and JOON T. PARK[†]

Since the first report of a complex involving a direct metal-to-C_{60} bond, $(Ph_3P)_2Pt(\eta^2\text{-}C_{60})$,[1] numerous studies have established that the fullerene C_{60} acts as a moderately electronegative alkene in coordinating to electron-rich metal centers.[2] In many cases the C_{60} ligand is subject to relatively facile displacement when the complex is in solution; however, the zerovalent, octahedral complexes $M(CO)_3(dppe)(C_{60})$ [M = Mo, W; dppe = 1,2-bis(diphenylphosphino)ethane] display outstanding stability even under severe conditions.[3] The overall time needed to prepare these complexes from commercially available $M(CO)_6$ is dramatically reduced by adopting a biphasic procedure for the synthesis of the precursor $M(CO)_4(dppe)$, which was first described for the preparation of $Mo(CO)_4(dppe)$.[4] Here details are presented for the biphasic synthesis of $W(CO)_4(dppe)$[5] and for its use in the preparation of $W(CO)_3(dppe)(C_{60})$.

[*]Department of Chemistry, University of Illinois at Urbana-Champaign, Urbana, IL 61801.

[†]Department of Chemistry and School of Molecular Science, Korea Advanced Institute of Science and Technology, Daejon, 305-701, Korea.

Materials

$W(CO)_6$ (Strem), dppe (Arapahoe), and [n-Bu_4N]Br (Aldrich) are reagent-grade and used as received. C_{60} (98%) is available from Southern Chemical Group. Other solvents and reagents are reagent-grade and used as received.

■ **Caution.** *These procedures must be performed in a good fume hood as toxic carbon monoxide is evolved.*

A. TETRACARBONYL-1,2-BIS(DIPHENYLPHOSPHINO)ETHANE TUNGSTEN(0), W(CO)$_4$(dppe)

$$W(CO)_6 + \text{dppe} \xrightarrow[\text{KI}]{[n\text{-Bu}_4\text{N}][\text{OH}]} W(CO)_4(\text{dppe}) + 2\ CO$$

Procedure

$W(CO)_6$ (367 mg, 1.043 mmol), dppe (420 mg, 1.054 mmol), and toluene (20 mL are added to a 50-mL round-bottomed flask containing a magnetic stirring bar. [n-Bu_4N]Br (20 mg, 0.062 mmol), KI (10 mg, 0.061 mmol), and 50% NaOH solution (2.0 mL) are added to this solution, and the flask is fitted with a small condenser. The mixture is deoxygenated with nitrogen bubbling for 5 min, and then the flask is heated to 90°C under positive pressure of N_2 for 3 h with stirring. The flask is cooled to room temperature, and the reaction mixture is separated with a separatory funnel. The organic layer is washed once with water (10 mL), then dried over anhydrous $MgSO_4$. The filtered solution is then concentrated to ~5 mL on a rotary evaporator. An equal volume of absolute ethanol is added, and the flask is cooled first in an ice bath and then placed in a freezer at −15°C for 2 h. The white crystals are collected with a small Hirsch funnel and air-dried. Yield: 521 mg (72%).

Anal. Calcd. C, 51.85; H, 3.48. Found: C, 51.63; H, 3.27.

Properties

The product is a white crystalline solid, mp 204–206°C (lit.[6] 206–208°C). The compound can be recrystallized from acetone/methanol. IR (CH_2Cl_2): 2017 (m), 1901 (s), 1877 (m, sh) cm^{-1}. ^{31}P NMR (202 MHz, $CDCl_3$, 25°C): δ 41.271 ($J_{P\text{–}W}$ = 230.6 Hz).

B. TRICARBONYL-1,2-BIS(DIPHENYLPHOSPHINO)ETHANE (FULLERENE-60)TUNGSTEN(0), $W(CO)_3(dppe)C_{60}$

$$W(CO)_4(dppe) + C_{60} \xrightarrow{h\nu} CO + W(CO)_3(dppe)(C_{60})$$

In a 100-mL Pyrex Schlenk tube C_{60} (12.7 mg, 0.0175 mmol) and $W(CO)_4$ (dppe) (12.6 mg, 0.0182 mmol) are dissolved in 50 mL toluene under a nitrogen atmosphere. The solution is irradiated with a sunlamp (GE 275 W) placed close to the tube. Periodically a sample of the solution is taken, the solvent evaporated, and the residue redissolved in CH_2Cl_2 for IR analysis. Prominent bands at 2001 (m), 1937 (m), and 1884 (s) cm^{-1} due to the products appear, and the specific band at 2017 (m) due to the starting material diminishes. After ~5 h total of irradiation, the dark green solution is evaporated, and the residue is separated by thin-layer chromatography (Merck, Kieselgel 60 F-254, 0.25 mm) with 1 : 1 hexane/dichloromethane as eluting medium. Trailing a faint purple band due to C_{60} is a bright green band ($R_f \sim 0.5$) and a broader green-brown band ($R_f \sim 0.3$). The former band is cut from the plate, the product extracted with CH_2Cl_2, and the solution evaporated to dryness. Yield: forest green $W(CO)_3$ (dppe)(C_{60}), 5.1 mg (21%).

Anal. Calcd. for $C_{89}H_{24}O_3P_2W$: C, 77.07; H, 1.74. Found: C, 76.70; H, 1.84.

[*Note*: IR analysis of the product may indicate that it is contaminated with the $W(CO)_4$(dppe) starting material (ν_{CO} at 2017 cm^{-1}), which travels as a colorless band on the TLC plate just behind the leading green band. A subsequent elution with 1 : 1 hexane/CS_2 provides a clean separation. The checkers used 5 : 1 CS_2/CH_2Cl_2 to separate the C_{60} adducts ($R_f \sim 0.8$, ~ 0.45), and then observed the second band to be contaminated with starting material.]

Properties

$W(CO)_3$ (dppe)(C_{60}): IR (CH_2Cl_2), ν_{CO} 2002 (m), 1936 (m), 1884 (s) cm^{-1}. ^{31}P NMR (121.5 MHz, C_7D_8): 38.165 ppm (d), 42.281 ppm (d), $J_{P-P} = 4$Hz. MS(FAB-): m/e 1386 (M^-). The crystal structure has been determined.[3] The gray-green second band is due to $C_{60}\{W(CO)_3$ (dppe)$\}_2$: IR (CH_2Cl_2), ν_{CO} 2000 (m), 1930 (m), 1882 (s) cm^{-1}. ^{31}P NMR studies suggest that this compound is a mixture of positional isomers, as shown quantitatively for the Mo analog.[7]

References

1. P. J. Fagan, J. C. Calabrese, and B. Malone, *Science* **252**, 1160 (1991).
2. M. M. Olmstead and A. L. Balch, *Chem. Rev.* **98**, 2123 (1998).

3. H. Hsu, Y. Du, T. E. Albrecht-Schmitt, S. R. Wilson, and J. R. Shapley, *Organometallics* **17**, 1756 (1998).
4. J. A. Howell, *J. Chem. Educ.* **74**, 577 (1997).
5. K. R. Birdwhistell, *Inorg. Synth.* **29**, 141 (1992).
6. S. O. Grim, W. L. Briggs, R. C. Barth, C. A. Tolman, and J. P. Jesson, *Inorg. Chem.* **13**, 1095 (1974).
7. H. F. Hsu and J. R. Shapley, *Proc. Electrochem. Soc.* **10**, 1087, (1995).

24. DODECACARBONYLTRIRUTHENIUM, $Ru_3(CO)_{12}$

Submitted by MATTHIEU FAURÉ,* CATHERINE SACCAVINI,* and GUY LAVIGNE*
Checked by ALEJANDRO TRUJILLO† and BEATRIZ OELCKERS†

Recent years have seen a dramatic increase in the practical use of $Ru_3(CO)_{12}$ as an efficient catalyst precursor for achieving selective carbon–carbon bond formation and cleavage in a number of important organic transformations.[1,2] Representative examples include Murai's catalytic C—H/olefin coupling reactions,[2] the hydroamination of alkynes,[3] the hydroesterificationof olefins,[4] and the Pauson–Khand reaction.[5] Thus, there is a real need for a rapid access to this basic compound. Numerous procedures have been proposed for the preparation of $Ru_3(CO)_{12}$.[6,7] High-pressure methods are generally well adapted to large-scale preparation, but require specific high-pressure equipment that is seldom available in the laboratory.

The present synthetic strategy rests on the fundamental principle of Hieber's "base reaction,"[8] which has only recently (as of 2003) been applied to carbonylhaloruthenium(II) complexes.[9–11] The one-pot procedure proposed here is ideally suited for the rapid conversion (3–4 h) under 1 atm CO of moderate quantities of commercially available $RuCl_3 \cdot {}_3H_2O$ into pure $Ru_3(CO)_{12}$ in yields exceeding 90%. It combines the advantages of simplicity, rapidity, high efficiency, and insensitivity to moisture, thereby allowing direct use of all commercial reagents as received.

The overall synthesis involves two consecutive steps. The first one is a CO-induced reductive carbonylation of $RuCl_3 \cdot 3H_2O$, producing in situ an equilibrium mixture of two carbonylchlororuthenium(II) complexes, $[Ru(CO)_2Cl_2]_n$ and $[Ru(CO)_3Cl_2]_2$.[12]

$$RuCl_3 \cdot n\ H_2O \xrightarrow[\text{2-ethoxyethanol}]{\text{CO}} [Ru(CO)_2Cl_2]_n \underset{}{\overset{\text{CO}}{\rightleftarrows}} [Ru(CO)_3Cl_2]_2$$

*Laboratoire de Chimie de Coordination, 205 route de Narbonne, 31077 Toulouse Cedex, France.
†Instituto de Química, Universidad Catolica de Valparaíso, Avda; Brasil 2950, Valparaiso, Chile.

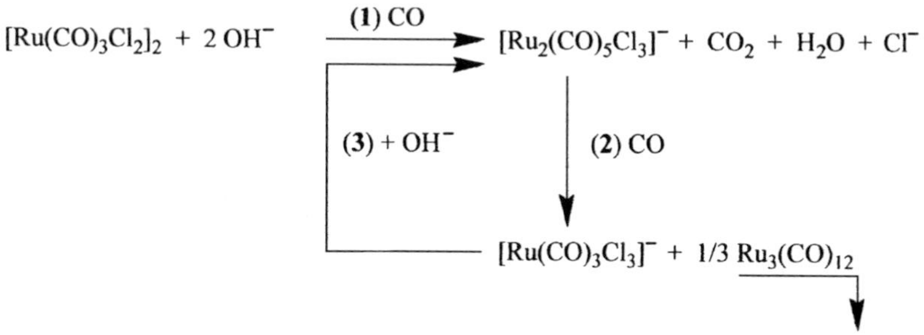

$$[Ru(CO)_3Cl_2]_2 + 2\,OH^- \xrightarrow{(1)\,CO} [Ru_2(CO)_5Cl_3]^- + CO_2 + H_2O + Cl^-$$

$$[Ru_2(CO)_5Cl_3]^- \xrightarrow{(2)\,CO} [Ru(CO)_3Cl_3]^- + 1/3\,Ru_3(CO)_{12}\downarrow$$

$$[Ru(CO)_3Cl_3]^- \xrightarrow{(3)\,+\,OH^-} [Ru_2(CO)_5Cl_3]^-$$

Scheme 1. Mechanism for production of $Ru_3(CO)_{12}$ with KOH and CO at 75°C.

As indicated here, there is no need to use distilled solvents since the ruthenium salt is hydrated and brings water to the reaction anyway. Yet, water might become a problem since it is susceptible to react (albeit only at high temperature) with the elusive transient monocarbonyl species formed in the early stage of the carbonylation of $RuCl_3$, to produce irreversibly a very stable green monocarbonyl aqua complex, previously identified by Halpern.[13] The formation of this undesirable green product is systematically avoided in the present procedure by performing the initial carbonylation under mild conditions (80°C during the first 45 min), at which time the transient mono-carbonyl intermediate reacts much faster with CO than with water. During this early stage, the excess of water is also evacuated along with the fast CO stream before the reflux temperature of 2-ethoxyethanol is reached. Adopting this stepwise heating procedure allows a significant reduction of the time required to obtain the characteristic yellow solution.

The second step involves the reaction cascade shown on Scheme 1, which is triggered by addition of KOH (2 equiv per Ru atom) to the solution shown above under CO at 75°C.[9,10] The observed reduction of $Ru^{(II)}$ to $Ru^{(0)}$ can be rationalized in terms of the following elementary steps:

1. Generation of an elusive bimetallic dianionic bishydroxycarbonyl adduct "$[Ru(CO)_2\{C(O)OH\}Cl_2]_2^{2-}$" which decarboxylates to produce in situ the polymeric Ru(I) polyanion $\{[Ru_2(CO)_4(\mu\text{-}CO)Cl_3]^-\}_n$ of known structure.[10]
2. Fast CO-induced disproportionation of the latter[10] to give equimolar amounts of $Ru^{(0)}$, recovered as insoluble $Ru_3(CO)_{12}$ and $Ru^{(II)}$, obtained under the form of $K[Ru(CO)_3Cl_3]$.
3. Automatic recycling of the newly formed soluble $Ru^{(II)}$ species $K[Ru(CO)_3Cl_3]$, which enters a new reduction/disproportionation cycle on reaction with the remaining hydroxide ions until total consumption of

both reagents. The total time required for this synthesis is about 3–4 h, namely, at least 1 h 45 min for the first step, 45 min for the second step, and 30 min for slow crystallization.

Chemicals

$RuCl_3 \cdot 3H_2O$ (Johnson Matthey), 2-ethoxyethanol (Fluka) taken directly from the bottle; KOH pellets (RIEDEL, or any other label); CO gas cylinder (Air Liquide, N 20 grade).

Apparatus

Heating magnetic stirrer and oil bath; large (olive shaped) magnetic stirbar; three-necked round-bottomed 500-mL flask equipped with a reflux condenser, the upper part of which is connected to a gas bubbler and a hose sending toxic vapors to the top of the hood. One lateral neck of the flask is equipped with a CO gas inlet consisting of a glass bubbler (steel needles are not convenient) of sufficiently large diameter (2–3 mm) to avoid clogging by accumulation of solid $Ru_3(CO)_{12}$. The second lateral neck is used to introduce KOH pellets in the second reaction step.

■ **Caution.** *CO gas and phosgene (which may be formed as a volatile byproduct in the first step) are highly toxic. Appropriate personal protection must be adopted and all manipulations must be carried out in a highly efficient fume hood.*

Procedure

1. Reduction of Ru(III) to Ru(II). Into the flask are introduced 5 g ($\sim$19.1 mmol) of $RuCl_3$ $3H_2O$ and 250 mL of 2-ethoxyethanol. The solution is first deaerated by bubbling nitrogen or by stirring under reduced atmosphere for a few minutes. The flask is then connected to the reflux condenser, whereupon the CO gas inlet is introduced. A fast CO stream ($\sim$2 bubbles per second) and a vigorous agitation are needed in this first reduction step. The temperature is first raised to 80°C for 45 min, during which time the solution progressively turns blood red. It is then increased up to 135°C (reflux) for 30–45 min, until a perfectly limpid golden yellow solution is obtained. At this stage, the temperature of the bath is allowed to cool down to 75°C and regulated prior to the beginning of the second step. IR spectrum of the yellow solution: ν_{CO}, 2135.5 (m), 2063.6 (s), 1992.9 (m) cm^{-1}. This three-band pattern corresponds to the overlapping of the characteristic bands of the polymeric $\{Ru(CO)_2Cl_2\}_n$ [IR ν_{CO}, 2063 (s), 1993 (s) cm^{-1}] and dimeric $\{Ru(CO)_3Cl_2\}_2$ [ν_{CO}, 2135 (s), 2065 (s) cm^{-1}].

2. *Reduction of Ru(II) to Ru(0).* A strict temperature control is required here; the temperature of the bath should be stabilized around 75–80°C and should remain below 85°C (see explanation below). The rate of CO bubbling may be reduced to ~1 bubble per second. KOH pellets (2.4 g) ~38.2 mmol, assuming 10% water by weight is then added directly into the solution (at 75°C) via the lateral neck of the flask (cautiously opened under the hood for just a few seconds). A progressive darkening of the solution is then observed over the next 15 min, after which time the orange crystalline $Ru_3(CO)_{12}$ begins to appear on the walls of the glassware. (*Note*: After 20 minutes, spectacular fumes, expelled from the exploding CO bubbles, may spontaneously appear above the solution, thus forming a kind of dense fog that tends to deposit traces of an orange sublimate on the walls of the reflux condenser.) After a total time of 45 min, the heater is stopped, but CO bubbling and a moderate magnetic stirring rate are maintained. Keeping the flask in the oil bath at this stage ensures a very slow rate of cooling during the crystallization of $Ru_3(CO)_{12}$, which occurs particularly efficiently between 70 and 50°C and is accompanied by a significant lightening of the initial brown color. A limpid and almost colorless solution is obtained at the end. Since $Ru_3(CO)_{12}$ is totally insoluble in 2-ethoxyethanol, its characteristic IR absorption bands [ν_{CO}, 2060 (vs), 2030 (s), 2011 m cm^{-1}] appear neither at the intermediate stage nor in the final solution spectrum, which generally shows evidence for only small amounts of the anionic Ru(II) complex $[Ru(CO)_3Cl_3]$- [ν_{CO}, 2126 (m); 2048 (s) cm^{-1}] and possibly also trace amounts of a soluble minor complex previously identified as the known dianionic Ru(I) oxo derivative $[Ru_4(\mu_4\text{-}O)(\mu\text{-}Cl)_4(CO)_{10}]^{2-}$ [ν_{CO}, 2014 (s) 1939 (m), 1733 (w) cm^{-1}], reflecting the presence of traces of water.[10]

Crystals of $Ru_3(CO)_{12}$ are perfectly air-stable. They can be easily recovered by filtration on a frit after cautiously venting the solution under nitrogen for a few minutes to evacuate CO. They generally need to be washed with alcohol and/or water to remove traces of KCl, the only sideproduct of this reaction. Alternatively, a rapid chromatographic workup of the crude product $Ru_3(CO)_{12}$ (solubilized in 300–400 mL of dichloromethane) on a silicagel column, using dichloromethane as the eluent, is the best way to eliminate KCl, which is retained at the top of the column. After solvent evaporation under reduced pressure, the crystals are subsequently dried under vacuum (~3.9 g, 6.1 mmol, 96%).

[*Note*: A precise adjustment of the amount of KOH is required in the present procedure. While OH^- undergoes highly preferential nucleophilic attack onto the electrophilic carbonyls of $K[Ru(CO)_3Cl_3]$ even in the presence of $Ru_3(CO)_{12}$ (this is also favored by the insolubility of the latter in 2-ethoxyethanol below a critical temperature of 80°C), any excess of OH^- that would be still present after total consumption of Ru(II) is potentially able to react with $Ru_3(CO)_{12}$ to produce the soluble anionic complex $[Ru_3(\mu\text{-}H)(CO)_{11}]^-$ (characteristic violet; IR ν_{CO}, 2016 (vs), 1989 (s), 1953 (mw), 1732 (w), cm^{-1}.[14]] In cases where this incidentally happens, recovery of $Ru_3(CO)_{12}$ is still possible, since $[Ru_3(\mu\text{-H})(CO)_{11}]^-$ is a hydride transfer agent whose reaction with water in the presence of CO leads back to $Ru_3(CO)_{12}$, according to the water–gas shift reaction[15]

$$[Ru_3(\mu\text{-H})(CO)_{11}]^- + H_2O + CO \longrightarrow Ru_3(CO)_{12} + H_2 + OH^-$$

Water is inevitably present in the second reaction step, since it is produced during the reduction of Ru(II) to Ru(I), thereby allowing this reaction to take place, albeit at slow rate. Effectively, dark solutions incidentally obtained at the end of the abovementioned preparation and reflecting the presence of $[Ru_3(\mu\text{-H})(CO)_{11}]^-$ become clearer on prolonged treatment overnight with CO at 25°C, with concomitant recovery of $Ru_3(CO)_{12}$.

Properties

$Ru_3(CO)_{12}$ is a yellow-orange crystalline solid that can be handled in air but slowly decomposes under ambient conditions. The compound is readily soluble in most organic solvents. IR (hexane): ν_{CO}, 2061 (s), 2031 (s), 2011 (m) cm^{-1}.[6]

References

1. T. Mitsudo and T. Kondo, *Synth. Lett.* 309 (2001) and refs. therein.
2. F. Kakiuchi and S. Murai, *Acc. Chem. Res.* **35**, 826 (2002) and refs. therein.
3. (a) T. Kondo, A. Tanaka, S. Kotachi, and Y. Watanabe, *Chem. Commun.* 413 (1995); (b) M. Tokunaga and Y. Wakatsuki, *Angew. Chem., Int. Ed. Engl.* **38**, 3222 (1999); (c) Y. Uchimaru, *Chem. Commun.* 1133 (1999); (d) T. Kondo, T. Okada, T. Suzuki, and T. Mitsudo, *J. Organomet. Chem.* **622**, 149 (2001).
4. S. Ko, Y. Na, and S. Chang, *J. Am. Chem. Soc.* **124**, 751 (2002) and refs. therein.
5. (a) T. Kondo, N. Suzuki, T. Okada, and T. Mitsudo, *J. Am. Chem. Soc.* **119**, 6187 (1997); (b) S.-K. Kang, K.-J. Kim, and Y.-T. Hong, *Angew. Chem., Int. Ed. Engl.* **41**, 1584, (2002).
6. M. I. Bruce, C. M. Jensen, and N. L. Jones, *Inorg. Synth.* **26**, 259 (1989) and refs. therein.
7. E. Lucenti, E. Cariati, C. Dragonetti, and D. Roberto, *J. Organomet. Chem.* **669**, 44 (2003).
8. A. F. Hill, *Angew. Chem., Int. Ed. Engl.* **39**, 130, (2000) and refs. therein.
9. M. Faure, L. Maurette, B. Donnadieu, and G. Lavigne, *Angew. Chem., Int. Ed. Engl.* **38**, 518, (1999); L. Maurette, B. Donnadieu, and G. Lavigne, *Angew. Chem., Int. Ed. Engl.* **38**, 3707 (1999).
10. G. Lavigne, *Eur. J. Inorg. Chem.* 917 (1999) and refs. therein.
11. M. Fauré, C. Saccavini, and Guy Lavigne, *Chen. Commun.* 1578 (2003).
12. (a) M. I. Bruce, and F. G. A. Stone, *J. Chem. Soc. A* 1238 (1967); (b) N. Lugan, G. Lavigne, J.-M. Soulié, S. Fabre, P. Kalck, J.-Y. Saillard, and J. F. Halet, *Organometallics* **14**, 1712 (1995) and refs. therein.
13. J. Halpern, B. R. James, and L. W. Kemp, *J. Am. Chem. Soc.* **88**, 5142 (1966).
14. C. R. Eady, J. J. Guy, B. F. G. Johnson, J. Lewis, M. C. Malatesta, and G. M. Sheldrick, *J. Chem. Soc., Dalton Trans.* 1358 (1978).

15. (a) M. W. Payne, D. L. Leussing, and S. G. Shore, *J. Am. Chem. Soc.* **109**, 617 (1987); (b) M. W. Payne, D. L. Leussing, and S. G. Shore, *Organometallics* 10, 574 (1991) and refs. therein.

25. DICARBONYLBIS(HEXAFLUOROACETYLACETONATO) RUTHENIUM, $Ru(CO)_2(hfac)_2$

Submitted by YUN CHI,* CHAO-SHIUAN LIU,* YING-HUI LAI,* and ARTHUR J. CARTY†
Checked by DAVID HOTH‡ and JIM D. ATWOOD‡

$$Ru_3(CO)_{12} + 6\ \text{hfacH} \rightarrow 3\ Ru(CO)_2(\text{hfac})_2 + 6\ CO + 3\ H_2$$

There has been considerable interest in deposition of highly conductive Ru and RuO_2 thin films as metallization materials for integrated circuits.[1,2] Various organometallic ruthenium complexes have been proposed as possible CVD (chemical vapor deposition) source reagents for these thin films. The first set of compounds comprises ruthenocene[3] and alkyl substituted derivatives such as $Ru(C_5H_4Et)_2$. The latter possesses a low melting point and appreciable volatility and has been tested for possible scaleup application.[4] A second group of possible organometallic precursors includes tris-β-diketonate complexes such as $Ru(acac)_3$, $Ru(tfac)_3$, and $Ru(tmhd)_3$, (tfac = 1,1,1-trifluoro-2,4-pentanedionate and tmhd = 2,2,6,6-tetramethyl-3,5-heptanedionate).[5] The tris-β-diketonate complexes $Ru(acac)_3$ and $Ru(tmhd)_3$ can be purchased from commercial chemical suppliers and thus have been extensively examined in CVD studies. However, high deposition rates using these tris-β-diketonate precursors are difficult to obtain, as these compounds are not very volatile. A third group includes alkenyl and alkene (ruthenium) complexes, such as bis(2,4-dimethylpentadienyl)ruthenium[6] $(\eta^6\text{-}C_6H_6)Ru(\eta^4\text{-}C_6H_8)$ (C_6H_8 = 1,3-cyclohexadiene), and $Ru(C_3H_5)_2(COD)$ (COD = 1,5-cyclooctadiene).[7] However, the preparation of these complexes requires multistep manipulations under an inert atmosphere, which makes handling of these materials more difficult.

The fourth type of source reagent, which is far more volatile than the previous groups, consists of the binary carbonyls $Ru(CO)_5$ and $Ru_3(CO)_{12}$[8] as well as

*Department of Chemistry, National Tsing Hua University, Hsinchu, Taiwan, 30013, Republic of China.
†Steacie Institute for Molecular Sciences, National Research Council Canada, Ottawa, Ontario K1A OR6, Canada.
‡Department of Chemistry, University of Buffalo, SUNY, Buffalo, NY 14260.

corresponding diene or alkyne-stabilized derivative complexes, such as $Ru(CO)_3(C_6H_8)$[9] (C_6H_8 = cyclohexadiene) and $Ru(CO)_4(hfb)$[10] (hfb = hexa-hexafluoro-2-butyne). Very pure metal films have been prepared from $Ru_3(CO)_{12}$, which has a relatively low vapor pressure, but the more volatile alternative, $Ru(CO)_5$, has poor stability, which limits its use as a CVD precursor. A recently discovered source reagents consists of a series of Ru(II) complexes with the general formula $Ru(CO)_2(diketonate)_2$. These complexes can be prepared in excellent yields via the direct reaction of an excess of a β-diketone, such as hexafluoroacetylacetone (hfacH), 2,2,6,6-tetramethyl-3,5-heptanedione (tmhdH) or others, with $Ru_3(CO)_{12}$ in a sealed stainless-steel autoclave at elevated temperature. This modification of the literature method for $Ru(CO)_2(acac)_2$[11] offers the possibility of incorporating widely utilized, commercially available β-diketones as starting reagents.[12]

Procedure

■ **Caution.** *This reaction evolves toxic carbon monoxide. All manipulations should be performed in a well-ventilated fume hood. Adequate shielding must be provided for the pressure vessel.*

A sample of $Ru_3(CO)_{12}$ (Strem; 0.5 g, 0.78 mmol), 6.5 equiv of (hfac)H (Aldrich; 0.72 mL, 5.1 mmol), and 50 mL of anhydrous pentane together with a stirring bar are added into a 160-mL stainless-steel autoclave. The autoclave is sealed and then heated under stirring to 170°C for 18 h. The autoclave is heated using a silicon oil bath, and the temperature readout is obtained from a thermal couple inserted into the autoclave. (*Note*: If internal temperature measurement is not possible, the silicon oil bath is set to a higher temperatures of 180–200°C, as such a temperature will not cause product decomposition and the higher temperature will guarantee completion of the reaction.) After cooling of the autoclave to room temperature, the pentane solution is transferred from the reactor and filtered to remove a small amount of insoluble, black precipitate that may be present. [*Note*: This precipitate is obtained if impure $Ru_3(CO)_{12}$ is used, as a small amount of black impurity will be always presented in a commercial sample of $Ru_3(CO)_{12}$ after long period of storage in air.] The filtrate is then concentrated on a rotary evaporator maintained at a pressure of 225 torr, and the temperature of the water bath is kept within 10–20°C. Prolonged exposure of the solution to reduced pressure will cause serious loss of the volatile product. The resulting solid residue is dissolved in a minimum amount of hot methanol (2–5 mL). Cooling the methanol solution to −20°C produces yellow crystals of $Ru(CO)_2(hfac)_2$, which are collected by filtration (1.27 g, 92%) and air-dried. Further purification of this compound is accomplished by sublimation at room temperature and a pressure of 300 mtorr, using a dry-ice cold finger.

Anal. Calcd. for $C_{12}H_2F_{12}O_6Ru$: C, 25.23; H, 0.35. Found: C, 25.45; H, 0.40.

Properties

The product is stable for an indefinite period in air under ambient conditions, mp 64–66°C. It is highly soluble in chlorinated hydrocarbon solvents, THF, and acetone, but is less soluble in methanol, hexane, and toluene. The molecular structure shows an octahedral geometry with two cis carbonyl ligands. FAB (fast-atom bombardment) mass spectrometry shows a parent molecular ion at m/z 572 (M^+). IR spectroscopy indicates two CO stretching bands at 2092 (s) and 2036 (vs) cm^{-1} in cyclohexane. ^1H NMR ($CDCl_3$, 298 K): δ 6.34 (s, 2H, CH). ^{13}C NMR ($CDCl_3$, 298 K): δ 192.5 (2C, CO), 179.1 (2C, CO, $^2J_{CF} = 36$ Hz), 176.9 (2C, CO, $^2J_{CF} = 36$ Hz), 116.7 (2C, CF_3, $^1J_{CF} = 285$ Hz), 115.9 (2C, CF_3, $^1J_{CF} = 285$ Hz), 91.8 (2C, CH). ^{19}F ($CDCl_3$, 298 K): δ -74.15 (s, 6F), -75.06 (s, 6F).

Deposition of ruthenium thin films has been achieved using this CVD source reagent.[12] For the experiments conducted under H_2, a large amount of the free ligand (hfac)H is obtained as a volatile byproduct, suggesting that the deposition of ruthenium is proceeding by the following pathway:

$$Ru(CO)_2(hfac)_2 + H_2 \rightarrow Ru(s) + 2\ CO + 2\ hfacH$$

For the deposition reactions carried out under an Ar atmosphere, up to 41% of the dark red tris-β-diketonate complex $Ru(hfac)_3$ is observed as a volatile byproduct. This finding suggests that the operation of a parallel pathway, which involves self-disproportionation of the source reagent, is necessary to account for the observed film growth:

$$3\ Ru(CO)_2(hfac)_2 \rightarrow Ru(s) + 6\ CO + 2\ Ru(hfac)_3$$

References

1. (a) G. S. Sandhu, *Thin Solid Films* **320**, 1 (1998); (b) T. Aoyama, M. Kiyotoshi, S. Yamazaki, and K. Eguchi, *Jpn. J. Appl. Phys.* **38**, 2194, (1999); (c) E. S. Choi, J. C. Lee, J. S. Hwang, J. B. Park, and S. G. Yoon, *J. Electrochem. Soc.* **146**, 4189 (1999).
2. S. Yamamichi, P. Y. Lesaicherre, H. Yamaguchi, K. Takemura, S. Sone, H. Yabuta, K. Sato, T. Tamura, K. Nakajima, S. Ohnishi, K. Tokashiki, Y. Hayashi, Y. Kato, Y. Miyasaka, M. Yoshida, and H. Ono, *IEEE Trans. Electron Devices* **44**, 1076 (1997).
3. (a) D. E. Trent, B. Paris, and H. H. Krause, *Inorg. Chem.* **3**, 1057 (1964); (b) M. L. Green, M. E. Gross, L. E. Papa, K. J. Schnoes, and D. Brasen, *J. Electrochem. Soc.* **132**, 2677 (1985); (c) J. Si, and S. B. Desu, *J. Mater. Res.* **8**, 2644 (1993); (d) W. C. Shin and S. G. Yoon, *J. Electrochem. Soc.* **144**, 1055 (1997).
4. (a) T. Aoyama and K. Eguchi, *Jpn. J. Appl. Phys.* **38**, L1134 (1999); (b) T. Aoyama, M. Kiyotoshi, S. Yamazaki, and K. Eguchi, *Jpn. J. Appl. Phys.* **38**, 2194 (1999).

5. (a) P. Hones, T. Gerfin, and M. Graetzel, *Appl. Phys. Lett.* **67**, 3078 (1995); (b) G. R. Bai, A. Wang, C. M. Foster, and J. Vetrone, *Thin Solid Films* **310**, 75 (1997); (c) J. M. Lee, J. C. Shin, C. S. Hwang, H. J. Kim, and C. G. Suk, *J. Vac. Sci. Technol.* **16**, 2768 (1998); (d) J. H. Lee, J. Y. Kim, and S. W. Rhee, *Electrochem. Solid-State Lett.* **2**, 622 (1999).
6. L. Meda, R. C. Breitkopf, T. E. Haas, and R. U. Kirss, *Mater. Res. Soc. Symp. Proc.* **495**, 75 (1998).
7. D. Barreca, A. Buchberger, S. Daolio, L. E. Depero, M. Fabrizio, F. Morandini, G. A. Rizzi, L. Sangaletti, and E. Tondello, *Langmuir* **15**, 4537 (1999).
8. (a) C. J. Smart, A. Gulhati, and S. K. Reynolds, *Mater. Res. Soc. Symp. Proc.* **363**, 207 (1995); (b) E. P. Boyd, D. R. Ketchum, H. Deng, and S. G. Shore, *Chem. Mater.* **9**, 1154 (1997).
9. S. Uhlenbrock, and B. A. Vaarstra, U.S. Patent 5,962,716 (1999).
10. (a) Y. Senzaki, W. L. Gladfelter, and F. B. McCormick, *Chem. Mater.* **5**, 1715 (1993); (b) Y. Senzaki, D. Colombo, W. L. Gladfelter, and F. B. McCormick, *Proc. Electrochem. Soc.* **97–25**, 933 (1997).
11. F. Calderazzo, C. Floriani, R. Henzi, and F. L'Eplattenier, *J. Chem. Soc.* 1378 (1969).
12. (a) F.-J. Lee, Y. Chi, P.-F. Hsu, T.-Y. Chou, C.-S. Liu, S.-M. Peng, and G.-H. Lee, *Chem. Vap. Deposition* **7**, 99 (2001). (b) Y.-H. Lai, T.-Y. Chou, Y.-H. Song, C.-S. Liu, Y. Chi, A. J. Carty, S.-M. Peng, and G.-H. Lee, *Chem. Mater.* **15**, 2454 (2003).

26. CARBONYL COMPLEXES OF RUTHENIUM(II) FROM DIMETHYLFORMAMIDE

Submitted by P. SERP,[*] M. HERNANDEZ,[†] and P. KALCK[*]
Checked by JOSHUA H. HOYNE[‡] and JOHN R. SHAPLEY[‡]

Ruthenium(II) carbonyl halides are useful precursors for access to a wide family of complexes containing donor ligands. The usual starting material is "hydrated ruthenium trichloride," and its direct carbonylation in solution leads to mixtures of ruthenium chlorocarbonyl species that have been used for the synthesis of many ruthenium(II) compounds.[1,2] The preparation of the complexes $\{RuCl_2(CO)_2\}_n$ (or a solvated form),[3] $[(CO)_3RuCl_3]^-$,[4] $\{(CO)_3RuCl_2\}_2$,[5] and $[(CO)_4RuCl_2]^{2-}$,[6] by carbonylation at high temperature or under CO pressure of $RuCl_3 \cdot xH_2O$ in alcoholic media, have been also reported.

Here we describe the easy synthesis of $[Ru(CO)_2Cl_3(DMF)]^-$.[7] This compound is produced by the reaction of $RuCl_3 \cdot xH_2O$ with the carbon monoxide generated in situ by the decarbonylation of the solvent, dimethylformamide, in the presence of dry HCl. As an illustration, this intermediate is then used to prepare a triphenylphosphine derivative, namely, *cis*, *cis*, *trans*-$Ru(CO)_2Cl_2(PPh_3)_2$.[8]

*Laboratoire de Catalyse Chimie Fine et Polymères, Ecole Nationale Supérieure d' Ingénieurs en Arts Chimiques et Technologies, 118 Route de Narbonne, Toulouse 31077, France.
†Departamento de Química Inorganica, Facultad de Ciencias, Universidad de Valladolid, Calle Prado de la Magdalena, 47005 Valladolid, Spain.
‡Department of Chemistry, University of Illinois at Urbana—Champaign, Urbana, IL 61801.

Starting Materials

Dimethylformamide (99.8%), ruthenium chloride hydrate ($RuCl_3 \cdot xH_2O$, 35–40% Ru), hydrogen chloride (4 M solution in 1,4-dioxane), and triphenylphosphine (99%) are purchased from Aldrich and used as received. DMF is conveniently dried with activated 4-Å molecular sieves. As $RuCl_3 \cdot xH_2O$ is a hygroscopic salt, it should be handled only briefly under ambient conditions.

■ **Caution.** *All reactions must be carried out in a well-ventilated hood. Carbon monoxide is released during the course of the reactions; it is a highly toxic gas. The HCl solutions cause severe burning. Protective gloves and safety glasses should be worn.*

A. BIS(TRIPHENYLPHOSPHORANYLIDENE)AMMONIUM DICARBONYLTRICHLORO(DIMETHYLFORMAMIDE) RUTHENATE(II), [PPN][Ru(CO)$_2$Cl$_3$(DMF)]

$$RuCl_3 \cdot xH_2O \xrightarrow[160^\circ C]{DMF-HCl\ (diox)} [NH_2Me_2][Ru(CO)_2Cl_3(DMF)] + [NH_2Me_2]Cl$$

$$[NH_2Me_2][Ru(CO)_2Cl_3(DMF)] + [PPN]Cl \xrightarrow[25^\circ C]{DMF} [PPN][Ru(CO)_2Cl_3(DMF)] + [NH_2Me_2]Cl$$

Procedure

A 100-mL Schlenk tube is purged with N_2 and charged with a Teflon-coated magnetic stirring bar, $RuCl_3 \cdot xH_2O$ (0.4 g, 1.53 mmol), 15 mL of dry DMF and anhydrous 4 M HCl in dioxane* (1 mL, 4 mmol). The Schlenk tube is stoppered (with no refluxing condenser system), keeping it connected to the nitrogen line so that a constant pressure of N_2 is maintained in the reaction vessel. The Schlenk tube is then put into a hot oil bath at 160°C for 60 min; the solution turns progressively from red to orange. [*Note*: Monitoring of the reaction indicates the formation of the chlorocarbonyl anion $[Ru(CO)_2Cl_3(DMF)]^-$ with ν_{CO} bands at 2042 and 1964 cm.$^{-1}$†] After cooling the tube to ~60°C or less, [PPN]Cl (0.88 g, 1.53 mmol) is added. The resulting solution is concentrated in vacuo

*Checkers prepared this solution from benzoyl chloride plus methanol (1 equivalent) in dry dioxane.
†The checkers found that additional aliquots of HCl/dioxane (1–3 mL) may be necessary to complete the reaction.

at ~70–80°C to a volume of ~5 mL, and dropwise addition of cold (20°C) distilled water (20 mL) precipitates the complex as a yellow powder.*

The product is collected by filtration using a Büchner funnel, washed with water (100 mL), and dried under vacuum. Recrystallization from acetone/heptane mixture gives yellow crystals. Yield: 0.85 g (64% based on $RuCl_3 \cdot 3H_2O$).

Anal. Calcd. for $C_{41}H_{37}Cl_3N_2O_3P_2Ru$: C, 56.27; H, 4.26; N, 3.20; O, 5.48. Found: C, 56.62, H, 4.27; N 3.05, O 5.34.

Properties

$[PPN][Ru(CO)_2Cl_3(DMF)]$ is a yellow crystalline solid that is relatively stable to air and moisture; it is better stored under nitrogen for long periods of time. This compound is soluble in dimethylformamide, acetone, and chlorinated solvents in which a slow decoordination of the DMF ligand has been observed. IR† (KBr pellets) ν_{CO}: 2045, 1970, 1633 (coordinated DMF) cm^{-1}. ^1H NMR ($CDCl_3$): δ 8.52 (s, 1 H, $HCON(CH_3)_2$), 2.91 (s, 6 H, $HCON(CH_3)_2$). MS (FAB-, CH_3CN): m/z 300 (100%) $[RuCl_3(CO)(DMF)]^-$.

B. *cis, cis, trans*-DICARBONYLDICHLOROBIS (TRIPHENYLPHOSPHINE) RUTHENIUM(II), $Ru(CO)_2Cl_2(PPh_3)_2$

$$[NH_2Me_2][Ru(CO)_2Cl_3(DMF)] + 2\ PPh_3 \xrightarrow[140°C]{DMF} Ru(CO)_2Cl_2(PPh_3)_2 + [NH_2Me_2]Cl$$

Procedure

The DMF solution containing the chlorocarbonyl anion $[Ru(CO)_2Cl_3(DMF)]^-$ is prepared as described in Section 26.A, starting from 0.4 g of $RuCl_3 \cdot xH_2O$. This solution is cooled to 140°C, and 2 equiv of PPh_3 (0.8 g, 3.05 mmol) are added. The solution is maintained at this temperature (~60 min) until its infrared spectrum shows exclusively the ν_{CO} bands at 2058 and 1994 cm^{-1}, characteristic of *cis, cis, trans*-$Ru(CO)_2Cl_2(PPh_3)_2$. The yellow solution is then allowed to cool to room temperature, and addition of 50 mL of water precipitates a pale yellow solid. The product is collected by filtration on a Büchner funnel, washed with

*The checkers occasionally encountered a microcrystalline precipitate that was difficult to collect by filtration. An alternative workup procedure is to dissolve the concentrated reaction solution in CH_2Cl_2 (100 mL), wash this solution with water (30 mL) in a separatory funnel, dry ($MgSO_4$), and evaporate to obtain yellow solid product.

†The checkers observed IR ν_{CO} at 2051 and 1975 cm^{-1} in CH_2Cl_2.

distilled water (100 mL), and dried in vacuo. Yield: 0.7 g (64% based on $RuCl_3 \cdot 3H_2O$).

Anal. Calcd. for $C_{38}H_{30}C_2O_2P_2Ru$: C, 60.65; H, 4.02; O, 4.25. Found: C 60.85; H, 4.15, O, 4.94.

Properties

The complex *cis, cis, trans*-$Ru(CO)_2Cl_2(PPh_3)_2$ is a yellow powder that is stable to air and moisture. IR* (KBr): ν_{CO} 2054, 1992 cm^{-1}. $^{31}P\{^1H\}$ NMR ($CDCl_3$): 17.2 (s) ppm.

References

1. J. Chatt, B. L. Shaw, and A. E. Field, *J. Chem. Soc.* 3466 (1964).
2. J. Halpern, B. R. James, and A. L. W. Kemp, *J. Am. Chem. Soc.* **88**, 5142 (1966).
3. R. Colton and R.H. Farthing, *Aust. J. Chem.* **20**, 1283 (1967).
4. M. L. Berch and A. Davison, *J. Inorg. Nucl. Chem.* **35**, 3763 (1973).
5. M. I. Bruce and F. G. A. Stone, *J. Chem. Soc. A* 1238 (1967).
6. V. F. Borbat, N. M. Sinitsyn, and L. I. Selina, *Zh. Neorg. Khim.* **21**, 751 (1976).
7. P. Serp, M. Hernandez, B. Richard, and P. Kalck, *Eur. J. Inorg. Chem.* 2327 (2001).
8. T. A. Stephenson and G. Wilkinson, *J. Inorg. Nucl. Chem.* **28**, 945 (1966).

27. RHODIUM(I) AND IRIDIUM(I) CARBONYL COMPLEXES

Submitted by P. SERP,[†] M. HERNANDEZ,[‡], B. RICHARD,[†] and P. KALCK[†]
Checked by DAESONG CHONG[§] and MARCETTA Y. DARENSBOURG[§]

Most of the procedures followed to prepare carbonyl containing rhodium(I) or iridium(I) complexes use carbon monoxide as a reducing agent and a ligand. The compound $\{Rh(\mu\text{-}Cl)(CO)_2\}_2$ has been prepared by passing CO slowly over solid $RhCl_3 \cdot 3H_2O$, with the complex subliming from the reactor.[1] By carbonylation of $IrCl_3 \cdot {}_3\delta \rightarrow 3H_2O$ or $\{Ir(\mu\text{-}Cl)(1,5\text{-}C_8H_{12})\}_2$ solutions, an insoluble blue-gray solid is obtained;[2] it is necessary to use dry $IrCl_3$ to obtain $\{IrCl_{1.1}(CO)_3\}_n$ by passing CO over the solid.[3] Sometimes, other strategies can be used to prepare

*The checkers observed IR ν_{CO} at 2058 and 1996 cm^{-1} in CH_2Cl_2.

†Laboratoire de Catalyse Chimie Fine et Polymères, Ecole Nationale Supérieure d' Ingénieurs en Arts Chimiques et Technologiques, 118 Route de Narbonne, Toulouse 31077, France.
‡Departamento de Química Inorganica, Facultad de Ciencias, Universidad de Valladolid, Calle Prado de la Magdalena, 47 005 Valladolid, Spain.
§Department of Chemistry, Texas A&M University, College Station, TX 77843.

the desired carbonyl compound. For instance, in triphenylphosphine/absolute ethanol mixtures, $RhCl_3 \cdot 3H_2O$ reacts quickly with formaldehyde to give $RhCl(CO)(PPh_3)_2$.[4] The complex $IrCl(CO)(PPh_3)_2$ has been synthesized conveniently from $IrCl_3 \cdot xH_2O$ and triphenylphosphine in high-boiling-point solvents such as 2-(2-methoxyethoxy)ethanol[5,6] or dimethylformamide.[7] Dimethylformamide can be decarbonylated in acidic medium to provide an easy source of carbon monoxide.[8] In addition, it is a polar solvent that easily dissolves the rhodium or iridium halide starting salts and, as a high-boiling liquid, it allows a very fast (minutes) carbonylation reaction at its boiling point of 153°C. We describe here the syntheses of the halocarbonyl anions $[Rh(CO)_2Cl_2]^-$ and $[Ir(CO)_2I_2]^-$ from which various complexes can be prepared by simple addition of a ligand, with yields generally higher than 70%.

■ **Caution.** *All reactions must be carried out in a well-ventilated hood. Carbon monoxide is a highly toxic gas. Both 2-methylpropanethiol and DMF have unpleasant odors.*

Materials

Dimethylformamide (99.8%), 2-methylpropanethiol (99%), triphenylphosphine (99%), and triphenylphosphite (99+%) were purchased from Aldrich and used as received. Rhodium trichloride hydrate was a generous loan from Engelhard-CLAL, and iridium iodide was purchased from Johnson Matthey as a mixture of IrI_3 and IrI_4 of several formula $IrI_{3.4}$. The trisodium salt of tris(3-sulfonatophenyl)phospine (TPPTS) was a gift from Hoechst.

A. DMF SOLUTION OF DIMETHYLAMMONIUM DICARBONYLDICHLORORHODATE, $[NH_2Me_2][Rh(CO)_2Cl_2]$

$$RhCl_3 \cdot 3H_2O + 2\ HC(O)NMe_2 \rightarrow [NH_2Me_2][Rh(CO)_2Cl_2] + [NH_2Me_2]Cl$$

Procedure

In a 100-mL Schlenk tube equipped with a magnetic stirrer, 0.35 mL (19.5 mmol) of water is added at room temperature to a suspension of $RhCl_3 \cdot 3H_2O$ (1.0 g, 3.8 mmol) in 15 mL of DMF. The Schlenk tube is stoppered (with no refluxing condenser system), keeping it connected to the nitrogen line so that a constant pressure of N_2 is maintained in the reaction vessel. The Schlenk tube is then placed in a hot oil bath (160°C) for 15 min, and the solution turns progressively from purple brown to light orange. The reaction is quenched by removing the tube from the oil bath as soon as the light orange color appears, since reduction to metallic rhodium can occur at longer reaction times.

Properties

The chlorocarbonyl anion displays two carbonyl stretching IR bands (ν_{CO}: 2063 and 1984 cm^{-1}). The solutions of dimethylamonium dicarbonyldichlororhodate are stable under nitrogen for several days and can be used to prepare several rhodium(I) carbonyl complexes.[8]

B. DICARBONYLBIS(TRIPHENYLPHOSPHINE)BIS (μ-2-METHYLPROPANETHIOLATO)DIRHODIUM, $\{Rh(\mu\text{-}S^tBu)(CO)(PPh_3)\}_2$

$$2[NH_2Me_2][Rh(CO)_2Cl_2] + 2\ {}^tBuSH \rightarrow \{Rh(\mu\text{-}S^tBu)(CO)_2\}_2 + 2HCl + 2[NH_2Me_2]Cl$$

$$\{Rh(\mu\text{-}S^tBu)(CO)_2\}_2 + 2\ PPh_3 \rightarrow \{Rh(\mu\text{-}S^tBu)(CO)(PPh_3)\}_2 + 2CO$$

Procedure

In a 100-mL Schlenk tube equipped with a magnetic stirrer, 10 equiv of 2-methylpropanethiol (3 mL, 33 mmol) are added at room temperature to the $[NH_2Me_2][Rh(CO)_2Cl_2]$ solution prepared as described in Section 27.A. After 30 min, the solution turns from light orange to brown, and an IR spectrum of the solution shows four ν_{CO} bands at 2062, 2050, 1999, and 1984 cm^{-1} characteristic of $\{Rh(\mu\text{-}S^tBu)(CO)_2\}_2$. Then, a 1.5 molar equivalent of triphenylphosphine per rhodium is added (1.5 g, 5.7 mmol). An immediate CO evolution is observed, and the reaction is completed within 5 min. The IR spectrum of the solution shows a single ν_{CO} band at 1977 cm^{-1}. Precipitation of the complex is achieved by adding slowly 50 mL of cold (20°C) distilled water. The product is collected by filtration on a Büchner funnel, washed with 150 mL of water, then dried under vacuum. Yield: 1.51 g (82%).

Anal. Calcd. for $C_{46}H_{48}O_2P_2Rh_2S_2$: C, 57.26; H, 4.97; O, 3.32; S, 6.64. Found: C, 56.81; H, 5.51; O, 3.20; S 6.38.

Properties

The product is air- and moisture-stable for extended periods of time. It is quite soluble in most of the organic solvents, such as toluene, dichloromethane, and various alcohols. IR (KBr) ν_{CO}: 1964, 1948 cm^{-1}. $^{31}P\{^1H\}$ NMR ($CDCl_3$, 101.256 MHz, 298 K): 37.5 ppm (d, J_{Rh-P} = 151 Hz). The complex $\{Rh(\mu\text{-}S^tBu)(CO)(PPh_3)\}_2$ is an effective catalyst for hydroformylation reactions.[9]

C. *trans*-CARBONYLCHLOROBIS(TRIS(3-SULFONATOPHENYL) PHOSPHINE)RHODIUM (SODIUM SALT), $Na_6[Rh(CO)Cl\{P(C_6H_4SO_3)_3\}_2]$

$$[NH_2Me_2][Rh(CO)_2Cl_2] + 2\ Na_3[P(C_6H_4SO_3)_3] \rightarrow Na_6[Rh(CO)Cl\{P(C_6H_4SO_3)_3\}_2] + CO + [NH_2Me_2]Cl$$

Procedure

In a 100-mL Schlenk tube equipped with a magnetic stirrer, 1.95 equiv (4.83 g) of the sodium salt of tris(3-sulfonatophenyl)phosphine (TPPTS) dissolved in a minimum amount of water (ca. 5 mL) are added at room temperature to the solution of $[NH_2Me_2][Rh(CO)_2Cl_2]$ prepared as described in Section 27.A. The reaction is fast, with significant CO evolution. After 15 min, an IR spectrum of the solution shows a single ν_{CO} band at 1982 cm^{-1}. The yellow product is precipitated by adding 50 mL of cold ethanol, then collected by filtration on a Büchner funnel, washed with 150 mL of ethanol, and dried under vacuum. Yield: 3.81 g (71%).

Anal. Calcd. for $C_{37}H_{36}ClNa_6O_{25}P_2S_6Rh$: C, 31.49; H, 2.57; O, 28.34; Found: C, 32.12; H, 2.63; O, 29.68.

Properties

The hygroscopic yellow solid product is air-stable indefinitely and shows high solubility in water. IR (KBr) ν_{CO}: 1968 cm^{-1}. $^{31}P\{^1H\}$ NMR (D_2O, 101.256 MHz, 298 K): 34.2 ppm (d, J_{Rh-P} = 126 Hz).

D. BIS(TRIPHENYLPHOSPHORANYLIDENE)AMMONIUM DICARBONYLDIIODOIRIDATE, $[PPN][Ir(CO)_2I_2]$

$$IrI_{3.4} + HC(O)NMe + CO + H_2O \rightarrow [NH_2Me_2][Ir(CO)_2I_2]\delta + CO_2$$

$$[NH_2Me_2][Ir(CO)_2I_2] + [PPN]Cl \rightarrow [PPN][Ir(CO)_2I_2] + [NH_2Me_2]Cl$$

Procedure

In a 500-mL two-necked flask fitted with a gas inlet adapter, a solution is prepared under nitrogen from 7.81 g (12.5 mmol) of $IrI_{3.4}$*, 280 mL of DMF, and 3 mL of water. The mixture is heated at reflux temperature for 5 h under vigorous CO bubbling. During this time the color of the solution turns from brown to yellow. The reaction mixture is cooled to room temperature, and the CO bubbling is replaced by a nitrogen flow. Occasionally, a small amount of black precipitate is formed; if this happens, the solution is filtered under nitrogen using a cannula/filter paper system. Then, 7.16 g (12.5 mmol) of [PPN]Cl is added to the solution, and the mixture is stirred for 15 min. Precipitation of $[PPN][IrI_2(CO)_2]$ is achieved by adding slowly 300 mL of distilled water under nitrogen. The yellow solid is collected on a sintered-glass frit, washed with 3 × 100 mL of distilled water, and dried under vacuum. Yield: 10.0 g (77%).

Anal. Calcd. for $C_{38}H_{30}I_2NO_2P_2$ Ir: C 43.95, H 2.88, N 1.35; Found: C 44.23, H 2.88, N 1.38.

Properties

The yellow iodocarbonyl complex is air-stable. IR (CH_2Cl_2) ν_{CO}: 2046, 1968 cm^{-1}. ^{13}C NMR ($CDCl_3$, 50.323 MHz, 298 K): δ_{CO} 169.9 (s). Addition of phosphines to the intermediate solution containing $[NH_2Me_2][IrI_2(CO)_2]$ yields the corresponding $[IrI(CO)(PR_3)_2]$ complexes. The $[IrI_2(CO)_2]^-$ anion is an active catalyst for the carbonylation of methanol.

E. TETRACARBONYLBIS(μ-2-METHYLPROPANETHIOLATO) DIIRIDIUM, $\{Ir(\mu\text{-}S^tBu)(CO)_2\}_2$

$$2[NH_2Me_2][Ir(CO)_2I_2] + 2\ {}^tBuSH \rightarrow \{Ir(\mu\text{-}S^tBu)(CO)_2\}_2 + 2\ HI + 2[NH_2Me_2]I$$

A solution is prepared as described in Section 27.D from 2 g (3.2 mmol) of $IrI_{3.4}$, 100 mL of DMF, and 0.5 mL of water, which is refluxed for 5 h under vigorous CO bubbling, then cooled to room temperature, and the CO flow is replaced by a

*The iodide content of commercial samples is variable and can affect the time needed for reduction, and is generally 3.4 per iridium.

nitrogen flow. Any black precipitate is removed by filtration with a cannula/filter paper system; then, the solution is concentrated to 5 mL at 110°C under vacuum. After cooling to room temperature, 1 mL (11 mmol) of 2-methyl-2-propanethiol is added under nitrogen, and the solution is stirred for 90 min. The desired complex is precipitated by the addition of 300 mL of distilled water at room temperature under nitrogen. The deep brown precipitate is collected on a Büchner funnel, washed with 3 × 100 mL of distilled water, and dried under vacuum. Yield 0.75 g (62%).

Anal. Calcd. for $C_{12}H_{18}O_4S_2Ir_2$: C, 22.4; S, 9.4; H, 2.7; Found: C, 22.7; S, 9.3; H, 2.8.

Properties

The crystals of $\{Ir(\mu\text{-}S^tBu)(CO)_2\}_2$ are red or black (two allotropic forms). The compound is stable to air and moisture but is preferably stored under nitrogen for extended periods. This complex is volatile and sublimes at 120°C under reduced pressure (0.1 torr). IR (*n*-hexane) ν_{CO}: at 2040, 1986, and 1963 cm^{-1}. ^1H NMR (CD_2Cl_2, 298 K): δ 3.3 (s). ^{13}C NMR (CD_2Cl_2, 50.323 MHz, 298 K): δ 175.0 (s, CO), 50.2 (s, CMe_3), and 34.3 (s, CH_3). MS-EI (70 eV, m/z): 674 $[Ir_2(\mu\text{-}SC_4H_9)_2(CO)_4]^+$, 618 $[Ir_2(\mu\text{-}SC_4H_9)_2(CO)_2)]^+$, 590 $[Ir_2(\mu\text{-}SC_4H_9)_2(CO)]^+$, and 562 $[Ir_2(\mu\text{-}SC_4H_9)_2]^+$. The volatile precursor has been used for iridium organometallic chemical vapor deposition.[10] An alternative synthesis of the compound by carbonylation of $\{Ir(1,5\text{-COD})(\mu\text{-}S^tBu)\}_2$ has been described.[11]

References

1. J. A. McCleverty and G. Wilkinson, *Inorg. Synth.* **8**, 211 (1966).
2. D. De Montauzon and R. Poilblanc, unpublished results.
3. P. Ginsberg, J. W. Koepke, and C. R. Sprinkle, *Inorg. Synth.* **19**, 18 (1979).
4. D. Evans, J. A. Osborn, and G. Wilkinson, *Inorg. Synth.* **11**, 99 (1968).
5. L. Vaska and J. W. DiLuzio, *J. Am. Chem. Soc.* **83**, 2784 (1961).
6. K. Vrieze, *Inorg. Synth.* **11**, 102 (1968).
7. J. P. Collman, C. T. Sears, and M. Kubota, *Inorg. Synth.* **11**, 103 (1968).
8. P. Serp, M. Hernandez, B. Richard, and P. Kalck, *Eur. J. Inorg. Chem.* 2327 (2001).
9. P. Kalck, *Polyhedron* **7**, 2441 (1988).
10. P. Serp, R. Reurer, P. Kalck, H. Gomes, J. L. Faria, and J. L. Figuelredo, *Chem. Vap. Depos.* **7**, 59 (2001).
11. D. De Montauzon and R. Poilblanc, *Inorg. Synth.* **20**, 237 (1980).

28. USEFUL MONONUCLEAR RHODIUM(I) AND IRIDIUM(I) COMPOUNDS

Submitted by MARC A. F. HERNANDEZ-GRUEL,* JESÚS J. PÉREZ-TORRENTE,* MIGUEL A. CIRIANO,* and LUIS A. ORO*
Checked by DON KROGSTAD†

Simple diolefin and carbonyl rhodium(I) and iridium(I) compounds are useful starting materials for the preparation of catalyst precursors and polynuclear complexes or clusters. Of particular relevance are the 2,4-pentanedionate (acetylacetonate) derivatives, since they provide the synthon ML_2^+ by protonation even with weak acids (pK_a Hacac = 9.0). These metal fragments with two accessible coordination sites in cis positions can either exist as the solvates in a donor solvent or can be captured by anionic ligands.[1] The "acac method" as a way of synthesis of related square planar gold(III) compounds has been reviewed.[2] In addition, dicarbonyl-2,4-pentanodionatorhodium is a catalyst precursor for the reduction of carbon monoxide to ethylene glycol[3] and for the hydroformylation and hydrosilylation of alkenes.[4] A second group of useful complexes are the anionic dihalocarbonyl complexes of rhodium and iridium. The anions $[Rh(CO)_2X_2]^-$ (X = Cl, Br, I) were prepared by Vallarino starting from di(μ-chloro)tetracarbonyldirhodium(I) and the hydrohalic acid.[5]

They have also been prepared from the salts $MCl_3 \cdot nH_2O$ (M = Rh, Ir) and formic acid in the presence of the corresponding hydrohalic acid.[6] Dihalocarbonylrhodate anions have a relevant role in the conversion of methanol to acetic acid (Monsanto process).[7] The anions $[Ir(CO)_2X_2]^-$ (X = Cl, Br, I) were first obtained by Malatesta[8] by reducing carbonyliridates with zinc and later by carbonylation of $IrCl_3 \cdot nH_2O$ in 2-methoxyethanol.[9] The following syntheses are uncomplicated and fast, use the readily available rhodium and iridium compounds $RhCl_3 \cdot 3H_2O$ and $\{Ir(COD)(\mu\text{-}Cl)\}_2$,[10] and provide the products cleanly and in good yields.

Materials

The precious-metal salts are obtained from Johnson Mathey Co. The commercial salts are sold on the basis of metal contained and may not be quite stoichiometric, such as for $RhCl_3 \cdot 3H_2O$. Other starting materials and solvents are reagent-grade samples used as received from commercial sources.

*Departamento de Química Inorgánica, Instituto de Ciencia de Materiales de Aragón, Universidad de Zaragoza-CSIC, E-50009 Zaragoza, Spain.
†Department of Chemistry, Moorhead State University, Moorhead, MN 56563.

A. DICARBONYL-2,4-PENTANEDIONATORHODIUM(I), $Rh(CO)_2(acac)$

$$RhCl_3 \cdot 3H_2O + 2\ HC(O)NMe_2 \rightarrow [NH_2Me_2][Rh(CO)_2Cl_2] + [NH_2Me_2]Cl$$

$$[NH_2Me_2][RhCl_2(CO)_2] + CH_2(COMe)_2 + HC(O)NMe_2 \rightarrow Rh(CO)_2(acac) + CO + 2[NH_2Me_2]Cl$$

Acetylacetonatodicarbonylrhodium has been prepared by two alternative routes, either from $\{Rh(CO)_2(\mu\text{-}Cl)\}_2$ and barium carbonate in the presence of acetylacetone[11] or by reaction of $RhCl_3 \cdot 3H_2O$ with *N,N*-dimethylformamide, also when acetylacetone is present.[12] The second method provides the compound in a single-pot reaction from the Rh(III) starting material. The anion $[Rh(CO)_2Cl_2]^-$ is an intermediate in this synthesis[13] that can be isolated, in the absence of acetylacetone, by addition of a large cation.

■ **Caution.** *This synthesis should be performed in a well-ventilated hood, since carbon monoxide is released during the reaction. Carbon monoxide is a highly toxic, colorless, and odorless gas.*

Procedure

The reaction is carried out in a 1-L two-necked, round-bottomed flask equipped with a stirring bar, a gas inlet, and a reflux condenser connected to a bubbler. After charging the flask with 10 g (0.38 mol) of $RhCl_3 \cdot 3H_2O$ and 200 mL of dimethylformamide, the condenser is fitted, the reaction vessel is purged with argon (or oxygen-free nitrogen), and the suspension is gradually heated until the solvent starts to reflux. Then 20 mL (1.94 mol) of acetylacetone are introduced through the top of the condenser against a slow flow of the inert gas. Release of gas (CO) is immediately observed, and the mixture turns color from deep red to deep orange in approximately 10 min. Heating to cause reflux is continued for a total of 30 min, and then the flask is allowed to cool to room temperature. The solution is transferred open to the air into a 2-L beaker, and 1 L of distilled water is added. The fine purple precipitate formed is collected by filtration with a fritted-glass funnel, washed with 10×100 mL of distilled water, and then with 2×15 mL of cold methanol, and finally dried by suction with air. The crude solid is transferred to a flask and extracted with 200 mL of dichloromethane. Anhydrous magnesium sulfate is added to dry the extract, and then the mixture is filtered through Celite. The cake of $MgSO_4$ and Celite is washed with 50 mL of dichoromethane, collecting the washing liquid with the extract to obtain a clear golden yellow filtrate. Evaporation of the solvent under vacuum to dryness gives the product as green-bronze needles, which are washed with cold methanol (2 × 10 mL), separated by filtration, and vacuum-dried. The product is sufficiently pure for most purposes. (**Note**: The synthesis can be carried out on a scale of two to ten grams with a proportionate adjustment in the amounts of chemicals). The compound can be purified further by sublimation at 90°C/0.1 torr. Yield: 8.3 g (85%).

Anal. Calcd. for $C_7H_7O_4Rh$: C, 32.58; H, 2.73. Found: C, 32.78; H, 2.56.

Properties

$Rh(CO)_2(acac)$ is a dichroic solid, appearing either as dark green neddles with a bronze shine or a purple microcrystalline solid. It is readily soluble in moderately polar organic solvents to give yellow solutions, and it is slightly soluble in cold methanol or cold alkanes. The solid is stable in air and can be stored in air for long periods in a closed vessel. The IR spectrum in dichoromethane shows ν_{CO} absorptions at 2086 (s), 2013 (s), and 1984 (w) cm^{-1}, whereas four bands are observed at 2084 (s), 2067 (w), 2015 (s), and 1985 (w) cm^{-1} in hexanes. ^1H NMR spectrum ($CDCl_3$, 25°C, 300 MHz) δ 2.04 (s, 6H, CH_3), 5.58 (s, 1H, CH). $^{13}C\{^1H\}$ NMR spectrum ($CDCl_3$, 25°C, 75.5 MHz) δ 26.9 (s, CH_3), 101.7 (s, *C*H), 183.8 (d, $J_{Rh-C} = 73$ Hz, Rh—*C*O), 187.4 (s, *C*O). One or both carbonyl ligands can be replaced by phosphorus donor ligands.[14]

B. BENZYLTRIPHENYLPHOSPHONIUM DICARBONYLDICHLORORHODATE, $[PPh_3CH_2Ph][Rh(CO)_2Cl_2\}$

$$Rh(CO)_2(acac) + HCl + [PPh_3CH_2Ph]Cl \rightarrow [PPh_3CH_2Ph][Rh(CO)_2Cl_2] + Hacac$$

Procedure

A one-necked 250 mL round-bottomed flask equipped with a stirring bar is charged with a mixture of $Rh(CO)_2(acac)$ (1.0 g, 3.88 mmol), $[PPh_3CH_2Ph]Cl$ (1.51 g, 3.88 mmol) and 10 mL of methanol. Addition of 20 drops (~0.7 mL, 8.4 mmol) of 37% HCl in water to the resulting suspension gives a clear yellow solution, which is stirred for 15 min at room temperature. Addition of 150 mL of cold diethyl ether with a strong stirring causes precipitation of the product as a yellow microcrystalline solid. (*Note*: Sometimes a yellow oil is obtained after adding diethyl ether; scratching the walls of the flask with a glass rod leads to the precipitation of the product). The solid product is collected on a sintered-glass filter, washed with 2 × 10 mL of diethyl ether, and dried on the filter by suction with air. Typical yields are 1.8–2.0 g (80–90%). The product is stored under argon or nitrogen, since it shows slight decomposition after some months in the air.

Anal. Calcd. for $C_{27}H_{22}Cl_2O_2PRh$: C, 55.60; H, 3.80. Found: C, 55.73; H, 3.62.

Properties

The product is a yellow crystalline solid that is soluble in polar solvents such as acetone, dichloromethane, chloroform, methanol, and nitromethane and insoluble in diethyl ether, hexanes, and toluene. IR spectrum (CH_2Cl_2): 2070 (s) and 1992 (s) cm^{-1}. ^1H NMR spectrum ($CDCl_3$, 25°C, 300 MHz): δ 4.83 (d, $^2J_{P\text{–}H} = 14.1$ Hz, 2H, CH_2Ph), 6.93–7.22 (m, 5H, CH_2*Ph*), 7.51–7.77 (m, 15H, P*Ph*$_3$). $^{13}C\{^1H\}$ NMR spectrum ($CDCl_3$, 25°C, 75.5 MHz): δ 30.9 (d, $^1J_{P\text{–}C} = 47$ Hz, CH_2Ph), 116.6–135.3 (*C* aromatic), 181.5 (d, $J_{Rh\text{–}C} = 72$ Hz, Rh—*C*O). $^{31}P\{^1H\}$ NMR spectrum ($CDCl_3$, 25°C, 121.5 MHz): 23.4(s) ppm.

C. BENZYLTRIPHENYLPHOSPHONIUM (1,5-CYCLOOCTADIENE) DICHLOROIRIDATE, $[PPh_3CH_2Ph][Ir(COD)Cl_2]$

$$\{Ir(COD)(\mu\text{-}Cl)\}_2 + 2[PPh_3CH_2Ph]Cl \rightarrow 2\ [PPh_3CH_2Ph][Ir(COD)Cl_2]$$

Procedure

In a 50-mL Schlenk tube fitted with a stirring bar, 5 mL of deoxygenated dichloromethane is added to a finely divided mixture of $\{Ir(COD)(\mu\text{-}Cl)\}_2$ (1.0 g, 1.49 mmol) and $[PPh_3CH_2Ph]Cl$ (1.17 g, 2.98 mmol) under an inert-gas atmosphere. All the solids dissolve in 5 min at room temperature to give a deep yellow solution. Then 40 mL of ice-cold deoxygenated diethyl ether is added with strong stirring, and the title compound precipitates immediately as a bright yellow microcrystalline solid. The stirring is continued for 5 min, and then the solid is quickly filtered in air on a sintered-glass filter, washed with 3 × 5 mL of diethyl ether, and vacuum-dried. The isolated solid is stored under argon or nitrogen at room temperature. Yield: 2.12 g (98%).

Anal. Calcd. for $C_{33}H_{34}PCl_2Ir$: C, 54.69; H, 4.73. Found: C, 54.65; H, 4.55.

Properties

$[PPh_2CH_2Ph][Ir(COD)Cl_2]$ is a yellow microcrystalline solid. It is very soluble in methanol, but these solutions decompose quickly. It is soluble in dichloromethane and chloroform, less soluble in acetone, and insoluble in diethyl ether, hexanes, and toluene. ^{1}H NMR spectrum ($CDCl_3$, 25°C, 300 MHz): δ 1.15 (m, 4H, CH_2), 1.93 (m br, 4H, CH_2), 3.79 (br s, 4H, =CH), 5.02 (d, $^{2}J_{P-H}$ = 14.1, 2H, CH_2Ph), 6.95–7.21 (m, 5H, CH_2Ph), 7.55–7.75 (m, 15H, PPh_3). ^{13}C{^{1}H} NMR spectrum ($CDCl_3$, 25°C, 75.5 MHz): δ 30.9 (d, $^{1}J_{P-C}$ = 47 Hz, CH_2Ph), 31.8 (s, CH_2), 59.1 (s, =CH), 117.4 (d, $^{1}J_{P-C}$ = 86 Hz PPh_3), 126.8 (d, $^{2}J_{P-C}$ = 8 Hz, CH_2*Ph*), 128.3 (d, $^{5}J_{P-C}$ = 4 Hz, CH_2*Ph*), 128.6 (d, $^{4}J_{P-C}$ = 3 Hz, CH_2*Ph*), 130.1 (d, $^{2}J_{P-C}$ = 12 Hz, PPh_3), 131.3 (d, $^{3}J_{P-C}$ = 5 Hz, CH_2*Ph*), 134.1 (d, $^{3}J_{P-C}$ = 10 Hz, P*Ph*$_3$), 134.9 (d, $^{4}J_{P-C}$ = 3 Hz, P*Ph*$_3$). ^{31}P{^{1}H} NMR spectrum ($CDCl_3$, 25°C, 121.5 MHz): 23.1(s) ppm.

D. BENZYLTRIPHENYLPHOSPHONIUM DICARBONYLDICHLOROIRIDATE, [PPh_3CH_2Ph][$Ir(CO)_2Cl_2$]

$$[PPh_3CH_2Ph][Ir(COD)Cl_2] + 2\ CO \rightarrow [PPh_3CH_2Ph][Ir(CO)_2Cl_2] + COD$$

■ **Caution.** *This preparation should be performed in a well-ventilated hood, since CO is used for the reaction. Carbon monoxide is a highly toxic, colorless, odorless gas.*

Procedure

A deep yellow dichloromethane solution of [PPh_3CH_2Ph][Ir(COD)Cl_2] is prepared as described in Section 28.C starting from {Ir(COD)(μ-Cl)}$_2$ (1.0 g, 1.49 mmol) in a 50-mL Schlenk flask. The flask is connected to a cylinder of carbon monoxide, evacuated, and filled with this gas, and the solution is stirred under a CO atmosphere at room temperature for 15 min to give a pale yellow-green solution. The connection to the CO reservoir is removed, and then 30 mL of ice-cold diethyl ether is immediately added with strong stirring. Scratching of the walls with a glass rod helps to precipitate an off-white solid. The stirring is continued for 5 min to complete the precipitation. The solid is filtered quickly on a fritted-glass funnel in air, washed with 3 × 10 mL of diethyl ether, and vacuum-dried. The isolated solid is stored under argon or nitrogen at room temperature. Yield: 1.80–1.90 g (90–95%).

Anal. Calcd. for $C_{27}H_{22}Cl_2O_2PIr$: C, 48.22; H, 3.30. Found: C, 48.52; H, 3.89.

Properties

[PPh_3CH_2Ph][$Ir(CO)_2Cl_2$] is a pale yellow solid that is soluble in dichloromethane, methanol, acetone, and chloroform and insoluble in diethyl ether and hexanes. The IR spectrum in dichloromethane shows two strong ν_{CO} bands at 2056 and 1973 cm^{-1}. ^{1}H NMR spectrum ($CDCl_3$, 25°C, 300 MHz): δ 4.90 (d, $^{2}J_{P-H}$ = 13.8, 2H, CH_2Ph), 6.94–7.23 (m, 5H, CH_2Ph), 7.53–7.72 (m, 15H, PPh_3). $^{13}C\{^{1}H\}$ NMR spectrum ($CDCl_3$, 25°C, 75.5 MHz): δ 30.9 (d, $^{1}J_{P-C}$ = 47.8 Hz, CH_2Ph), 116.7–135.3 (HC aromatic), 167.8 (s, Ir—*CO*). $^{31}P\{^{1}H\}$ NMR spectrum ($CDCl_3$, 25°C, 121.5 MHz: 23.4 (s) ppm.

References

1. P. R. Sharp, in *Comprehensive Organometallic Chemistry II*, E. W. Abel, F. G. A. Stone, and G. Wilkinson, eds., Pergamon, Oxford, Vol. 8, Chapter 3, 1995; M. A. Casado, J. J. Pérez-Torrente, M. A. Ciriano, A. J. Edwards, and F. J. Lahoz, L. A. Oro, *Organometallics* **18**, 5299 (1999).
2. J. Vicente and M. T. Chicote, *Coord. Chem. Rev.* **193–195**, 1143 (1999).
3. E. Watanabe, K. Muruyama, Y. Hara, Y. Kobayashi, K. Wada, and T. Onada, *J. Chem. Soc., Chem. Commun.* 227 (1986).
4. C. D. Frohning, and Ch. W. Kochpaintner, in *Applied Homogeneous Catalysis with Organometallic Compounds*, B. Cornils and W. A. Herrmann, eds. Wiley-VCH, Weinheim, 2000, p. 28.
5. L. M. Vallarino, *Inorg. Chem.* **4**, 161 (1965).
6. M. J. Cleare and W. P. Griffith, *J. Chem. Soc. A* 2788 (1970).
7. A. Haines, B. E. Mann, G. E. Morris, and P. M. Maitlis, *J. Am. Chem. Soc.* **115**, 4093 (1993).
8. L. Malatesta, L. Naldini, and F. Cariati, *J. Chem. Soc.* 961 (1964); L. Malatesta and F. Canziani, *J. Inorg. Nucl. Chem.* **19**, 81 (1961).
9. D. Foster, *Inorg. Nucl. Chem. Lett.* **5**, 433 (1969).
10. J. L. Herde, J. C. Lambert, and C. V. Senoff, *Inorg. Synth.* **15**, 18 (1974).
11. F. Bonati and G. Wilkinson, *J. Chem Soc.* 3156 (1964).
12. Y. S. Varshavskii and T. G. Cherkasova, *Russ. J. Inorg. Chem., Int. Ed. Engl.* **12**, 899 (1967).
13. P. Serp, M. Hernandez, and Ph. Kalck. *C. R. Acad. Sci. Paris, Ser II c* **2**, 267 (1999).
14. J. G. Leipoldt, L. C. D. Bok, J. S. van Vollenhoven, and A. I. Pieterse, *J. Inorg. Nucl. Chem.* **40**, 61 (1978); A. M. Trzeciak and J. J. Ziólkowski, *Inorg. Chim. Acta Lett.* **64**, L267 (1982).

Chapter Four

CYANIDE COMPOUNDS

29. HEXACYANOMETALLATES AS TEMPLATES FOR DISCRETE PENTANUCLEAR AND HEPTANUCLEAR BIMETALLIC CLUSTERS

Submitted by LEONE SPICCIA,* KEITH S. MURRAY,* and JACQUI F. YOUNG*
Checked by TALAL MALLAH†

Hexacyanometallates are excellent building blocks for constructing extended-array multimetallic assemblies with novel structural features and physicochemical properties. Recognition of the propensity of cyanide groups to bridge metal centers has led to the application of hexacyanometallates as templates for the synthesis of discrete heteropolynuclear complexes with high-spin ground states,[1–7] with target applications as "single molecule magnets," as well as heterometallic coordination polymers with one-, two- or three-dimensional extended-array structures capable of exhibiting long-range magnetic ordering.[8]

One synthetic strategy to prepare discrete heteropolynuclear assemblies has involved *N*-capping of the cyanides on hexacyanometallates, such as $[Fe(CN)_6]^{3-/4-}$ and $[Cr(CN)_6]^{3-}$, with appropriate metal complexes. Strongly coordinating multidentate ligands are introduced on the second metal center so as to reduce the number of sites available for bridge formation to the hexacyanometallate. Displacement of weakly binding ligands by the cyano groups

* School of Chemistry, Monash University, PO Box 23, Victoria 3800, Australia.
† Laboratoire de Chimie Inorganique, CNRS 8613, Université Pars-Sud, 91405 Orsay, France.

Inorganic Syntheses, Volume 34, edited by John R. Shapley
ISBN 0-471-64750-0

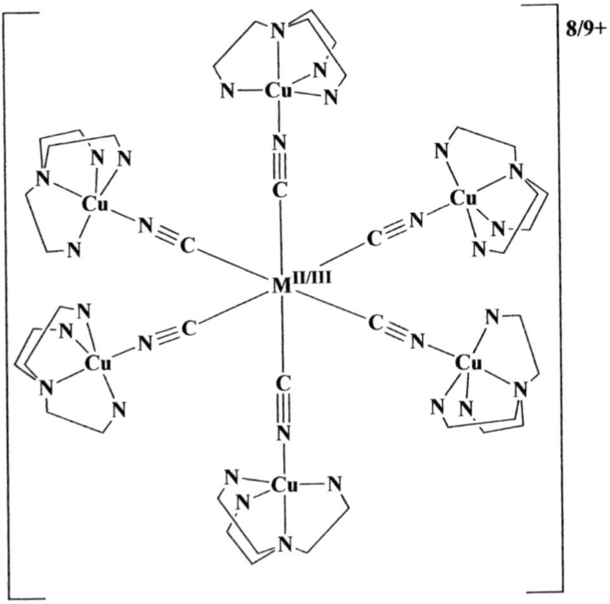

Scheme 1. Generation of five-coordinate Cu^{II} geometries. Around a hexacyanometallate core (adapted from Ref. 4 with permission from the American Chemical Society).

forms the heterometallic complexes. For metal ions that form six coordinate complexes (e.g., Ni^{II}, $Mn^{II/III}$), heptanuclear cations have been formed by reacting the complexes of pentadentate ligands with the hexacyanometallate (usually in a 6 : 1 molar ratio). For example, the reaction of $[Cr(CN)_6]^{3-}$ or $[Fe(CN)_6]^{4-}$ with appropriate metal complexes generated $[\{Ni(L^1)(CN)\}_6Cr][ClO_4]_9$,[1] $[\{Mn(L^3)CN\}_6Cr][Cr(CN)_6](ClO_4)_6$,[2] $[\{Mn(L^3)CN\}_6Fe][ClO_4]_8$,[2] and $[\{Ni(L^2)CN\}_6Fe][ClO_4]_8$[2] ($L^1$ = tetraethylenepentamine; L^2 = *N*,*N*,*N'*-(tris(2-pyridylmethyl)-*N'*-methyl-ethane)-1,2-diamine; L^3 = 1,4-bis(2-pyridylmethyl)-1,4,7-triazacyclononane). Heptanuclear clusters have also been formed through the use of Cu^{II} complexes of branched tetradentate ligands, since such a combination can generate five-coordinate Cu^{II} geometries, such as $[\{Cu(L^4)CN\}_6Fe][ClO_4]_8 \cdot 3H_2O$,[3] $[\{Cu(L^5)CN\}_6Fe][ClO_4]_8 10H_2O$,[4] and $[\{Cu(L^5)CN\}_6Fe][Fe(CN)_6]_2[ClO_4]_2 \cdot 18H_2O$,[4] ($L^4$ = tris(2-pyridylmethyl)amine, L^5 = tris(2-aminoethyl)amine (see Scheme 1). The isolation of $[M_4Co_4(CN)_{12}(L^6)_8]^{12+}$,[5] an octanuclear cluster with a cubic arrangement of metal ions, produced by reacting $[(L^6)M(OH_2)_3]^{3+}$ (M = Cr or Co) with $[(L^6)Co(CN)_3]$, and $[\{(L^7)_2Ni\}_3\{Fe(CN)_6\}_2]$,[6] a neutral pentanuclear cluster prepared from $[(L^7)_2Ni(OH_2)_2]^{2+}$ and $[Fe(CN)_6]^{3-}$ (L^6 = 1,4,7-triazacyclononane; L^7 = bis(pyrazolyl)methane), further highlights the importance of the choice of ligands. Reliable syntheses to one pentanuclear and three heptanuclear cyano-bridged bimetallic clusters are detailed below.

■ **Caution.** *Although no problems were encountered in the syntheses that follow, transition metal perchlorates are potentially explosive and should be prepared in small quantities. Due care must be taken when handling perchlorate and cyanide salts. All materials are commercial samples of reagent grade unless indicated otherwise.*

A. AQUATRIS(2-PYRIDYLMETHYL)AMINECOPPER(II) DIPERCHLORATE, $[Cu(tpa)(OH_2)][ClO_4]_2$

$$Cu(ClO_4)_2 \cdot 6H_2O + tpa \longrightarrow [Cu(tpa)(OH_2)][ClO_4]_2$$

Procedure

Tris(2-pyridylmethyl)amine (tpa) can be prepared by the method reported by Anderegg and Wenk[9] or by a modification reported recently by Canary et al.[10] The latter procedure avoids the isolation of a hydroperchlorate salt. Other reagents are available from commercial sources.

The Cu(II)-tpa complex is prepared by a literature method[11] as follows. Equimolar quantities of tpa (1.00 g, 3.45 mmol) and copper(II) perchlorate hexahydrate (1.28 g, 3.45 mmol) are added to 50 mL of acetonitrile in a 100-mL round-bottomed flask. The resulting solution is heated on a steam bath for a short time (5–10 min), after which the volume is reduced to about half, *in vacuo*. Crystals of $[Cu(tpa)(OH_2)][ClO_4]_2$ form on standing at room temperature. Yield: 1.63 g, 83%. Analytical data are in agreement with those reported in the literature.[11]

Properties

In the IR spectrum of the compound, a broad O—H stretch at 3418 cm^{-1} indicates the presence of water; bands at 1609, 1480, and 1439 cm^{-1} confirm the presence of the pyridyl rings of the ligand, and bands at 1103 and 626 cm^{-1} are characteristic of the perchlorate counterions. The UV–visible spectrum shows a broad band at 858 nm (201 $M^{-1}\ cm^{-1}$).

B. HEXAKIS{CYANOTRIS(2-PYRIDYLMETHYL)-AMINECOPPER(II)}IRON(II) OCTAPERCHLORATE TRIHYDRATE, $[\{Cu(tpa)(CN)\}_6Fe][ClO_4]_8 \cdot 3H_2O$[3]

$$6[Cu(tpa)(OH_2)][ClO_4]_2 + K_3[Fe^{III}(CN)_6] \longrightarrow [\{Cu(tpa)(CN)\}_6Fe^{II}][ClO_4]_8 \cdot 3H_2O$$

Procedure

A solution prepared by dissolving 0.96 g of $[Cu(tpa)(OH_2)][ClO_4]_2$ (1.7 mmol) in 20 mL of water is stirred, and the pH is adjusted to pH 7 using 0.5 M sodium hydroxide solution. A solution of 92 mg of $K_3[Fe(CN)_6]$ (0.28 mmol) dissolved in 8 mL of water is then added dropwise to the stirred solution. This resulted in the immediate formation of a purple precipitate that was collected by filtration; washed successively with cold water, ethanol, and ether; and then air-dried to afford a purple powder of $[\{Cu(tpa)(CN)\}_6Fe][ClO_4]_8 \cdot 3H_2O$ (Yield: 0.66 g, 74%). Small dark purple crystals suitable for X-ray structure determination were obtained by slow evaporation of a solution prepared by dissolving the purple powder in hot water.

Anal. Calcd. for $C_{114}H_{114}N_{30}Cl_8O_{35}Cu_6Fe$: C, 43.0; H, 3.6; N, 13.2. Found: C, 42.9; H, 3.6; N, 13.2%.

Properties

The encapsulation of a ferrocyanide core by six $[Cu(tpa)]^{2+}$ moieties forming a quasispherical heptanuclear complex is confirmed by X-ray crystallography.[3] The geometry of the hexacyanoferrate core is little perturbed from that in the free ion, with Fe—CN distances of 1.85–1.90 Å, and bond angles typical for octahedral geometry. These distances, Cu—NC distances of 1.92–1.95 Å and the almost linear Cu—N—C angles, result in Fe···Cu distances of 4.89–4.95 Å. The IR spectrum shows a strong band at 2109 cm^{-1} attributable to the stretching of the CN group in Fe^{II}—CN—Cu^{II} units. The UV–visible spectrum shows bands at 675 (sh) nm (820 M^{-1} cm^{-1}) and 843 nm (1370 M^{-1} cm^{-1}), which are typical of Cu^{II} in trigonal bipyramidal geometry, and a band at 515 nm (1200 M^{-1} cm^{-1}) due to a $Fe^{II} \rightarrow Cu^{II}$ metal-to-metal charge transfer (MMCT) transition. A small drop in the effective magnetic moment, observed as the temperature is lowered, is indicative of weak antiferromagnetic coupling.[3]

C. 1,4-BIS(2-PYRIDYLMETHYL)-1,4,7-TRIAZACYCLONONANE (dmptacn)[12]

Procedure

Samples of tacn · 3HCl (available from Sigma-Aldrich Fine Chemicals) (0.65 g, 2.44 mmol) and 2-picolylchloride hydrochloride (available from Sigma-Aldrich Fine Chemicals) (0.80 g, 4.88 mmol) are dissolved in 20 mL of water. The pH of the solution is adjusted to 9 by the addition of 2 M NaOH with stirring. After 3 days, the pH of the resulting pale orange solution falls to approximately 7. It is readjusted to 9, and the solution is stirred for an additional 3 days. The pH of the reaction mixture is increased to ~13 with 2 M NaOH, and the free ligand is extracted into chloroform (6 × 70 mL). The chloroform solution is dried over Na_2SO_4, filtered, and evaporated under reduced pressure to give a yellow oil (yield: 0.36 g, 50%). The oil is converted into the less hygroscopic tetrahydrobromide by treatment with hydrobromic acid.

Anal. Calcd. for $C_{18}H_{25}N_5 \cdot 4HBr$: C, 33.8; H, 4.6; N, 11.0. Found: C, 34.0; H, 5.3; N, 11.0%.

Properties

Free ligand: ^1H NMR spectrum ($CDCl_3$): δ 2.5–2.8 (m, 6H), 3.43 (bs, NH), 3.9 (s, 2H) $pyCH_2$, 7.11 (t, 1H), 7.48–7.65 (m, 2H), 8.5 (d, 1H). ^{13}C NMR spectrum ($CDCl_3$): δ 157.9, 148.9, 136.5, 122.6, and 122.2 ppm (pyridine ring carbons); 60.5 ppm ($pyCH_2$) 52.1, 48.8, and 44.9 (tacn ring carbons). Mass spectrum: m/z^+ 311.

D. AQUA(1,4-BIS(2-PYRIDYLMETHYL)-1,4,7-TRIAZACYCLONONANE)MANGANESE(II) DIPERCHLORATE, $[Mn(dmptacn)(OH_2)][ClO_4]_2$[13]

Procedure

$MnCl_2 \cdot 4H_2O$ (0.48 g, 2.4 mmol) is added to a solution of dmptacn (0.76 g, 2.4 mmol) in 125 mL of methanol, and the solution is stirred for 15 min. Then $NaClO_4$ (1.0 g) is added, and stirring is continued until a homogeneous solution is formed. Cooling of the solution to 4°C overnight gives a pale yellow precipitate that is collected by filtration and washed with ethanol. Yield: 1.03 g, 73%.

Anal. Calcd. for $C_{18}H_{27}N_5O_9Cl_2Mn$: C, 36.0; H, 4.8; N, 11.7. Found: C, 36.0; H, 4.6; N, 11.8%. Electron microprobe analysis: Mn : Cl ratio 1 : 2.

Properties

A single N—H stretch at 3326 cm^{-1} and bands due to the pyridyl rings at 1607, 1484, and 1441 cm^{-1} confirm the presence of the ligand in the complex. The magnetic moment of $\mu_{eff} = 5.92$ BM (Bohr magnetons) at 295 K falls in the expected range for high-spin Mn(II) complexes.

E. HEXAKIS{CYANO(1,4-BIS(2-PYRIDYLMETHYL)-1,4,7-TRIAZACYCLONONANE)MANGANESE(II)}IRON(II) OCTAPERCHLORATE PENTAHYDRATE, $[\{Mn(dmptacn)(CN)\}_6Fe][ClO_4]_8 \cdot 5H_2O$

$$6[Mn(dmptacn)(OH_2)][ClO_4]_2 + K_4[Fe^{II}(CN)_6] \longrightarrow [\{Mn(dmptacn)(CN)\}_6Fe^{II}][ClO_4]_8 \cdot 5H_2O$$

Procedure

An aqueous solution of $K_4[Fe(CN)_6]$ (62 mg, 0.19 mmol) in 10 mL of water is added dropwise to a stirred solution of $[Mn(dmptacn)(OH_2)][ClO_4]_2$ (0.58 g, 1.00 mmol) in 50 mL of water, resulting in a change from colorless to pale green. Once the addition is complete, the pale green solution is left to slowly evaporate at room temperature. Pale green needles of the heptanuclear complex (0.53 g, 96%) precipitate.

Anal. Calcd. for $C_{114}H_{160}N_{36}Cl_8O_{37}Mn_6Fe$: C, 41.5; H, 4.9; N, 15.3; $(ClO_4)]^-$, 24.2%. Found: C, 41.5; H, 4.5; N, 15.4; ClO_4^-, 24.7%.

Properties

The X-ray crystal structure of the complex confirms the formation of quasitoroidal heptanuclear $[\{Mn(dmptacn)CN\}_6Fe]^{8+}$ cations whose charge is balanced by perchlorate counteranions.[2b] The geometry around the ferrocyanide shows little deviation from that in the free ion, whereas the Mn^{II} centers are in a distorted trigonal prismatic environment. The complex exhibits weak ferromagnetic coupling between the paramagnetic Mn^{II} centers. The IR spectrum has a strong band at 2052 cm^{-1} attributed to the CN stretch of Fe^{II}—CN—Mn^{II} units.[2b]

F. HEXAKIS{CYANO(TRIS(2-AMINOETHYL)AMINE)COPPER(II)}-CHROMIUM(III) NONAPERCHLORATE HYDRATE, $[\{Cu(tren)(CN)\}_6Cr][ClO_4]_9 \cdot H_2O$

$$6[Cu(tren)(OH_2)][ClO_4]_2 + K_3[Cr^{III}(CN)_6] \longrightarrow [\{Cu(tren)(CN)\}_6Cr^{III}][ClO_4]_9 \cdot H_2O$$

Procedure

Potassium hexacyanochromate(III) is available by a literature method.[14] A solution of $[Cu(tren)(OH_2)][ClO_4]_2$ is prepared by mixing aqueous solutions of $Cu(ClO_4)_2 \cdot 6H_2O$ (2.60 g, 7.00 mmol) in 20 mL of water and tren (1.02g, 6.98 mmol) in 10 mL of water. The dropwise addition of a hexacyanochromate(III) solution, prepared by dissolving 0.38 g of $K_3[Cr(CN)_6]$ (1.16 mmol) in 15 mL of water, to the copper solution results in the immediate formation of a turquoise precipitate. Once the addition is complete, the precipitate is collected by filtration; washed successively with cold water, ethanol, and ether; and then air-dried to afford a blue powder of $[\{Cu(tren)(CN)\}_6Cr][ClO_4]_9 \cdot H_2O$ Yield: 1.50 g, 55%.

Anal. Calcd. for $C_{42}H_{110}N_{30}Cl_9O_{37}Cu_6Fe$: C, 21.2; H, 4.7; N, 17.7. Found: C, 21.3; H, 4.5; N, 17.7%.

Properties

A strong infrared band at 2178 cm^{-1} corresponds to the stretch of the cyano group located within Cr^{III}—CN—Cu^{II} units. Electronic transitions at 590 (sh) (780 M^{-1} cm^{-1}) and 775 nm (1140 M^{-1} cm^{-1}) are typical of Cu^{II} centers in trigonal bipyramidal geometry. Both the variable temperature and variable field magnetic behavior indicate ferromagnetic coupling between Cr^{III} and Cu^{II} mediated by the cyano bridges, and there is evidence of weak intercluster antiferromagnetic coupling at very low temperatures.

G. DIAQUABIS(BIS(1-PYRAZOLYL)METHANE)NICKEL(II) DINITRATE, $[Ni(bpm)_2(OH_2)_2](NO_3)_2$

Procedure

Bis(1-pyrazolyl)methane (bpm) can be prepared by the method of Elguero and co-workers,[15] or by the procedure reported more recently by Jameson and Castellano.[16] Solid bpm is added (2.96 g, 20 mmol) to a stirred solution of $[Ni(OH_2)_6](NO_3)_2$ (2.91 g, 10 mmol) dissolved in 100 mL of water. Total dissolution of the bpm occurs after 15 min, resulting in a deep blue solution. Slow evaporation of the solution over several days gives blue crystals of the product. The crystals are filtered, washed with a small amount of water and diethyl ether, and dried in air. Yield: 4.3 g, 84%.

Anal. Calcd. for $C_{14}H_{20}N_{10}O_8Ni$: C, 32.6; H, 3.9; N, 27.2. Found: C, 32.4; H, 4.1; N, 29.1.

Properties

Electronic spectral transitions at 367 (7 M^{-1} cm^{-1}), 596 (4 M^{-1} cm^{-1}), and 743 nm (0.6 M^{-1} cm^{-1}) are typical of octahedral Ni(II) and are assigned to transitions from $^3A_{2g}$ to $^3T_{1g}$ (P), $^3T_{1g}$ (F) and $^3T_{2g}$, respectively. The magnetic moment of 3.07 μ_B at 295 K is also typical of octahedral geometry. A powder XRD pattern is similar to that of the Cu(II) analog whose structure shows trans arrangement of bpm ligands.[17]

H. TRIS{BIS(BIS(1-PYRAZOLYL)METHANE)NICKEL(II)} BIS{HEXACYANOFERRATE(III)} HEPTAHYDRATE, $\{bpm)_2Ni\}_3\{Fe(CN)_6\}_2 \cdot 7H_2O$

$$3[(bpm)_2Ni(OH_2)_2](NO_3)_2 + 2\ K_3[Fe^{III}(CN)]_6 \longrightarrow \{(bpm)_2Ni\}_3\{Fe^{III}(CN)_6\}_2 \cdot 7H_2O$$

Procedure

A solution of $K_3[Fe(CN)_6]$ (0.33 g, 1 mmol) dissolved in 50 mL of water is added slowly to a vigorously stirred solution of $[(bpm)_2Ni(OH_2)_2](NO_3)_2$ (0.77 g, 1.5 mmol) in 50 mL of water. This leads to the immediate precipitation of a fine yellow solid. The suspension is centrifuged, and the solid is recovered by decantation of the supernatant. The solid residue is washed with water and recentrifuged twice, then left to dry in air overnight. Yield: 0.79 g, 98%. Crystal suitable for X-ray structure determination are obtained by diffusing a solution of $K_3[Fe(CN)]_6$ (0.08 g, 0.25 mmol) in 2-propanol/water (1 : 1 v/v; 30 mL of mixed solution) through the glass frit of a H-tube into a solution of $[(bpm)_2Ni(OH_2)_2]$ $(NO_3)_2$ (0.13 g, 0.25 mmol) in 30 mL of DMF. After several weeks in the dark, well-shaped yellow-orange crystals are obtained.

Anal. Calcd. for $C_{54}H_{62}N_{36}O_7Ni_3Fe_2$: C, 40.2; H, 3.9; N, 31.2. Found: C, 40.0; H, 3.9; N, 31.3.

Properties

The complex consists of pentanuclear clusters in which two $[Fe(CN)_6]^{3-}$ moieties are connected via three CN—Ni(bpm)$_2$—CN bridges in a *fac* arrangement for each Fe center.[6] Four coordination sites on each Ni(II) center are occupied by two cis-disposed bpm ligands, and the remaining two sites are occupied by the nitrogen atom of a cyano group from each $[Fe(CN)_6]^{3-}$. In the IR spectrum, strong ν_{CN} stretches are found at 2173, 2164, 2151, 2125, and 2116 cm^{-1}. Magnetic susceptibility and magnetization studies show that the cluster displays ferromagnetic coupling and that crystalline samples exhibit intercluster long-range magnetic ordering with $T_c = 23$ K.[6] Related Ni_3Fe_2 clusters do not show long-range orders.[18]

References

1. T. Mallah, C. Auberger, M. Verdaguer, and P. Veillet, *Chem. Commun.* 61 (1995); A. Scuiller, T. Mallah, M. Verdaguer, A. Nivorozkhin, J. Tholence, and P. Veillet, *New J. Chem.* **20**, 1 (1996); M.-A. Arrio, A. Scuiller, Ph. Saintctavit, Ch. Cartier de Moulin, T. Mallah, and M. Verdaguer, *J. Am. Chem. Soc.* **121**, 6414 (1999).
2. (a) R. J. Parker, L. Spiccia, K. J. Berry, G. D. Fallon, B. Moubaraki, and K. S. Murray, *Chem. Commun.* 333 (2001); (b) R. J. Parker, L. Spiccia, B. Moubaraki, K. S. Murray, D. C. R. Hockless, A. D. Rae, and A. C. Willis, *Inorg. Chem.* **41**, 2489 (2002).
3. R. J. Parker, D. C. R. Hockless, B. Moubaraki, K. S. Murray, and L. Spiccia, *Chem. Commun.* 2789, (1996).
4. R. J. Parker, L. Spiccia, S. R. Batten, J. D. Cashion, and G. D. Fallon, *Inorg. Chem.* **40**, 4696 (2001).
5. J. L. Heinrich, P. A. Berseth, and J. R. Long, *Chem. Commun.* 1232 (1998).
6. K. Van Langenberg, S. R. Batten, K. J. Berry, D. C. R. Hockless, B. Moubaraki, and K. S. Murray, *Inorg. Chem.* **36**, 5006 (1997).
7. Z. J. Zhong, H. Seino Y. Mizobe, M. Hidai, A. Fujishima, S. Ohkoshi, and K. Hashimoto, *J. Am. Chem. Soc.* **122**, 2952, (2000); Z. J. Zhong, H. Seino Y. Mizobe, M. Hidai, M. Verdaguer, S. Ohkoshi,

and K. Hashimoto, *Inorg. Chem.* **39**, 5095 (2000); J. Larianova, M. Gross, M. Pilkington, H. Audres, H. Stoeckli- Evans, H. U. Güdel, and S. Decurtins, *Angew. Chem., Int,. Ed. Engl.* **39**, 1605 (2000).
8. See, for example, S. Ferlay, T. Mallah, J. Vaissermann, F. Bartolome, P. Veillet, and M. Verdaguer, *Chem. Commun.* 2481 (1996); N. Re, E. Gallo, C. Floriani, H. Miyasaka, and N. Matsumoto, *Inorg. Chem.* **35**, 5964, 6004 (1996); H. Miyasaka, H. Okawa, A. Miyasaki, and T. Enoki, *Inorg. Chem.* **37**, 4878 (1998); E. Colacio, J. M. Dominguez-Vera, M., Ghazi, R. Kivekäs, F. Lloret, and J. M. Moreno, *Chem. Commun.* 987, (1999); E. Colacio, J. M. Dominguez-Vera, M. Ghazi, R. Kivekäs, J. M. Moreno, and A. Pajuner, *Dalton Trans.* 505 (2000); J. A. Smith, J.-R. Galan-Mascaros, R. Clerac, and K. R. Dunbar, *Chem. Commun.* 1077, (2000); S.-W. Zhang, D. G. -Fu, W.-Y. Sun, Z. Hu, K.-B. Yu, and W.-X. Tang, *Inorg. Chem.* **39**, 1142 (2000) and refs. therein.
9. G. Anderegg and F. Wenk, *Helv. Chim. Acta* **50**, 2330 (1967).
10. J. W. Canary, Y. Wang, and R. Roy Jr., *Inorg. Synth.* **32**, 70 (1998).
11. R. R. Jacobson, Z. Tyeklar K. D. Karlin, and J. Zubieta, *Inorg. Chem.* **30**, 2035 (1991).
12. G. A. McLachlan, G. D. Fallon, R. L. Martin, B. Moubaraki, K. S. Murray, and L. Spiccia, *Inorg. Chem.* **33**, 4663 (1994).
13. G. D. Fallon, G. A. McLachlan, B. Moubaraki, K. S. Murray, L. O'Brien, and L. Spiccia, *J. Chem. Soc., Dalton Trans.* 2765 (1997).
14. J. H. Bigelow, *Inorg. Synth.* **2**, 203 (1947).
15. S. Juliá, P. Sala, J. del Mazo, M. Sancho, C. Ochoa, J. Elguero, J.-P. Fayet, and M.-C. Vertut, *J. Heterocyclic Chem.* **19**, 1141 (1982).
16. D. L. Jameson and R. K. Castellano, *Inorg. Synth.* **32**, 51 (1998).
17. K. Van Langenberg, B. Moubaraki, K. S. Murray, and E. R. T. Tiekink, *Z. Kristallogr.* NCS **218**, 345 (2003).
18. C. P. Berlinguette, J. R. Galan-Mascaos, and K. R. Dunbar, *Inorg. Chem.* **42**, 3416 (2003).

30. A MIXED-VALENCE HEPTANUCLEAR IRON COMPLEX

Submitted by GUILLAUME ROGEZ,[*] ARNAUD MARVILLIERS,[*] and TALAL MALLAH[*]
Checked by BOUJEMAA MOUBARAKI[†] and LEONE SPICCIA[†]

The reaction of hexacyanometalates with metal complexes chelated by pentadentate ligands may afford polynuclear complexes.[1] The presence of the pentadentate ligand precludes the polymerization that leads to extended systems. The preparation of a representative heptanuclear, mixed-valance iron complex,[2] $[Fe^{II}(CNFe^{III}(salmeten))_6]Cl_2 \cdot 6H_2O$, is detailed herein.

A. BIS(3-SALICYLIDENEAMINOPROPYL) METHYLAMINE (salmetenH$_2$)

$$(H_2N(CH_2)_3)_2NCH_3 + 2\ 2\text{-}HOC_6H_4CHO \xrightarrow{(abs.\ C_2H_5OH)} (2\text{-}HOC_6H_4CH{-}N(CH_2)_3)_2NCH_3 + 2H_2O$$

[*]Laboratoire de Chimie Inorganique, CNRS 8613, Université Paris-Sud, 91405 Orsay, France.
[†]School of Chemistry, Monash University, P.O. Box 23, Victoria 3800, Australia.

Procedure

Salicylaldehyde (0.1 mol) is mixed with 50 mL of absolute ethanol in a 250-mL round-bottomed flask containing a stirring bar. To this solution bis(3-aminopropyl)methylamine (0.05 mol) mixed with 50 mL of absolute ethanol is added dropwise (10 min) through a pressure-equalizing dropping funnel. The mixture is stirred for 15 min, and then the solution is evaporated under vacuum. A yellow oil is obtained with a yield of almost 100%.

Properties

^{1}H NMR (200 MHz, $CDCl_3$, 20°C, TMS), δ 8.31 (s, 2H), 7.24 (m, 4H), 6.85 (m, 4H), 3.6 (t, $J = 6.84$ Hz, 4H), 2.41(t, $J = 7.08$ Hz, 4H), 2.20 (s, 3H), 1.83 (q, $J = 6.96$ Hz, 4H).

B. CHLOROBIS(3-SALICYLIDENOAMINOPROPYL)-METHYLAMINEIRON(III), Fe(salmeten)Cl

$$\mathrm{FeCl_3 \cdot 6H_2O + salmetenH_2 + 2NEt_3 \xrightarrow{C_2H_5OH} Fe(salmeten)Cl + 6H_2O + 2[NHEt_3]Cl}$$

Procedure

$FeCl_3 \cdot 6H_2O$ (0.02 mol, 5.4 g) is dissolved in 50 mL of ethanol, and this solution is added dropwise to 30 mL of an ethanolic solution of bis(3-salicylideneamino-propyl)methylamine (0.022 mol, 7.77 g) contained in a 150-mL round-bottomed flask. Triethylamine (0.044 mol, 6.2 mL) mixed with 10 mL of ethanol is added dropwise to the preceding solution. The mixture is heated at 60°C for 20 min. The black precipitate formed is filtered with a sintered glass, washed 5 times with 10-mL aliquots of absolute ethanol, and dried under vacuum overnight. Yield: 6.6 g, 75%.

Recrystallization is carried out by preparing a saturated solution of the black precipitate (6.6 g is dissolved in about 150 mL of acetonitrile). The mixture is heated to reflux for 15 min, then rapidly filtered through a glass frit. The filtrate is cooled in an ice bath. The remaining blackbrown precipitate on the frit is taken up once more in 150 mL of acetonitrile, and the mixture is heated to reflux for 15 min, then filtered rapidly, and the filtrate is allowed to cool in an ice bath. If the filtered solid still contains black material, then the same operations are carried out until the remaining filtered solid is only brown. This brown precipitate can be discarded; it consists mainly of $Fe(OH)_3$. The filtrates are combined and in the ice bath for at least 30 min to give black needles. The black needles are collected by suction. The remaining solution is reduced in volume on a rotatory evaporator by half, then cooled in an ice bath for 30 min, and finally filtered to give more black needles. The black needles are collected and dried under vacuum for 1 h. Yield: 5.3 g, 60%.

Anal. Calcd. for $C_{21}H_{25}N_3O_2ClFe$: C, 56.96; H, 5.65; N, 9.49; Cl, 8.06; Fe, 12.51. Found: C, 56.79; H, 5.68; N, 9.41, Cl, 8.01; Fe, 12.60.

Properties

The complex is air-stable; it is almost black in the solid state and violet in solution. It is soluble in most polar organic solvents, slightly soluble in THF, and insoluble in ether. The complex is paramagnetic ($\chi_M T = 4.38$ cm^3 mol^{-1} K^{-1} between 300 and 10 K). IR in KBr: 1619 (s), 1541 (s), 1467 (s), 1444 (s), 1401 (m), 1303 (s), 1199 (m), 1147 (m), 1124 (w), 760 (s), 597 (m), 441 (w) cm^{-1}. UV–vis (CH_2Cl_2, λ_{max} in nm, ε in M^{-1} cm^{-1}): 514 (1940), 436 (sh), 327 (9190). Mössbauer ($T = 77$ K), shift in mm/s: −0.500, 0.917.

C. HEXAKIS {CYANO(SALMETEN)IRON(III)}IRON(II) CHLORIDE HEXAHYDRATE, [Fe(II)(μ-CN)Fe(III)(salmeten)}$_6$]$Cl_2 \cdot 6H_2O$

$$6\ \mathrm{Fe(salmeten)Cl} + \mathrm{K_4[Fe(CN)_6]} \xrightarrow{\mathrm{CH_3OH/H_2O}} [\mathrm{Fe\{CNFe(salmeten)\}_6}]\mathrm{Cl_2} \cdot 6\mathrm{H_2O} + 4\ \mathrm{KCl}$$

Procedure

$K_4[Fe(CN)_6] \cdot 3H_2O$ (211 mg, 5×10^{-4} mol) dissolved in 40 mL of a 4/1 methanol/water mixture is added dropwise to a 50-mL methanolic solution of [Fe(salmeten)Cl] (1.327 g, 3×10^{-3} mol) while stirring. The microcrystalline powder formed is filtered, washed with cold methanol (3 × 20 mL) and then water (3 × 10 mL), and finally dried overnight under vacuum. Yield: 1.18 g, 83%.

Anal. Calcd. for $C_{132}H_{162}N_{24}O_{18}C_{12}Fe_7$: C, 55.92; H, 5.76; N, 11.86; Cl, 2.50; Fe, 13.79. Found: C, 55.89; H, 5.79; N, 11.93; Cl, 2.50; Fe, 13.65.

Properties

The complex is air-stable; it has a dark blue color in the solid state and is blue-violet in CH_2Cl_2 solution. It is soluble in chloroform, slightly soluble in methanol, and insoluble in acetonitrile and ether. The complex is paramagnetic

($\chi_M T = 26.1$ cm^3 mol^{-1} K^{-1} between 300 and 40 K, $\chi_M T = 35$ cm^3 mol^{-1} K^{-1} at $T = 2$ K). IR in KBr: 2079 (s), 2066 (sh), 1622 (s), 1542 (m), 1467 (m), 1444 (m), 1302 (m), 1148 (w), 1124 (w), 759 (m), 592 (w), 432 (w) cm^{-1}. UV–vis (CH_2Cl_2, λ_{max} in nm, ε in M^{-1} cm^{-1}): 689 (sh), 581 (18,500), 436 (sh), 330 (55,320). Mössbauer ($T = 77$ K), shift in mm s^{-1}: -0.159, 0.797 with calculated fit to $\nu = 0.95$ mm s^{-1} (Fe(III) and $\nu = 0.337$ mm s^{-1}, $\delta Q = 0.923$ mm s^{-1} [6 Fe (II)].

References

1. T. Mallah, C. Auberger, M. Verdaguer, and P. Veillet, *Chem. Commun.* 61 (1995).
2. G. Rogez, A. Marvilliers, E. Rivière, J.-P. Audière, F. Lloret, F. Varret, A. Goujon, N. Menendez, J.-J. Girerd, and T. Mallah, *Angew. Chem.* **39**, 2885 (2000).

31. POTASSIUM HEXACYANOCHROMATE(III) AND ITS ^{13}C-ENRICHED ANALOG

Submitted by VALÉRIE MARVAUD,* TALAL MALLAH,† and MICHEL VERDAGUER*
Checked by MARGRET BINER‡ and SILVIO DECURTINS‡

$$2\ CrCl_3 \cdot 6\ H_2O + Zn/HCl \rightarrow 2\ CrCl_2 \cdot 6H_2O + ZnCl_2$$

$$2\ CrCl_2 \cdot 6\ H_2O + 4\ Na[O_2CCH_3] \rightarrow Cr_2(O_2CCH_3)_4 \cdot 2\ H_2O + 4\ NaCl + 4\ H_2O$$

$$Cr_2(O_2CCH_3)_4 \cdot 2\ H_2O + 12\ KCN \rightarrow 2K_4[Cr(CN)_6] + 4\ K[O_2CCH_3] + 2\ H_2O$$

$$4\ K_4[Cr(CN)_6] + O_2 + 2H_2O \rightarrow 4\ K_3[Cr(CN)_6] + 4\ KOH$$

A new direction was opened in cyanide chemistry in the mid-1990s when it was realized that high-nuclearity clusters could be formed from hexacyanometalates.[1] Such an investigation is related to the so-called bottom–up approach of nano-magnetism and allows the formation of high-spin molecules or even single-molecule magnets. Potassium hexacyanochromate(III) is a useful precursor in

*Laboratoire de Chimie Inorganique et Matériaux Moléculaires, CNRS 7071, Université Pierre et Marie Curie, 75 252 Paris, France.
†Laboratoire de Chimie Inorganique, CNRS 8613, Université Paris-Sud, 91405 Orsay, France.
‡Departement für Chemie und Biochemie, Universität Bern, Bern CH-3012, Switzerland.

the formation of polynuclear, CN-bridged complexes. The salt has been prepared by reduction of potassium dichromate with sulfur dioxide[2] or alcohol in hydrochloric acid,[3] leading to the formation of chromium hydroxide. After acidification, evaporation, and addition to a boiling solution of potassium cyanide, $K_3[Cr(CN)_6]$, is separated on cooling. A more convenient synthetic route presented here starts from chromium(III) chloride hexahydrate $CrCl_3 \cdot 6H_2O$, which is reduced before being converted into the Cr(II) acetate complex, $Cr_2(O_2CCH_3)_4 \cdot 2H_2O$, following a procedure adapted from the literature.[4,5] The subsequent addition of an excess of potassium cyanide to the chromous(II) acetate compound under inert atmosphere affords $K_4[Cr^{II}(CN)_6]$, which is easily oxidised by air to give the target complex.

Procedure

Chromium(III) chloride hexahydrate (30 g, 0.113 mol) is dissolved in 40 mL of distilled water and introduced to a Schlenk flask containing 40 g of zinc (20 mesh). The solution is carefully deoxygenated under vacuum and then saturated with argon. With the help of a cannula and a pressure gradient, 75 mL of deoxygenated hydrochloric acid (35%) is slowly added to the constantly stirred mixture.

■ **Caution.** *The reaction is exothermic; the apparatus should be chilled in an ice-water bath.*

The resulting deep blue solution is periodically exposed to vacuum to eliminate the excess of hydrogen generated during the reaction. After 1 h of stirring, when the reduction is completed and the hydrogen bubbling stopped, the solution is separated from the zinc powder and transferred with a filter canula to a deoxygenated solution of sodium acetate dissolved in a minimum of warm distilled water (100 g of $Na[O_2CCH_3]$, 1.22 mole, in 100 mL of H_2O). The resulting mixture is stirred for 15 min and left standing for an additional 30 min. Under argon, the precipitate formed is separated from the supernatant solution by using a filter canula. The deep red solid chromous acetate $Cr_2(O_2CCH_3)_4 \cdot 2H_2O$ is successively washed with deoxygenated water (100 mL) and oxygen-free ethanol (100 mL) and then dried for several hours under vacuum.

The freshly prepared chromous(II) acetate is suspended in 50 mL of distilled water previously deoxygenated and saturated with argon. To the resulting mixture is added, under inert atmosphere, an aqueous solution of potassium cyanide (82 g, 1.26 mol in 200 mL of water). A deep green precipitate of $K_4[Cr^{(II)}(CN)_6]$ is formed.

■ **Caution.** *Potassium cyanide is an extremely toxic material that must be handled with care.*

The mixture is poured into 1.5 L of methanol and then filtered. The precipitate is mixed with 300 mL of methanol in a 1-L beaker, and the suspension is stirred for 3 h with a continuous stream of air. The yellow precipitate is filtered, washed with methanol, and dried by suction. For recrystallization the product is dissolved in a minimum of hot water, the solution is immediately filtered, and the filtrate is left standing for 3 days in a refrigerator to give large yellow crystals of $K_3[Cr(CN)_6]$ (15 g, 42% yield).

Anal. Calcd. for $C_6N_6CrK_3$: C, 22.15; N, 25.83. Found: C, 21.90; N, 25.54.

Properties

Potassium hexacyanochromate(III) is a yellow solid, highly soluble in water, that crystallizes as large square platelets. It crystallizes in the orthorhombic system, with the space group *Pcan*. The unit parameters are $a = 8.53$, $b = 10.60$, $c = 13.68$ Å.[6] Because of its disorder behavior, it presents a complex crystallographic problem.[7] The compound shows a ν_{CN} band at 2131 cm^{-1}. The molar extinction coefficients in aqueous solution of the two observable *d–d* bands at 376 nm ($^4A_{2g} \rightarrow {}^4T_2$) and 309 nm ($4A_{2g} \rightarrow T_{1g}$) are 93 and 62 L mol^{-1} cm^{-1}, respectively.

^{13}C-Enriched Compound, $K_3[Cr(^{13}CN)_6]$

The ^{13}C-enriched hexacyanochromate(III), $K_3[Cr(^{13}CN)_6]$, is obtained by following a similar synthetic route. When 0.230 g of chromous(II) acetate complex is used in the reaction with 1 g of ^{13}C-enriched potassium cyanide, then 0.198 g of $K_3[Cr(^{13}CN)_6]$ is obtained (50% yield). The ^{13}C-enriched hexacyanochromate(III) displays a ν_{CN} band at 2085 cm^{-1} in the IR spectrum.

References

1. T. Mallah, C. Auberger, M. Verdaguer, and P. Veillet, *J. Chem. Soc. Chem. Commun* 61 (1995).
2. J. Bigelow, *Inorg. Synth.* **8**, 203 (1966).
3. F. V. D. Cruser and E. H. Miller, *J. Am. Chem. Soc.* **28**, 1133 (1906).
4. J. H. Balthus and J. C. Bailar, Jr. *Inorg. Synth.* **1**, 122 (1939).
5. M. R. Hatfield and A. Witkowska, *Inorg. Synth.* **6**, 144 (1960).
6. S. Jagner, E. L. Jungström, and N.-G. Vannerberg, *Acta Chem. Scand. A* **28**, 623 (1974).
7. A. G. Sharpe, *The Chemistry of Cyano Complexes of the Transition Metals*, Academic Press, New York, 1976, p. 46 and refs. therein.

32. TWO POLYNUCLEAR NICKEL(II)–HEXACYANOCHROMIUM(III) COMPLEXES

Submitted by VALÉRIE MARVAUD,* ARIANE SCUILLER,* and MICHEL VERDAGUER*
Checked by MARGARET BINERT† and SILVIO DECURTINS†

Salts of hexacyanochromate(III) are suitable Lewis base precursors for forming polynuclear complexes when combined with mononuclear transition metal compounds able to act as Lewis acids. A rational synthetic strategy is to use a pentadentate polyamine ligand for the mononuclear reactant, which is then limited to only one labile position able to link the cyanide.[1,2] The *trans*-hexacyano-based trinuclear complex $[Cr(CN)_4\{CNNi(tetren)\}_2]Cl$, denoted hereafter as $CrNi_2$, is obtained by the direct reaction between mononuclear (tetren)nickel(II) chloride (tetren = tetraethylenepentamine) generated in situ and potassium hexacyanochromate(III). The presence of chloride ions in solution allows the selective formation of the trinuclear complex, as $CrNi_2$. However, the heptanuclear complex, $[Cr\{CNNi(tetren)\}_6](ClO_4)_9$, hereafter denoted as $CrNi_6$, is obtained when nickel(II) perchlorate is used and all chloride is removed from the system.

A. *trans*-BIS(TETRENNICKEL(II))HEXACYANOCHROMIUM(III) CHLORIDE, $[Cr(CN)_4\{CNNi(tetren)\}_2]Cl$

$$tetren \cdot 5HCl + NaOH \rightarrow tetren + 5\ NaCl + 5\ H_2O$$

$$tetren + [Ni(H_2O)_6]Cl_2 \rightarrow [Ni(tetren)Cl]Cl + 6\ H_2O$$

$$K_3[Cr(CN)_6] + 2[Ni(tetren)Cl]Cl \rightarrow [Cr(CN)_4\{CNNi(tetren)\}_2]Cl + 3\ KCl$$

Procedure

To a solution of tetraethylenepentamine pentahydrochloride (tetren·5HCl, Aldrich) (2.24 g 6.0 mmol) dissolved in 20 mL of water is added 1.21 g of sodium hydroxide (30 mmol, 5 equiv). A solution of nickel(II) chloride hexahydrate in 10 mL of water (0.95 g, 4.0 mmol) is then slowly added. The mixture is stirred for 20 min before adding dropwise a solution of potassium hexacyanochromate(III) (0.325 g, 1 mmol) in a minimum of water. The solution is left

*Laboratoire de Chimie Inorganique et Matériaux Moléculaires, CNRS 7071, Université Pierre et Marie Curie, 75 252 Paris Cedex 05, France.
†Departement für Chemie und Biochemie, Universität Bern, Friestrasse 3, Bern Ch-3012, Switzerland.

standing for a few days. The brown-red hexagonal crystals that are formed are collected, washed with ethanol, and air-dried (0.553 g, 74% yield).

Anal. Calcd. for $[C_{22}H_{46}CrNi_2N_{16}Cl](H_2O)_7$: C, 30.54; H, 6.93; N, 25.90; Cr, 6.01; Ni, 13.57; Cl, 4.10. Found: C, 30.51; H, 6.85; N, 25.78; Cr, 5.92; Ni, 13.43; Cl, 4.08.

Properties

The trinuclear $CrNi_2$ complex is obtained as brown-red crystals, soluble in water and in aqueous acetonitrile. The IR spectrum shows three different ν_{CN} bands of equivalent intensity at 2118, 2130, and 2144 cm^{-1}. The latter one corresponds to the bridging cyano group between chromium and nickel transition metal cations. The two other bands are attributed to singly coordinated cyano ligands; one of them is involved in hydrogen bonding as shown by the X-ray crystallographic structure. The product crystallises in the orthorhombic system, space group $Pbc2_1$, with parameters $a = 8.55$, $b = 16.71$ and $c = 28.49$ Å. The magnetic properties are consistent with ferromagnetic interactions between the spin carriers [Cr(III), d^3, $S = \frac{3}{2}$; Ni(II), d^8, $S = 1$] leading to a spin $S = \frac{7}{2}$ ground state.[3]

B. HEXAKIS(TETRENNICKEL(II))HEXACYANOCHROMIUM(III) PERCHLORATE, $[Cr\{CNNi(tetren)\}_6][ClO_4]_9$

$$\text{tetren} \cdot 5HCl + NaOH \rightarrow \text{tetren} + 5\ NaCl + 5\ H_2O$$

$$\text{tetren} + Ni(ClO_4)_2 \rightarrow [Ni(\text{tetren})(H_2O)](ClO_4)_2$$

$$NaCl + AgClO_4 \rightarrow AgCl + NaClO_4$$

$$K_3[Cr(CN)_6] + 6[Ni(\text{tetren})(H_2O)](ClO_4)_2 \rightarrow [Cr\{CNNi(\text{tetren})\}_6][ClO_4]_9 + 3KClO_4$$

■ **Caution.** *Perchlorate salts are potentially explosive and should be handled with care.*

Procedure

To a solution of tetraethylene pentamine pentahydrochloride (tetren.5HCl, Aldrich) (1.12 g, 3.0 mmol) dissolved in 20 mL of water is added 600 mg of NaOH (15 mmol, 5 equiv) followed by nickel(II) perchlorate (755 mg, 2.06 mmol) dissolved in 10 mL of water. The mixture is stirred for 20 min. Then silver perchlorate (3.12 g, 15 mmol, 5 equiv) is added in order to eliminate the chloride ions. The solution is left stirring for 1 h in the dark, then the solution is filtered through a glass frit to separate silver chloride. Then 20 mL of acetonitrile is added to the deep blue filtrate. A solution of potassium hexacyanochromate(III) (0.110 g, 0.33 mmol) in a minimum of water is added drop by drop, and the solution is left standing for a few days. The red-purple octahedral crystals that form are collected, washed with water, and air-dried (0.540 mg, 63% yield).

Anal. Calcd. for $[C_{54}H_{138}CrNi_6N_{36}Cl_9O_{36}](H_2O)_6$: C, 24.03; H, 5.60; N, 18.68; Cr, 1.93; Ni, 13.05; Cl, 11.82. Found: C, 23.72; H, 5.39; N, 18.68; Cr, 1.90; Ni, 13.30; Cl, 12.31.

Properties

$[Cr\{CNNi(tetren)\}_6](ClO_4)_9$ is a red-purple solid, soluble in water–acetonitrile mixtures where the undissociated heptanuclear entity can be identified by electrospray ionization mass spectrometry. IR spectroscopy shows the presence of a unique ν_{CN} band at 2146 cm^{-1} of the bridging cyano ligand and an intense band of the perchlorate anion at $\sigma = 1090$ cm^{-1}. The IR spectrum of the ^{13}C-enriched compound prepared from $K_3[Cr(^{13}CN)_6]$ displays a ν_{CN} band at 2103 cm^{-1}. The compound crystallizes in a trigonal system, space group $R\bar{3}$, with parameters $a = b = 15.274$ Å and $c = 41.549$ Å. The magnetic properties indicate a ferromagnetic interaction between the spin carriers (Cr(III), d^3, $S = \frac{3}{2}$; Ni(II), d^8, $S = 1$) leading to a spin $S = \frac{15}{2}$ ground state.

References

1. T. Mallah, C. Auberger, M. Verdaguer, and P. Veillet, *J. Chem. Soc. Chem. Commun.* 61 (1995).
2. M. Verdaguer, A. Bleuzen, V. Marvaud, J. Vaissermann, M. Seuleiman, C. Desplanches, A. Scuiller, C. Train, R. Garde, and G. Gelly, *Coord. Chem. Rev.* **190**, 1023 (1999).
3. A. Scuiller, V. Marvaud, J. Baisseremann, I. Rosenman, and M. Verdaguer, *Mol. Cryst. Liq. Cryst.* **335**, 453 (1999).

33. TRIMETHYLTRIAZACYCLONONANECHROMIUM(III) COMPLEXES AND A CHROMIUM(III)–NICKEL(II)–CYANIDE CLUSTER WITH A FACE-CENTERED CUBIC GEOMETRY

Submitted by MATTHEW P. SHORES,* POLLY A. BERSETH,* and JEFFREY R. LONG*

Checked by VALÉRIE MARVAUD,† RAQUEL GARDE,† and MICHEL VERDAGUER†

The discovery that certain metalloxo clusters can sustain a remanent magnetization prompted interest in the synthesis of new "single-molecule magnets" as potential media for high-density information storage.[1] Unfortunately, the broad

*Department of Chemistry, University of California, Berkeley, CA 94720.
†Laboratoire de Chimie Inorganique et Materiaux Moleculaires, CNRS 7071, Université Pierre et Marie Curie, 75252 Paris, France.

structural variability inherent to metalloxo cluster systems severely hampers the design of species with higher magnetic moment reversal barriers. Alternative cluster systems that might permit better control over magnetic properties such as the spin and anisotropy associated with the ground state are therefore sought.

One such system derives from transition metal–cyanide cluster chemistry.[2–6] The preference of cyanide for adopting a linear bridging geometry allows a more directed approach to synthesis, as well as a means of predicting the nature of the pairwise magnetic exchange interactions.[7] Indeed, recognition of these factors has enabled the design of Prussian blue–type solids exhibiting bulk magnetic ordering above room temperature.[8] By using appropriate capping ligands on the metal complex precursors, analogous reactions can direct the formation of molecular metal–cyanide clusters instead of extended solids. Such synthetic control has been demonstrated with the assembly of a cubic $[(tacn)_8Co_8(CN)_{12}]^{12+}$ (tacn = 1,4,7-triazacyclononane) cluster[2,9] from $[Co(tacn)(CN)_3$ and $[(tacn)Co(H_2O)_3]^{3+}$, and the use of a capping ligand on just one of the reactant species can lead to even larger metal–cyanide clusters.[3,4] For example, the reaction of $[(Me_3tacn)Cr(CN)_3]$ (Me_3tacn = *N,N′,N″*-trimethyl-1,4,7-triazacyclononane) with $[Ni(H_2O)_6]^{2+}$ generates a 14-metal $[(Me_3tacn)_8Cr_8Ni_6(CN)_{24}]^{12+}$ cluster that displays the unique face-centered cubic geometry depicted in Fig. 1.[3]

Herein, we report the detailed syntheses of $[(Me_3tacn)_8Cr_8Ni_6(CN)_{24}]Br_{12}\cdot 25H_2O$ and a sequence of mononuclear chromium(III) precursors,

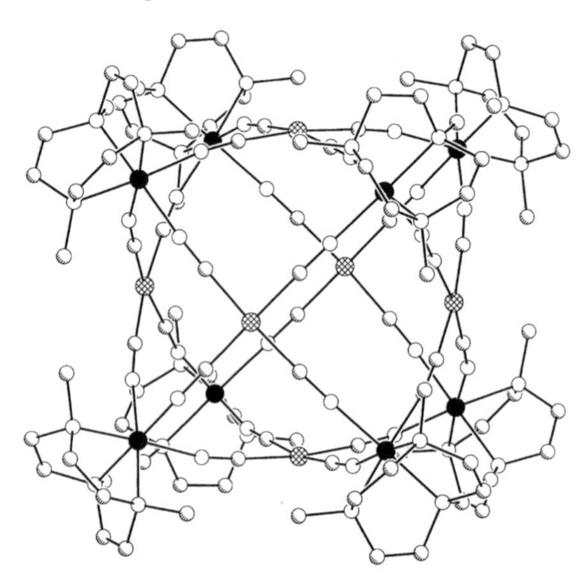

Figure 1. Structure of the face-centered cubic cluster $[(Me_3tacn)_8Cr_8Ni_6(CN)_{24}]^{12+}$. Black, crosshatched, shaded, and white spheres represent Cr, Ni, C, and N atoms, respectively; H atoms are omitted for clarity.

$(Me_3tacn)CrX_3$ (X = Cl, CF_3SO_3, CN). The preparation of the trichloride complex is adapted from previous reports,[10,11] while the preparations for the other precursors represent modifications of those described for the analogous (tacn)-containing complexes.[12]

■ **Caution.** *The reagents trimethyltriazacyclonane and trifluoromethane sulfonic acid are harmful by inhalation and by contact with skin. Potassium cyanide is highly toxic and produces volatile, poisonous hydrogen cyanide on protonation. All operations must be conducted in a well-ventilated fume hood. Wear appropriate protective clothing, gloves, and eye protection.*

A. (*N*,*N*′,*N*″-TRIMETHYL-1,4,7-TRIAZACYCLONONANE)-TRICHLOROCHROMIUM(III), $(Me_3tacn)CrCl_3$

$$CrCl_3 \cdot 6H_2O + Me_3tacn \rightarrow Cr(Me_3tacn)Cl_3 + 6H_2O$$

Procedure

A 10-mL round-bottomed flask is charged with a Teflon-coated magnetic stirring bar, dimethylsulfoxide (3 mL, commercial, dried over 3-Å sieves), and solid $CrCl_3 \cdot 6H_2O$ (3.76 g, 14.1 mmol, Fisher). A small portion of methanol is used to wash all the $CrCl_3 \cdot 6H_2O$ into the flask. The resulting green solution is stirred and heated to 120°C for one hour to drive off the water and methanol, forming a deep purple solution. Approximately 30 granules of Zn (30 mesh, Fisher) are added to the solution, and, with the solution maintained at a temperature of 70°C, Me_3tacn (2.41 g, 14.0 mmol, Unilever*) is added dropwise over a 5-min period. The resulting dark green mixture must be stirred vigorously to prevent solidification. Small aliquots of methanol can be added to wash in all the Me_3tacn as well as break up the mixture. The mixture is heated at 120°C for 2 h, then cooled to room temperature. The reaction mixture is transferred to a 250-mL beaker, mixed with water (100 mL, distilled and deionized), and cooled at 5°C in a refrigerator for 4 h. The mixture is filtered, washed with water (50 mL) and diethyl ether (2 × 20 mL), and dried in air to afford $(Me_3tacn)CrCl_3$ (2.65 g, 8.04 mmol) as a green solid.† Yield: 57%.

*The authors gratefully acknowledge Dr. C. Crawford and Unilever for a donation of crude Me_3tacn, which was purified by vacuum distillation (36°C, ~80 mtorr) prior to use. For a preparation of Me_3tacn, see Ref. 11. The checkers used commercial material available from Strem Chemicals.

†In the event that only a small amount of green precipitate is obtained, the product can be obtained from the purple filtrate by reducing it to dryness using a rotary evaporator, heating the residue in vacuo at 120°C for 4 h, and washing the resulting solid twice with 50 mL of water.

Anal. Calcd. for $C_9H_{21}Cl_3CrN_3$: C, 32.79; H, 6.42; N, 12.75. Found: C, 32.69; H, 6.48; N, 12.67.

Properties

The compound $(Me_3tacn)CrCl_3$ is a green solid that is air- and moisture-stable at room temperature. It is soluble in dimethylformamide and dimethylsulfoxide, but insoluble in water and most other common solvents. Its UV–visible absorption spectrum (DMF) shows maxima (ε) at 468 (132) and 634 (111) nm.

B. (*N,N′,N″*-TRIMETHYL-1,4,7-TRIAZACYCLONONANE) TRIS(TRIFLUOROMETHANE SULFONATO) CHROMIUM(III), $(Me_3tacn)Cr(CF_3SO_3)_3$

$$(Me_3tacn)CrCl_3 + 3CF_3SO_3H \rightarrow (Me_3tacn)Cr(CF_3SO_3)_3 + 3HCl$$

Procedure

A 150-mL Schlenk flask is charged with a Teflon-coated magnetic stirring bar and $(Me_3tacn)CrCrCl_3$ (2.60 g, 7.89 mmol). Under a flow of dinitrogen and while stirring, trifluoromethanesulfonic acid (15 mL, 170 mmol, Acros, 99%) is slowly added, producing a thick purple solution and gaseous hydrochloric acid. The flask is fitted with an oil bubbler, and dinitrogen is passed over the solution while it is stirred and heated for 6 h at 50°C. The flask is cooled to room temperature, and diethyl ether (100 mL) is added to precipitate a purple solid. The solid is filtered on a fine sintered-glass funnel, washed with diethyl ether (4 × 25 mL), and dried in air briefly to afford $(Me_3tacn)Cr(CF_3SO_3)_3$ (5.28 g, 7.88 mmol) as a violet solid. Yield: 99%.

Anal. Calcd. for $C_{12}H_{21}CrF_9N_3O_9S_3$, 21.50; H, 3.16; N, 6.27; S, 14.34. Found: C, 21.22; H, 3.10; N, 6.20; S, 14.16.

Properties

The compound $(Me_3tacn)Cr(CF_3SO_3)_3$ is a violet solid when anhydrous. During prolonged exposure to air, its color deepens to dark purple as it picks up water. It is soluble in dimethylformamide and dimethylsulfoxide but insoluble in water and methanol. Its IR spectrum shows ν_{SO} as a broad, intense band at 1203 cm^{-1}. Its UV–visible absorption spectrum (DMF) shows maxima (ε) at 417 (35) and 566 (162) nm.

C. (*N*,*N*′,″-TRIMETHYL-1,4,7-TRIAZACYCLONONANE)-TRICYANOCHROMIUM(III), $(Me_3tacn)Cr(CN)_3$

$$(Me_3tacn)Cr(CF_3SO_3)_3 + 3KCN \rightarrow (Me_3tacn)Cr(CN)_3 + 3K[CF_3SO_3]$$

Procedure

A 150-mL Schlenk flask fitted with a gas bubbler is charged with a Teflon-coated magnetic stirring bar, dimethylsulfoxide (40 mL, commercial, dried over 3-Å sieves), $(Me_3tacn)Cr(CF_3SO_3)_3$ (5.25 g, 7.83 mmol), and KCN (10.25 g, 157 mmol, Mallinckrodt, ground with mortar and pestle). The purple mixture is stirred and heated under a dinitrogen atmosphere at 110°C for 24 h, whereupon a color change to orange-brown is observed. The mixture is cooled to room temperature and transferred to a 250-mL beaker containing 150 mL of dichloromethane to complete precipitation of the raw product. The mixture is filtered and washed with dichloromethane (2 × 15 mL) to obtain a yellow solid and a brown filtrate. The filtrate is discarded. To separate the product from unreacted KCN, a yellow solution is extracted from the solid with hot (~50°C) dimethylformamide (6 × 50 mL). The solvent is removed using a rotary evaporator (rotovap), and the resulting yellow solid is washed with diethyl ether (50 mL). The product is redissolved in water (40 mL), the volume reduced with heat to 25 mL, and the solution cooled in a refrigerator at 5°C to precipitate a first crop of $(Me_3tacn)Cr(CN)_3$ (1.29 g, 4.28 mmol) as yellow crystals. The precipitate is washed with acetone (5 mL) and diethyl ether (5 mL) and dried in air. Concentration of the mother liquor to 8 mL and cooling as above yields a second crop of yellow solid (0.328 g, 1.09 mmol), which is washed similarly. Further concentration of the mother liquor to 1 mL yields a third crop (0.136 g, 0.452 mmol) on cooling. Total yield: 74% (1.76 g, 5.83 mmol).

Anal. Calcd. for $C_{12}H_{21}CrN_6$: C, 47.83; H, 7.02; N, 27.89. Found: C, 47.79; H, 7.30; N, 27.79.

Properties

The compound $(Me_3tacn)Cr(CN)_3$ is a yellow microcrystalline solid that is air- and moisture-stable at room temperature. It is soluble in polar solvents such as water, methanol, acetonitrile, dimethylformamide, and dimethylsulfoxide. It is insoluble in common organic solvents such as ether, dichloromethane, and acetone. Its IR spectrum shows ν_{CN} as a sharp but very weak band at 2132 cm^{-1}. Its UV–visible absorption spectrum (H_2O) shows maxima (ε) at 339 (52) and 425 (43) nm. At 295 K, the compound exhibits an effective magnetic moment of $\mu_{eff} = 3.85\ \mu_B$. Single crystals suitable for X-ray analysis can be grown by slow evaporation of an aqueous solution of the complex. It crystallizes as the monohydrate in orthorhombic space group $P2_12_12_1$ with unit cell dimensions $a = 8.3292$ (2) Å, $b = 13.6144$ (4) Å, $c = 13.6916$ (4) Å, $V = 1552.59$ (7) Å^3, and Z = 4.

D. [OCTAKIS{*N*,*N*′,*N*″-TRIMETHYL-1,4,7-TRIAZACYCLONONANE)-CHROMIUM(III)} HEXAKIS{TETRACYANONICKEL(II)}] BROMIDE HYDRATE, $[Cr_8Ni_6(Me_3tacn)_8(CN)_{24}]Br_{12} \cdot 25H_2O$

$$8(Me_3tacn)Cr(CN)_3 + 6NiBr_2 \cdot 6H_2O \rightarrow (Me_3tacn)_8[Cr_8Ni_6(CN)_{24}]Br_{12} \cdot 25H_2O$$

Procedure

A 50-mL round-bottomed flask is charged with a Teflon-coated stirring bar, $(Me_3tacn)Cr(CN)_3$ (150 mg, 0.50 mmol), and water (15 mL). To the stirred mixture is added solid $NiBr_2 \cdot 6H_2O$ (130 mg, 0.38 mmol, Aldrich). Within 5 min, an orange solution forms. The solution is stirred and heated under reflux for 10 h, then concentrated to 2 mL, and allowed to cool to room temperature and stand for 12 h. A first crop of orange crystals precipitates from the solution. The crystals are filtered, washed with ethanol (5 mL), acetone (5 mL), and diethyl ether (5 mL), and then dried in air to give $(Me_3tacn)_8Cr_8Ni_6(CN)_{24}]Br_{12} \cdot 25H_2O$ as an orange microcrystalline powder (98 mg, 24 μmol). The mother liquor is further concentrated with heat to 0.5 mL to yield a second crop of crystals, which are washed as described above to give more orange powder (86 mg, 21 μmol). Total yield: 67% (180 mg, 44 μmol).

Anal. Calcd. for $C_{96}H_{218}Br_{12}Cr_8N_{48}Ni_6O_{25}$: C, 27.64; H, 5.27; N, 16.11. Found: C, 27.89; H, 5.54; N, 15.90.

Properties

The compound $[(Me_3tacn)_8Cr_8Ni_6(CN)_{24}]Br_{12} \cdot 25H_2O$ is an orange microcrystalline solid that is stable under ambient conditions. The water content of this compound varies somewhat with humidity, and was confirmed by thermogravimetric analysis. It is soluble in water and methanol and insoluble in common organic solvents such as ether, acetonitrile, and acetone. Its IR spectrum shows ν_{CN} as a strong band at 2148 cm^{-1} with shoulders at 2175 and 2111 cm^{-1}. Its UV–visible absorption spectrum (H_2O) shows a maximum at 463 nm

($\varepsilon = 52\ \mathrm{L\ mol^{-1}\ cm^{-1}}$). At 280 K, the compound exhibits an effective magnetic moment of $\mu_{\mathrm{eff}} = 11.3\ \mu_{\mathrm{B}}$. Single crystals suitable for X-ray analysis can be grown by slow evaporation of an aqueous solution of the compound. It crystallizes with approximately 20 additional solvate water molecules in trigonal space group $R\bar{3}$ with unit cell dimensions $a = 19.4604$ (3) Å, $c = 41.7172$ (1) Å, $V = 13682.0$ (3) Å^3, and $Z = 3$.

References

1. R. Sessoli, H.-L. Tsai, A. R. Schake, S. Wang, J. B. Vincent, K. Folting, D. Gatteschi, G. Christou, and D. N. Hendrickson, *J. Am. Chem. Soc.* **115**, 1804 (1993); R. Sessoli, D. Gatteschi, A. Caneschi, and M. Novak, *Nature* **365**, 141 (1993); H. J. Eppley, H.-L. Tsai, N. de Vries, K. Folting, G. Christou, and D. N. Hendrickson, *J. Am. Chem. Soc.* **117**, 301 (1995); S. M. J. Aubin, M. W. Wemple, D. M. Adams, H.-L. Tsai, G. Christou, and D. N. Hendrickson, *J. Am. Chem. Soc.* **118**, 7746 (1996); A.-L. Barra, P. Debrunner, D. Gatteschi, C. E. Schultz, and R. Sessoli, *Europhys. Lett.* **35**, 193 (1996); S. L. Castro, Z. Sun, C. M. Grant, J. C. Bolinger, D. N. Hendrickson, and G. Christou, *J. Am. Chem. Soc.* **120**, 2365 (1998); E. K. Brechin, J. Yoo, M. Nakano, J. C. Huffman, D. N. Hendrickson, and G. Christou, *Chem. Commun.* 783 (1999); A. L. Barra, A. Caneschi, A. Cornia, F. Fabrizi de Biani, D. Gatteschi, C. Sangregoria, R. Sessoli, and L. Sorace, *J. Am. Chem. Soc.* **121**, 5302 (1999); S. M. J. Aubin, Z. Sun, L. Pardi, J. Krzystek, K. Folting, L.-C. Brunel, A. L. Rheingold, G. Christou, and D. N. Hendrickson, *Inorg. Chem.* **38**, 5329 (1999).
2. J. L. Heinrich, P. A. Berseth, and J. R. Long, *Chem. Commun.* 1231 (1998).
3. P. A. Berseth, J. J. Sokol, M. P. Shores, J. L. Heinrich, and J. R. Long, *J. Am. Chem. Soc.* **122**, 9655 (2000).
4. J. J. Sokol, M. P. Shores, and J. R. Long, *Angew. Chem., Int. Ed.* **40**, 236 (2001); J. L. Heinrich, J. J. Sokol, A. G. Hee, and J. R. Long, *J. Solid State Chem.* **159**, 293 (2001).
5. T. Mallah, C. Auberger, M. Verdaguer, and P. Veillet, *J. Chem. Soc., Chem. Commun.* 61 (1995); A. Scuiller, T. Mallah, M. Verdaguer, A. Nivorozkhin, J.-L. Tholence, and P. Veillet., *New J. Chem.* **20**, 1 (1996).
6. Z. J. Zhong, H. Seino, Y. Mizobe, M. Hidai, A. Fujishima, S. Ohkoshi, and K. Hashimoto, *J. Am. Chem. Soc.* **122**, 2952 (2000); J. Larionova, M. Gross, M. Pilkington, H. Andres, H. Stoeckli-Evans, H. U. Gudel, and S. Decurtins, *Angew. Chem. Int. Ed.* **39**, 1605 (2000).
7. W. R. Entley, C. R. Treadway, and G. S. Girolami, *Mol. Cryst. Liq. Cryst.* **273**, 153 (1995); H. Weihe and H. U. Güdel, *Comments Inorg. Chem.* **22**, 75 (2000).
8. V. Gadet, T. Mallah, I. Castro, and M. Verdaguer, *J. Am. Chem. Soc.* **114**, 9213 (1992); T. Mallah, S. Thiébaut, M. Verdaguer, and P. Veillet, *Science* **262**, 1554 (1993); W. R. Entley and G. S. Girolami, *Science* **268**, 397 (1995); S. Ferlay, T. Mallah, R. Ouahès, P. Veillet, and M. Verdaguer, *Nature* **378**, 701 (1995); K. R. Dunbar and R. A. Heintz, *Prog. Inorg. Chem.* **45**, 283 (1997) and refs. therein; E. Dujardin, S. Ferlay, X. Phan, C. Desplanches, C. Cartier dit Moulin, P.Sainctavit, F. Baudelet, E. Dartyge, P. Veillet, and M. Verdaguer, *J. Am. Chem. Soc.* **120**, 11347 (1998); S. M. Holmes and G. S. Girolami, *J. Am. Chem. Soc.* **121**, 5593 (1999); ø. Hatlevik, W. E. Buschmann, J. Zhang, J. L. Manson, and J. S. Miller, *Adv. Mater.* **11**, 914 (1999).
9. Analogous clusters containing cyclopentadienyl ligands have also been reported: K. K Klausmeyer, T. B. Rauchfuss, and S. R. Wilson, *Angew. Chem. Int. Ed.* **37**, 1694 (1998).
10. K. Wieghardt, W. Schmidt, H. Endres, and C. R. K. Wolfe, *Chem. Ber.* **112**, 2837 (1979).
11. K. Wieghardt, P. Chaudhuri, B. Nuber, and J. Weiss, *Inorg. Chem.* **21**, 3086 (1982).
12. C. K. Ryu, R. B. Lessard, D. Lynch, and J. F. Endicott, *J. Phys. Chem.* **93**, 1752 (1989).

34. CYANO-BRIDGED $M(II)_9M(V)_6$ MOLECULAR CLUSTERS, M(II) = Mn,Co,Ni; M(V) = Mo,W

Submitted by FEDERICA BONADIO,[*] JOULIA LARIONOVA,[†] MATHIAS GROSS,[*] MARGRET BINER,[*] HELEN STOECKLI-EVANS,[‡] SILVIO DECURTINS,[*] and MELANIE PILKINGTON[*]
Checked BY HOLMING F. YUEN,[§] MATTHEW P. SHORE,[§] and JEFFREY R. LONG[§]

The synthesis of molecular clusters containing paramagnetic transition metal ions and investigation of their magnetic properties represent significant challenges in the area of molecular magnetism.[1–6] These efforts are driven by the search for useful properties, in particular as certain clusters function as nanoscale magnets at cryogenic temperatures and exhibit magnetic bistability of purely molecular origin. Cyanometalate building blocks have proved to be a popular choice for the self-assembly of magnetic materials, since cyanide linkages facilitate the interactions between paramagnetic transition metal centers.[7,8] The self-assembly of octacyanometalate building blocks $[M(CN)_8]^{3-}$, M = Mo(V),W(V), together with an appropriate divalent metal salt (M(II) = Mn,Co,Ni) affords a series of μ-cyano-bridged molecular clusters.[9,10] The clusters are all isostructural and comprise 15 metal ions, namely, nine M(II) ions and six M(V) ions. The overall M(V)–CN–M(II) geometry is such that the atoms are linked to form an aesthetically pleasing topological pattern in which the polyhedron spanned by the peripheral metal ions is closest in geometry to a rhombic dodecahedron (see Fig. 1). This topology is reminiscent of the well-documented $[Mo_6X_8]^{4+}$ unit from molybdenum dichloride or dibromide.[11] The octacyanometalate precursors are prepared according to published procedures.[12] Details of the preparation of tetravalent $K_4[M(CN)_8]$ salts together with their subsequent oxidation and the isolation of pentavalent $[N(C_4H_9)_4]_3[M(CN)_8]$ salts are included here for completeness.

■ **Caution.** *Because of the toxicity of metal cyanide compounds, all reactions must be carried out in a well-ventilated fume hood, in particular avoiding exposure to acid.*

[*]Departement für Chemie und Biochemie, Universität Bern, 3012 Bern, Switzerland.
[†]Université de Montpellier II, CNRS 5636, 34095 Montpellier, France.
[‡]Institut de Chimie, Université de Neuchâtel, 2000 Neuchâtel, Switzerland.
[§]Department of Chemistry, University of California, Berkeley, CA 94720.

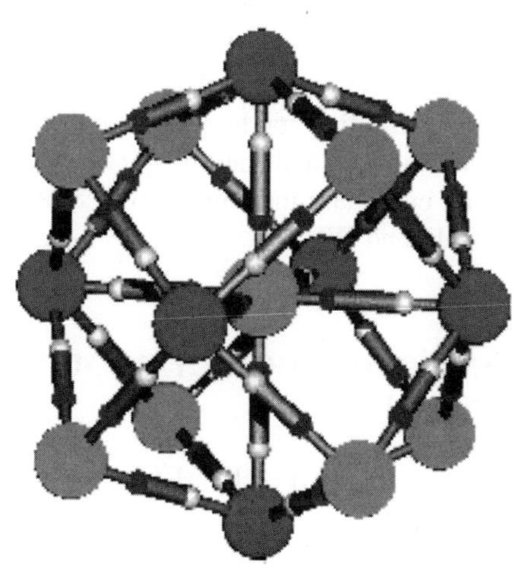

Figure 1. Representation of the $[M(II)_9(\mu\text{-}CN)_{30}M(V)_6]$ cluster core in an idealized O_h symmetry. The light spheres represent M(II) ions and the dark spheres, M(V) ions.

General Remarks

All starting materials and solvents can be purchased from Fluka or Aldrich and used without further purification. All experiments are performed under normal laboratory conditions. The most suitable method for isolation of the clusters is direct crystallization from the reaction mixtures prepared as indicated below. Since all compounds decompose with rapid loss of solvent, final yields are not given.

A. POTASSIUM OCTACYANOMETALATE(IV) HYDRATE, $K_4[M(CN)_8]\cdot 2H_2O$, M = Mo,W

$$Na_2[MO_4]\cdot 2H_2O + 8KCN + 2K[BH_4] \xrightarrow[H_2O]{HOAC} K_4[M(CN)_8]\cdot 2H_2O + 4K[BO_3H_2] + 2NaOH$$

Procedure

In a two-necked 250-mL flask, $Na_2[MO_4]2H_2O$ (18 g for W, 12 g for Mo, 0.05 mol) and KCN (81 g, 1.25 mol) are dissolved in H_2O (90 mL). A freshly prepared solution of $K[BH_4]$ (8 g, 0.15 mol) in H_2O (90 mL) is then added at room temperature with stirring, followed by the dropwise addition of acetic acid (99%, 60 mL) over a period of 1 h. The reaction mixture is then heated to 90°C for 30 min, then cooled to room temperature and stirred overnight. The flask is wrapped in aluminum foil to avoid exposure to light. The mixture is then poured into ethanol (500 mL), and the resulting dark brown precipitate is collected by filtration. Boiling water (300 mL) is then added, followed by decolorizing charcoal, and the resulting mixture is filtered through Celite. Ethanol (300 mL) is added to the filtrate, and the precipitate is collected by filtration, washed with ethanol and ether, and then air-dried to afford $K_4[M(CN)_8]\cdot 2H_2O$ in 70% yield (for both Mo and W).

B. TETRABUTYLAMMONIUM OCTACYANOMETALATE(V), $[N(C_4H_9)_4]_3[M(CN)_8]$, M = Mo,W

$$K_4[M(CN)_8] + [NH_4]_4[Ce(SO_4)_4] \rightarrow K_3[M(CN)_8] + K[NH_4]_4[Ce(SO_4)_4]$$

$$K_3[M(CN)_8] + 3Ag\ NO_3 \rightarrow Ag_3[M(CN)_8] + 3KNO_3$$

$$Ag_3[M(CN)_8] + 3H\ Cl \rightarrow H_3[M(CN)_8] + 3AgCl$$

$$H_3[M(CN)_8] + 3[N(C_4H_9)_4]Br \rightarrow [N(C_4H_9)_4]_3[M(CN)_8] + 3HBr$$

In a 200-mL conical flask $K_4[M(CN)_8]2H_2O$ (3 g for W, 2.5 g for Mo, 5.10 mmol) is dissolved in water (50 mL). To the resulting yellow solution an acidic solution (H_2SO_4) of $[NH_4]_4[Ce^{IV}(SO_4)_4]$ (50 mL, 0.10 M) is added dropwise over a period of 20 min. The reaction mixture is then stirred for an additional 20 min, after which a solution of $AgNO_3$ (2.7 g, 0.016 mol) in water (25 mL) is added in a dropwise manner to give an orange-brown precipitate. The precipitate is collected and washed with water. Exposure to light is avoided. Then HCl (450 mL, 0.05 M) is added to the crude product, resulting in the immediate precipitation of AgCl, which is removed via filtration through Celite. Solid $[N(C_4H_9)_4]Br$ (5 g, 0.015 mol) is then added to the filtrate, after which the product $[N(C_4H_9)_4]_3[M(CN)_8]$ precipitates as a yellow solid for Mo^V and a white solid for W^V. The solid is collected, washed thoroughly with water, and then dried under vacuo to afford pure $[N(C_4H_9)_4]_3[M(CN)_8]$ in 40% yield (for both Mo^V and W^V).

C. $[Mn\{Mn(MeOH)_3\}_8(\mu\text{-}CN)_{30}\{Mo(CN)_3\}_6]\cdot 5MeOH\cdot 2H_2O$

$$2[N(C_4H_9)_4]_3[Mo(CN)_8] + 3[Mn(H_2O)_6][NO_3]_2 + 29MeOH \rightarrow$$
$$Mn\{Mn(MeOH)_3\}_8(\mu\text{-}CN)_{30}\{Mo(CN)_3\}_6 \cdot 5MeOH \cdot 2H_2O$$

In a 20-mL conical flask $[N(Bu)_4]_3[Mo(CN)_8]$ (100 mg, 0.094 mmol) is dissolved in methanol (6 mL) at room temperature. A solution of $[Mn(H_2O)_6][NO_3]_2$ (36 g, 0.144 mmol) in methanol (6 mL) is then added at room temperature with stirring, to afford immediately a red-brown solution. This solution is transferred into a test tube that is placed directly inside a glass jar containing diethyl ether. The jar is sealed, and slow diffusion of diethyl ether into this solution over a period of 1–2 days affords dark red-brown needles. The mother liquid is then carefully removed with a glass pipette, leaving only the crystals, which are immediately washed with additional portions of diethyl ether to remove any remaining impurities. The crystals are stable in diethyl ether and can be stored as such in the refrigerator.

D. $[Ni\{Ni(MeOH)_3\}_8(\mu\text{-}CN)_{30}\{M^V(CN)_3\}_6]\cdot x MeOH$; M = Mo,W

In a 20-mL conical flask, $[N(Bu)_4]_3$ $[M(CN)_8]$ (100 mg for Mo, 107 mg for W, 0.096 mmol) is dissolved in methanol (6 mL). A solution of $[Ni(H_2O)_6][(NO_3]_2$ (42 mg, 0.144 mmol) in methanol (4 mL) is then added at room temperature with stirring, which immediately gives a yellow solution. Slow diffusion of diethyl ether into this solution as described above gives pale yellow crystals (where M = Mo) and yellow-brown crystals for the W analog. The crystals are isolated as described above and any precipitate is removed by washing with a mixture of diethyl ether/methanol (2 : 1). The crystals are stored cold in this solvent mixture.

E. $[Co\{Co(MeOH)_3\}_8(\mu\text{-}CN)_{30}\{M^V(CN)_3\}_6]\cdot x MeOH\cdot x H_2O$; M = Mo,W

Both compounds are prepared from methanolic solutions of $[N(Bu)_4]_3[M(CN)_8]$ (100 mg for Mo, 107 mg for W, 0.096 mmol) and $[Co(H_2O)_6][X]_2$ (X = ClO_4, BF_4) (42 mg, 0.144 mmol) following the procedure described above. Slow diffusion of diethyl ether into the resulting red solutions affords dark red crystals for M = Mo and red crystals for the W analog. Once again the crystals are washed and stored in diethyl ether.

Properties

Characterization by routine spectroscopic methods is difficult since the clusters are fairly light-sensitive, lose solvent readily, and decompose rapidly on standing in air. For this class of compounds characterization via elemental analysis is not a suitable method. For the Ni(II) compounds UV–vis spectra recorded in MeOH show bands at 379 and 347 nm for Ni_9Mo_6 and Ni_9W_6, respectively. (Compare for $[M(CN)_8]^{3-}$, 389 nm, M = Mo and 358 nm, M = W). In all other cases the clusters are not stable in solution. Because of the excellent quality of the single crystals obtained by the procedures described above, the exact molecular geometry of the compounds can be determined via low-temperature (223 K) single-crystal X-ray analysis. For reference, selected crystallographic information for all the clusters is tabulated as follows:

Parameter	Mn_9Mo_6	Ni_9Mo_6	Ni_9W_6	Co_9Mo_6	Co_9W_6
Crystal system	Monoclinic	Monoclinic	Monoclinic	Monoclinic	Monoclinic
Space group	$C2/c$	$C2/c$	$C2/c$	$C2/c$	$C2/c$
a(Å)	29.3847 (18)	28.4957 (18)	28.5278 (16)	28.5857 (18)	28.6588 (16)
c(Å)	32.8841 (11)	32.4279 (17)	32.4072 (17)	32.550 (2)	32.4911 (17)
β (degrees)	114.188 (6)	113.155 (6)	113.727 (6)	113.298 (6)	113.522 (6)
Volume (Å^3)	17,088.0 (15)	16,312.9 (16)	16,251 (2)	16,550 (17)	16,465 (2)
Z	4	4	4	4	4
$\rho_{calcd.}$(g/cm^3)	1.277	1.500	1.695	1.480	1.631

Additional crystallographic information for the $Mn^{II}{}_9Mo^{V}{}_6$ cluster is deposited in the CCDC.* A more detailed description of the structural topology can be found in the original publication.[9] All the clusters contain 15 paramagnetic metal centers and, as a consequence, display a diverse range of interesting magnetic properties.[8–10]

References

1. G. Aromi, M. J. Knapp, J. P. Claude, J. C. Huffman, D. N. Hendrickson, and G. Christou, *J. Am. Chem. Soc.* **120**, 5489 (1999).
2. Y. Pontillon, A. Caneschi, D. Gatteschi, R. Sessoli, E. Ressouche, J. Schweizer, and E. Lelievre-Berna, *J. Am. Chem. Soc.* **121**, 5301 (1999).
3. A. L. Barra, A. Caneschi, A. Cornia, F. Fabrizi de Biani, D. Gatteschi, D. Sanggregorio, R. Sessoli, and L. Sorace, *J. Am. Chem. Soc.* **121**, 5302 (1999).
4. S. L. Castro, Z. Sun, C. M. Grant, J. C. Bollinger, D. N. Hendrickson, and G. Christou, *J. Am. Chem. Soc.* **120**, 2365 (1998).
5. S. M. J. Aubini, N. R. Dilley, L. Pardi, J. Krzystek, M. W. Wemple, L. C. Brunel, M. B. Maple, G. Christou, and D. N. Hendrickson, *J. Am. Chem. Soc.* **120**, 4991 (1998).
6. L. Thomas, F. Lionti, R. Ballou, D. Gatteschi, R. Sessoli, and B. Barbara, *Nature* 383, 145 (1996).
7. O. Kahn, *Molecular Magnetism*, Wiley-VCH, New York, 1993.
8. M. Pilkington and S. Decurtins, *Chimia* **54**, 593 (2000).
9. J. Larionova, M. Gross, M. Pilkington, H.-P. Andres, H. Stoeckli-Evans, H. U. Güdel, and S. Decurtins, *Angew. Chem. Int. Ed.* **39**, 1605 (2000).
10. F. Bonadio, M. Gross, H. Stoeckli-Evans, and S. Decurtins, *Inorg. Chem.* **41**, 5891 (2000).
11. N. N. Greenwood and A. Earnshaw, *Chemistry of the Elements*, Wiley-VCH, Weinheim, 1990, 1314.
12. (a) J. G. Leipoldt, L. D. C. Bok, and P. J. Cillers, *Z. Anorg. Allg. Chem.* **407**, 350 (1974); (b) N. H. Furman and C. O. Miller, *Inorg. Synth.* **3**, 160 (1950); (c) S. S. Basson, L. D. C Bok, and S. R. Eiroder, *Z. Analyt. Chem.* **268**, 287 (1974); (d) B. J. Corden, J. A. Cunningham, and R. Eisenberg, *Inorg. Chem.* **9**, 356 (1970); (e) R. A. Pribush and R. D. Archer, *Inorg. Chem.* **13**, 2556 (1974).

*Crystallographic data for the material tabulated here has been deposited with the CCDC as supplementary publication CCDC-136762; CSD: WISVOJ. For the manganese–molybdenum cluster the checkers also obtained a second crystalline modification with an alternative unit cell, namely, $R3$, a = 17.376 (5) Å, c = 44.25 (2) Å, V = 11570 (5) Å^3, featuring a different packing arrangement of the cluster units.

35. CYANOMETAL-SUBSTITUTED DERIVATIVES OF THE Fe_4S_4 CLUSTER CORE

Submitted by RALF APPELT* and HEINRICH VAHRENKAMP*
**Checked by DONALD J. DARENSBOURG,† ANDREA L. PHELPS,†
and M. JASON ADAMS†**

Molecular magnetism is one of the challenging fields of modern inorganic chemistry,[1] and a wide variety of polynuclear metal complexes have been designed as magnetic materials. Among them, cyanide-bridged species play an important role, motivated by the ferromagnetism of materials with Prussian blue structures,[2] the amazingly high-molecular spins of complexes with $M(\mu\text{-CN—M}')x$ cores ($x = 4$–7),[3] and studies of electronic interactions along chains of M—CN—M′ units.[4,5,6] Here we report the preparations of two cyanometal-substituted derivatives of the Fe_4S_4 cluster and their precursors. Such octanuclear $Fe_4S_4(NC—M')_4$ complexes[7,8] have an unusual temperature-independent paramagnetism resulting from the electronic influence of the cyanometal "ligands" on the magnetic interactions in the $[Fe_4S_4]^{2+}$ core.[7]

■ **Caution.** *Metal cyanides are very toxic. All manipulations should be carried out in a well-ventilated fume hood. Any excess cyanides may be destroyed by the addition of a strong oxidizing agent such as hydrogen peroxide. (The pH must be higher than 7 to suppress the formation of HCN!)*

A. CYANO(η^5-CYCLOPENTADIENYL)-1,2-BIS (DIPHENYLPHOSPHINO)ETHANEIRON(II), CpFe(dppe)CN

$$\text{CpFe(dppe)Br} + \text{KCN} \rightarrow \text{CpFe(dppe)CN} + \text{KBr}$$

Procedure

CpFe(dppe)Br is prepared by the literature procedure.[9] The solvents (methanol, CH_2Cl_2, hexane) as well as potassium cyanide are obtained commercially and used as received.

■ **Caution.** *Unreacted KCN!*

*Institut für Anorganische und Analytische Chemie der Universität Freiburg, Albertstr. 21, D-79104 Freiburg, Germany.
†Department of Chemistry, Texas A&M University, College Station, TX 77843.

A 100-mL single-necked round-bottomed flask is charged with a solution of CpFe(dppe)Br (1.50 g, 2.6 mmol) in 50 mL of methanol. Potassium cyanide (0.68 g, 10.4 mmol) is added and the resulting solution stirred for 3 h at room temperature. The solvent is removed under reduced pressure and the resulting solid is extracted 2 times with 15 mL of dichloromethane. The yellow-brown solution is layered with 60 mL of hexane, and the mixture is kept at room temperature* for 2 days. After filtration, washing with hexane, and drying in vacuo, 1.07 g (74%) of CpFe(dppe)CN is obtained as yellow brown crystals.

Anal. Calcd. for $C_{32}H_{29}FeNP_2$: C, 70.47; H, 5.36; N, 2.56. Found: C, 70.30; H, 5.56; N, 2.36.

Properties

The yellow-brown crystals of CpFe(dppe)CN[10,11] have mp 188–190°C. The product is soluble in alcohol, acetonitrile, and trichloromethane; slightly soluble in benzene; but insoluble in hexane. The IR spectrum shows characteristic bands; in KBr: 2062 (m) (CN), 696 (s) (C_6H_5); in CH_2Cl_2: 2063 (CN). The ^1H NMR spectrum (90 MHz, $CDCl_3$) shows multiplets at δ 7.8–7.0 (20H, aryl-H), a triplet at δ 4.3 ($J = 1.5$ Hz, 5H, Cp) and a multiplet at δ 2.8–2.2 (4H, CH_2).

B. SODIUM CYANOPENTACARBONYLTUNGSTATE, $Na[W(CO)_5CN]$

$$W(CO)_6 + Na[N\{Si(CH_3)_3\}_2] \longrightarrow Na[W(CO)_5CN] + O[Si(CH_3)_3]_2$$

Procedure

All manipulations are performed under an atmosphere of dry nitrogen by using standard Schlenk and syringe techniques. The solvents toluene, benzene, and hexane are dried by reflux over the drying agent sodium and distilled prior to use. Sodium bis(trimethylsilyl)amide is prepared by the literature procedure.[12] $W(CO)_6$ is obtained commercially and used without further purification.

A filtered solution of sodium bis(trimethylsilyl) amide (5.8 g, 31.6 mmol) in 100 mL of toluene is placed in a pressure-equalized dropping funnel (200 mL). The mixture is added dropwise to a magnetically stirred suspension of $W(CO)_6$ (11.3 g, 32.1 mmol) in 100 mL of toluene, contained in a 500-mL Schlenk flask. The mixture is stirred for 5 days at room temperature. The precipitate is collected by filtration with a large-diameter D3 Schlenk sintered-glass funnel and washed first with 3 × 30 mL of benzene and then with 3 × 30 mL of hexane. The filtration and washing steps may take more than one day. To prevent the sintered-glass

*Checkers crystallized product at −10°C; yield 78%.

frit from becoming blocked, the filtration should be performed without suction. After extensive drying in vacuo at 50°C to ensure complete removal of unreacted tungsten hexacarbonyl, 9.2 g (77%) of $Na[W(CO)_5CN]$ is obtained as a white solid.

Anal. Calcd. for C_6NNaO_5W: C, 19.30; N, 3.80. Found: C, 19.10; N, 3.60.

Properties

$Na[W(CO)_5CN]$[13] is a white, very fluffy solid [mp 210°C (dec.)] with an intensely bitter taste that can be experienced merely from traces of dust present in the air by normal handling. The product is soluble in methanol, ethanol, water, and acetonitrile. The IR spectrum shows characteristic bands (cm^{-1}); in KBr: 2111 (w) (CN), 2056 (m), 1933 (s), and 1901 (s) (CO); in methanol: 2104 (w) (CN), 2057 (m), 1931 (s), 1903 (s) (CO).

C. BIS(TETRAPHENYLPHOSPHONIUM) TETRACHLOROTETRA (μ_3-THIO)TETRAFERRATE, $[PPh_4]_2[Fe_4S_4Cl_4]$

$$4FeCl_2 + 6\ KSPh + 2[PPh_4]Cl + 4S \rightarrow [PPh_4]_2[Fe_4S_4Cl_4] + 6KCl + 3\ PhSSPh$$

Procedure

All manipulations are performed under an atmosphere of dry nitrogen using standard Schlenk and syringe techniques. The solvents are dried by reflux over the appropriate drying agents (CH_3CN/CaH_2; diethyl ether/4-Å molecular sieve) and distilled prior to use. $FeCl_2$ is prepared by the literature procedure.[14]

A 100-mL Schlenk flask is charged with a mixture of anhydrous $FeCl_2$ (1.00 g, 7.89 mmol), KSPh (1.75 g, 11.83 mmol), $[PPh_4]Cl$ (1.48 g, 3.94 mmol), and sulfur (0.32 g, 9.86 mmol). Then 40 mL of acetonitrile is added with continuous stirring. The resulting reaction mixture is stirred at room temperature for 45 min. The solution is filtered to remove KCl and unreacted sulfur, and 100 mL of diethyl ether is added to the filtrate. On standing for some hours, the solution exhibits deposition of a black crystalline solid. The product is isolated by filtration, washed twice with diethyl ether, and dried in vacuo. Yield: 1.57 g (68%).

Anal. Calcd. for $C_{48}H_{40}Cl_4Fe_4P_2S_4$: C, 49.18; H, 3.44. Found: C, 48.62; H, 3.44.

Properties

$[PPh_4]_2[Fe_4S_4Cl_4]$in CH_2Cl_2 has three absorptions in the visible and near-IR range at 507 nm ($\varepsilon = 1800\ L \cdot mol^{-1}\ cm^{-1}$), 694 nm ($\varepsilon = 1500$ L

mol^{-1} cm^{-1}) and 1077 nm (ε = 400 L mol^{-1} cm^{-1}). The electrochemical properties of the product, as obtained by cyclic voltammetry (in CH_2Cl_2, reference Ag/AgCl), show a reversible reduction ($E_{1/2} = -0.87$ V) and a reversible oxidation ($E_{1/2} = 0.53$ V).

D. TETRAKIS{μ-CYANO(η^5-CYCLOPENTADIENYL)(1,2-BIS(DIPHENYLPHOSPHINO)ETHANE)IRON}TETRA(μ_3-THIO)TETRAIRON BIS(HEXAFLUOROPHOSPHATE), $[Fe_4S_4\{NCFe(dppe)Cp\}_4][PF_6]_2$

$$[PPh_4]_2[Fe_4S_4Cl_4] + 4\ CpFe(dppe)CN + 4\ [NH_4][PF_6] \rightarrow$$
$$[Fe_4S_4\{NCFe(dppe)Cp\}_4][PF_6]_2 + 4\ NH_4Cl + 2\ [PPh_4][PF_6]$$

Procedure

All manipulations are performed under an atmosphere of dry nitrogen using standard Schlenk and syringe techniques. The solvents are dried by reflux over the appropriate drying agents (CH_2Cl_2 and CH_3CN/CaH_2; hexane and benzene/sodium) and distilled prior to use.

A 50-mL Schlenk flask is charged with a mixture of $[PPh_4]_2[Fe_4S_4Cl_4]$ (0.060 g, 0.050 mmol) and CpFe(dppe)CN (0.133 g, 0.20 mmol). Then 20 mL of acetonitrile is added with continuous stirring. A brownish black solution is obtained. Then $[NH_4][PF_6]$ (60 mg, 0.36 mmol) is added to the solution. The color changes rapidly to dark blue-purple. After stirring overnight, the solution is filtered. The filtrate is evaporated to dryness, and the residue is extracted with 10 mL of CH_2Cl_2. Then 40 mL of hexane is added to the extract to precipitate the product. Filtration, washing with 20 mL of hexane, and drying in vacuo yields 0.035 g (25%)* of $[Fe_4S_4\{NCFe(dppe)Cp\}_4][PF_6]_2$ as a dark purple powder.

Anal. Calcd. for $C_{128}H_{116}F_{12}Fe_8N_4P_{10}S_4$: C, 52.99; H, 4.21; N, 1.92. Found: C, 54.56; H, 4.14; N, 1.98.

Properties

Crystalline $[Fe_4S_4\{NCFe(dppe)Cp\}_4][PF_6]_2$ is relatively stable [mp 87°C (dec.)], but it is very air-sensitive in solution. The IR spectrum shows characteristic bands (cm^{-1}); in KBr: 2016 (s) (CN), 839 (s) (PF_6). The electronic spectrum, measured in dichloromethane, shows a strong band at 526 nm (ε = 12,000 L mol^{-1} cm^{-1}), which is assigned to a metal–metal charge transfer (MMCT) via the bridging cyanides from the organometallic CpFe(dppe)CN units to the

*The checkers' yield was 66%.

cluster core. The electrochemical properties of the product, as obtained by cyclic voltammetry (in CH_2Cl_2, reference Ag/AgCl) show a reversible reduction ($E_{1/2} = -0.64$ V) and an irreversible oxidation ($E_{p(ox)} = 0.70$ V).

E. BIS(TETRAPHENYLPHOSPHONIUM) TETRAKIS {μ-CYANOPENTACARBONYLTUNGSTEN}TETRA (μ_3-THIO)TETRAFERRATE, $[PPh_4]_2[Fe_4S_4\{NCW(CO)_5\}_4]$

$$[PPh_4]_2[Fe_4S_4Cl_4] + 4\ Na[W(CO)_5CN] \rightarrow [PPh_4]_2[Fe_4S_4\{NCW(CO)_5\}_4] + 4NaCl$$

To a mixture of $[PPh_4]_2[Fe_4S_4Cl_4]$ (0.120 g, 0.104 mmol) and $Na[NCW(CO)_5]$ (0.180 g, 0.483 mmol), contained in a 50-mL Schlenk flask is added 15 mL of CH_2Cl_2 with stirring. A dark violet solution is formed. After stirring for 30 min, the solution is filtered to remove NaCl. The filtrate is layered with a mixture of 10 mL of benzene and 30 mL of hexane and kept at −27°C for 2 days. After filtration, washing with petroleum ether, and drying in vacuo, 0.233 g (92%)* of $[PPh_4]_2[Fe_4S_4\{NCW(CO)_5\}_4]$ is obtained as dark violet crystals; mp 128°C (dec.)].

Anal. Calcd. for $C_{72}H_{40}N_4Fe_4O_{20}P_2S_4W_4$: C, 35.59; H, 1.66; N, 2.31; Fe, 9.19. Found: C, 35.43; H, 1.86; N, 2.19; Fe, 9.34.

Properties

In the solid state $[Ph_4P]_2[Fe_4S_4\{NCW(CO)_5\}_4]$ is stable in air [mp 128°C (dec.)], but it decomposes immediately when dissolved in acetonitrile, acetone, or alcohol. The compound is soluble in chlorobenzene and diethyl ether, slightly soluble in benzene or toluene, but insoluble in petroleum ether. The IR spectrum shows characteristic bands (cm^{-1}); in KBr: 2034 (m) (CN), 2105 (w), 1981 (w) and 1916 (s) (CO); in CH_2Cl_2: 2037 (m) (CN), 2105 (w), 1981 (w), 1939 (s), and 1910 (shoulder) (CO).† The electronic spectrum, recorded in dichloromethane, shows a strong band at 506 nm ($\varepsilon = 18{,}000\ L\ mol^{-1}\ cm^{-1}$) that is assigned to a metal–metal charge transfer via the bridging cyanides from the organometallic $Na[NCW(CO)_5]$ units to the cluster core.

*The checkers' yield was 43%.

†This band assignment is consistent with the IR data of a series of related complexes.[8] However, the alternative assignment with the CN band at 2105 cm^{-1} and the highest CO at 2037 cm^{-1} is also realistic.

References

1. O. Kahn, *Molecular Magnetism*, VCH Publishers, New York/Weinheim, 1993.
2. W. R. Entley and G. Girolami, *Inorg. Chem.* **33**, 5165 (1994) and refs. therein.
3. T. Mallah, C. Auberger, M. Verdaguer, and P. Veillet, *J. Chem. Soc., Chem. Commun.* 61 (1995).
4. G. N. Richardson, U. Brand, and H. Vahrenkamp, *Inorg. Chem.* **38**, 3070 (1999).
5. A. Geiss and H. Vahrenkamp, *Inorg. Chem.* **39**, 4029 (2000).
6. H. Vahrenkamp, A. Geiss, and G. N. Richardson, *J. Chem. Soc., Dalton Trans.* 3643 (1997).
7. N. Zhu, J. Pebler, and H. Vahrenkamp, *Angew. Chem., Int. Ed. Engl.* **35**, 894 (1996).
8. N. Zhu, R. Appelt, and H. Vahrenkamp, *J. Organomet. Chem.* **565**, 187 (1998).
9. S. G. Davies, H. Felkin, and O. Watts, *Inorg. Synth.* **24**, 170 (1986).
10. P. M. Treichel and P. C. Molzahn, *Synth, React. Inorg. Met.-Org. Chem.* **9**, 21 (1979).
11. G. J. Bird and S. G. Davies, *J. Organomet. Chem.* **262**, 215 (1984).
12. C. R. Krüger and H. Niederprüm, *Inorg. Synth.* **8**, 15 (1966).
13. R. B. King, *Inorg. Chem.* **6**, 25 (1967).
14. P. Kavacic and N. O. Brace, *Inorg. Synth.* **6**, 172 (1960).
15. M. G. Kanatzidis, W. R. Dunham, W. R. Hagen, and D. Coucouvanis, *J. Chem. Soc., Chem. Commun.* 356 (1984).
16. D. Coucouvanis, A. Salifoglou, M. G. Kanatzidis, W. R. Dunham, A. Simopoulos, and A. Kostikas, *Inorg. Chem.* **27**, 4066 (1988).
17. (a) G. B. Wong, M. A. Bobrik, and R. H. Holm, *Inorg. Chem.* **17**, 578 (1978); (b) R. W. Johnson and R. H. Holm, *J. Am. Chem. Soc.* **100**, 5338 (1978).

36. TRICYANOMETALATE BUILDING BLOCKS AND ORGANOMETALLIC CYANIDE CAGES

Submitted by STEPHEN M. CONTAKES,[*] KEVIN K. KLAUSMEYER,[*] and THOMAS B. RAUCHFUSS[*]
Checked by DANIEL CARMONA[†], MARÍA PILAR LAMAT,[†] and LUIS A. ORO[†]

Tricyanometalate complexes are useful building blocks for the construction of molecular cage compounds[1–6] and coordination solids.[7,8] One class of tricyanometalates that has proved especially useful are those with the formula $[M(C_5R_5)(CN)_3]^-$, since the cyclopentadienyl coligand limits the degree of polymerization and facilitates the formation of molecular cages. The tricyanides of the monocyclopentadienyl complexes with middle transition series metals were not well studied until the early 1970s. Pauson and Dineen reported the preparation of $K[CpCo(CN)_3]$ and $[PPh_3CH_2Ph][CpCo(CN)_3]$ via the reaction of cyanide with $CpCoI_2(CO)$.[9] We have obtained the former compound using a

[*]School of Chemical Sciences, University of Illinois at Urbana—Champaign, Champaign, IL 61820.
[†]Departamento de Química Inorgánical, Instituto de Ciencia de Materiales de Aragón, Universidad de Zaragoza-Consejo Superior de Investigaciones Científicas, 50009 Zaragoza, Spain.

slightly modified procedure and have been able to convert it to the organic solvent soluble $[K(18\text{-crown-6})]^+$ salt. The related tricyanides $[Cp^*Rh(CN)_3]^-$, , $[Cp^*Ir(CN)_3]^-$, and $[Cp^*Ru(CN)_3]^{2-}$ are readily prepared from commercially available $[Cp^*MCl_2]_2$,[9] and the related $[CpFe(CN)_3]^{2-}$ can be obtained from the photolysis of $[CpFe(CO)(CN)_2]^-$.[10] Of these, $[CpCo(CN)_3]^-$, $[Cp^*Rh(CN)_3]^-$, and $[Cp^*Ir(CN)_3]^-$ are useful in the synthesis of molecular cages. Of particular note is the condensation between $[CpCo(CN)_3]^-$ and $[Cp^*Rh(NCMe)_3]^{2+}$[11] to give a molecular cube of formula $[\{CpCo(CN)_3\}_4\{Cp^*Rh\}_4]^{4+}$ or a molecular bowl of formula $[\{CpCo(CN)_3\}_4\{Cp^*Rh\}_3]^{2+}$, depending on the stoichiometic ratio of the reactants used.

■ **Caution.** *Because of the toxic nature of cyanide, special care should be exercised in these preparations. We make amylnitrite available in the laboratory as an antidote to acute cyanide poisoning. Organic soluble salts of cyanide (and potentially cyanometallates) are especially hazardous contact poisons because they are transported intraveneously through the skin.*

A. TETRAETHYLAMMONIUM TRICYANO(PENTAMETHYL CYCLOPENTADIENYL)RHODATE(III) MONOHYDRATE, $[NEt_4][Cp^*Rh(CN)_3]\cdot H_2O$

$$[Cp^*RhCl_2]_2 + 6KCN + 2NEt_4Cl \rightarrow 2NEt_4[Cp^*Rh(CN)_3] \cdot H_2O + 6KCl$$

Procedure

A 250-mL round-bottomed flask containing a stirring bar is covered with aluminum foil (to exclude light) and then charged with 0.30 g (0.485 mmol) of $(Cp^*RhCl_2)_2$ and 50 mL of H_2O. To the resulting slurry, 0.329 g (1.94 mmol) of $AgNO_3$ is added. Over the course of 40 min, the color of the solution becomes yellow and a white precipitate of AgCl appears. The slurry is filtered through Celite. To the filtered solution of $[Cp^*Rh(H_2O)_3]^{2+}$ is added 0.284 g (4.36 mmol) of KCN. This solution is allowed to stir overnight. The resulting colorless solution is treated with 0.161 g (0.970 mmol) of Et_4NCl. After a further 20 min, the solvent is evaporated. The pale yellow solid residue is extracted into 30 mL of CH_2Cl_2, and the solution is filtered through a medium-porosity frit to remove a white solid consisting of KCl, KCN, and KNO_3. The product, $[Et_4N][Cp^*Rh(CN)_3]$, is precipitated by the addition of 50 mL Et_2O to the filtered extract. The precipitate is collected by filtration, washed with 10 mL of Et_2O, and dried under vacuum. Yield: 0.401 g (93%).*

*The checkers' yield was 73–76%.

Anal. Calcd. for $C_{21}H_{37}N_4ORh$: C, 54.31; H, 8.03; N, 12.06. Found: C, 53.93; H, 8.20; N, 11.97.

Properties

The product is a pale yellow crystalline solid. It is air-stable both as a solid and in solutions. It is soluble in polar solvents such as water, acetone, and acetonitrile as well as CH_2Cl_2 and is insoluble in ether and hexanes. The ^1H NMR spectrum (CD_3CN) exhibits a triplet at δ 3.16 [$N(CH_2CH_3)_4$, 8H], a singlet at 1.86 (Cp^*, 15H), and a quartet at 1.20 [$N(CH_2CH_3)_4$, 12H]. The ^1H NMR spectrum in $CDCl_3$ is similar: δ 1.36, 1.93, and 3.40. The $^{13}C\{^1H\}$ NMR spectrum ($CDCl_3$) shows signals at δ 128.9 (d, $J_{C-Rh} = 51$ Hz, CN), 100.59 (d, $J_{C-Rh} =$ = 4.6 Hz, Me_5C_5), 52.7 [s, $(CH_3CH_2)_4N$], 9.9 [s, $(CH_3)_5C_5$] 7.7 [$(CH_3CH_2)_4N$]. The molecular ion is evident in the negative-ion FAB mass spectrum at m/z 316.0. The infared spectrum of the solid (KBr pellet) exhibits the typical two-band pattern for pseudo-C_{3v} symmetry at 2110 and 2122 cm^{-1} in the cyanide region.

B. TETRAETHYLAMMONIUM TRICYANO (PENTAMETHYLCYCLOPENTADIENYL)IRIDATE(III) MONOHYDRATE, $[NEt_4][Cp^*Ir(CN)_3]\cdot H_2O$

$$[Cp^*IrCl_2]_2 + 6KCN + 2NEt_4Cl \rightarrow 2[NEt_4][Cp^*Ir(CN)_3] \cdot H_2O + 6KCl$$

Procedure

A 250-mL round-bottomed flask containing a stirring bar is charged with 0.300 g (0.377 mmol) of $(Cp^*IrCl_2)_2$ and 100 mL of H_2O. To this slurry is added 0.150 g (2.30 mmol) of KCN. The resulting mixture is heated at reflux for 3 h under N_2 to give a clear, colorless solution. To the solution is added 0.150 g (0.905 mmol) of $[Et_4N]Cl$, and then the solvent is removed under vacuum. The pale yellow solid residue is extracted into 50 mL CH_2Cl_2, and the volume of the resulting colorless solution is reduced under vacuum to ~5 mL. The product, $[Et_4N][Cp^*Ir(CN)_3]$, is precipitated by the addition of 50 mL Et_2O. The precipitate is collected by filtration, washed with 10 mL of Et_2O, and dried under vacuum. Yield: 0.293 g (73%).*

Anal. Calcd. for $C_{21}H_{37}IrN_4O$: C, 45.55; H, 6.73; N, 10.12. Found: C, 45.63; H, 6.66; N, 10.22.

*The checkers' yield was 85–87%.

Properties

The product is a colorless crystalline solid. It is air stable both as a solid and in solutions. It is soluble in polar solvents such as water, acetone, and acetonitrile as well as CH_2Cl_2 and is insoluble in ether and hexanes. The ^1H NMR spectrum (CD_2Cl_2) exhibits a quartet at δ 3.35 [8H, $(CH_3C\mathit{H}_2)_4N$], a singlet at δ 1.99 [15H, $(C\mathit{H}_3)_5C_5$), and a triplet at δ 1.34 [12H, $(C\mathit{H}_3CH_2)_4N$]. The ^1H NMR spectrum in CD_3CN is similar, with signals at δ 3.15, 1.95, and 1.20, respectively. The cyanide region in the infrared spectrum of the solid (KBr pellet) shows a typical two-band pattern at 2107 and 2124 cm^{-1}.

C. (18-CROWN-6)POTASSIUM TRICYANO(CYCLOPENTADIENYL) COBALTATE(III), [K(18-CROWN-6)][CpCo(CN)$_3$]

$$CpCo(CO)_2 + I_2 \rightarrow CpCo(CO)I_2 + CO$$

$$CpCo(CO)I_2 + KCN \rightarrow K[CpCo(CN)_3] + CO + 2KI$$

$$K[CpCo(CN)_3] + 18\text{-crown-6} \rightarrow [K(18\text{-crown-6}][CpCo(CN)_3]$$

Procedure

■ **Caution.** *The procedure involves the formation of gaseous CO and should be conducted in a well-ventilated hood.*

The synthesis of $CpCo(CO)I_2$ is performed by using a modification of literature methods.[12] A tared syringe is used to weigh out 3.0 g (17.0 mmol) $CpCo(CO)_2$ in air. The compound is then dissolved in 50 mL of MeOH to give a red-brown solution. Then 4.23 g (17.0 mmol) I_2 is added to the stirred solution in 0.1–0.2 g portions over 20 min. The addition occurs with evolution of gas and formation of a black precipitate. The resulting mixture is stirred for 2 h, and then the solvent is removed under vacuum. The residue is extracted with four 100-mL portions of CH_2Cl_2 (or until further extracts are colorless), and the extracts are evaporated to dryness to give black microcrystals. The crystals are washed with two 10-mL portions of Et_2O and dried in air on the frit for 20 min. Yield: 6.1 g $CpCo(CO)I_2$ (88%).

Under a nitrogen atmosphere, a solution of 2.41 g (37 mmol) of KCN in 80 mL of MeOH is added to a solution of 5.0 g (12 mmol) of $CpCo(CO)I_2$ in 20 mL of MeOH, and the resulting solution is heated under reflux for 3 days. The reaction mixture is allowed to cool, and then it is filtered in air. The filtrate is evaporated to dryness and the residue is washed with 800 mL of MeCN in 50-mL portions until the washings contain no KI (as indicated by the failure to obtain a precipitate on addition of Et_2O to the washings). The solid $K[CpCo(CN)_3]$ is washed with two 10-mL portions of Et_2O and dried under vacuum for 2 h. Yield: 1.16 g (41%).

Then 1.16 g (4.81 mmol) of $K[CpCo(CN)_3]$ is suspended in a solution containing 1.28 g (4.84 mmol) of 18-crown-6 in 250 mL of MeCN, and the suspension is stirred for 3 h. During this time the $K[CpCo(CN)_3]$ dissolves to give a golden yellow solution. The solution is filtered to remove a small quantity of an insoluble brown impurity, and then the solvent is reduced in volume to ~100 mL under vacuum. Addition of 150 mL Et_2O precipitates the product as yellow microcrystals. The product is collected by filtration, washed with two 10-mL portions of Et_2O, and finally dried under vacuum for 1 h. Yield: 2.31 g (91% from $K[CpCo(CN)_3]$; 33% overall).*

Anal. Calcd. for $C_{20}H_{29}CoKN_3O_6$: C, 47.52; H, 5.78; N, 8.31. Found: C, 47.50; H, 5.82; N, 8.36.

Properties

The compound is a golden yellow microcrystalline solid. It is stable in air both as a solid and in solution. It is soluble in water, methanol, and acetonitrile and is insoluble in ether and hexanes. The ^1H NMR spectrum (CD_3CN) shows two singlets at δ 3.56 (24H, $[K(C_{12}H_{24}O_6)]$) and 5.23 (5H, C_5H_5). The infared spectrum of the solid (KBr) shows a single band in the cyanide region at 2119 cm^{-1}.

D. DODECACYANOTETRAKIS {CYCLOPENTADIENYL) COBALT(III)} TETRAKIS{PENTAMETHYL CYCLOPENTADIENYLRHODIUM(III)} TETRAKIS (HEXAFLUOROPHOSPHATE), $[(CpCo(CN)_3)_4(Cp^*Rh)_4](PF_6)_2$

$$4\ [K(18\text{-crown-6})][CpCo(CN)_3] + 4\ [Cp^*Rh(NCMe)_3](PF_6)_2 \rightarrow [(CpCo)_4(Cp^*Rh)_4(CN)_{12}](PF_6)_4 + 4\ [K(18\text{-crown-6})](PF)_6$$

Procedure

A solution of 50 mg (0.10 mmol) of $[K(18\text{-crown-6})][CpCo(CN)_3]$ in 15 mL of MeCN is added dropwise to a stirred solution of 65 mg (0.1 mmol)

*The checkers' yields were 88–90% and 25–30% overall.

$[Cp^*Rh(NCMe)_3](PF_6)_2$,[13] in 15 mL MeCN. The resulting solution is allowed to stand for 36 h, and then the product is precipitated as a yellow powder by addition of 100 mL of Et_2O. The product is collected by filtration, washed with two 10-mL portions of Et_2O, and dried under vacuum for 1 h. Yield: 58 mg (99%).*

Anal. Calcd. for $C_{72}H_{80}Co_4F_{24}N_{12}P_4Rh_4$: C, 36.95; H, 3.44; N, 7.18. Found: C, 37.29; H, 3.73; N, 7.24.

Properties

The compound, a golden yellow powder, is stable in air both as a solid and in solutions. It is soluble in acetonitrile and slightly so in dichloromethane. It is insoluble in water, ether, and hexanes. The ^1H NMR spectrum (CD_3CN) shows two singlets at δ 1.762 [60H, $C_5(CH_3)_5$] and 5.720 (20H, C_5H_5). The ESI-MS shows peaks at $m/z = 2195$ amu corresponding to $[(CpCo)_4(Cp^*Rh_4)(CN)_{12}]$-$[PF_6]_3^+$ and at $m/z = 1025$ amu (atomic mass units) due to $[(CpCo)_4(Cp^*Rh)_4(CN)_{12}][PF_6]_2$.$^{2+}$ The infared spectrum of the solid (KBr) shows a single band in the cyanide region at 2180 cm^{-1}.

References

1. K. K. Klausmeyer, T. B. Rauchfuss, and S. R. Wilson, *Angew. Chem., Int. Ed.* **37**, 1808 (1998).
2. K. K. Klausmeyer, S. R. Wilson, and T. B. Rauchfuss, *J. Am. Chem. Soc.* **121**, 2705 (1999).
3. S. M. Contakes, K. K. Klausmeyer, R. M. Milberg, S. R. Wilson, and T. B. Rauchfuss, *Organometallics* **17**, 3633 (1998).
4. S. M. Contakes, and T. B. Rauchfuss, *Angew. Chem., Int. Ed.* **38**, 1984 (2000).
5. S. M. Contakes and T. B. Rauchfuss, *Chem. Commun.* 553 (2001).
6. S. M. Contakes, M. L. Kuhlman, M. Ramesh, S. R. Wilson, and T. B. Rauchfuss, *Proc. Natl. Acad. Sci.* (USA) **99**, 4889 (2002).
7. S. M. Contakes, M. Schmidt, and T. B. Rauchfuss, *Chem. Commun.* 1183 (1999).
8. S. M. Contakes, K. K. Klausmeyer, and T. B. Rauchfuss, *Inorg. Chem.* **39**, 2069 (2000).
9. J. A. Dineen and P. L. Paulson, *J. Organometal. Chem.* **43**, 209 (1972).
10. D. J. Darensbourg, W. Z. Lee, M. J. Adams, D. L. Larkins, and J. H. Reibenspies, *Inorg. Chem.* **38**, 1378 (1999).
11. C. White, A. Yates, and P. M. Maitlis, *Inorg. Synth.* **29**, 228 (1992).
12. R. B. King, *Inorg. Chem.* **5**, 82 (1966).
13. C. White, S. J. Thompson, and P. M. Maitlis, *J. Chem. Soc., Dalton Trans.* 1654 (1977).

*The checkers' yield was 86–91%.

37. PHOSPHINE LIGAND DERIVATIVES OF DIAMOND-SHAPED, CYANIDE-BRIDGED IRON(II)–COPPER(I) COMPLEXES

Submitted by DONALD J. DARENSBOURG,* WAY-ZEN LEE,* and M. JASON ADAMS*
Checked by ARANZAZU MENDIA,† ELENA CERADA,† and MARIO LAGUNA†

Cyanide-containing anionic organometallic complexes have served as ligands to a variety of other metals, thereby affording one-, two-, and three-dimensional derivatives containing bridging cyanide groups.[1] These derivatives have proved very useful in studies ranging from the electronic communication between metal centers[2] to the understanding of cyanide's role as a respiratory inhibitor in heme–copper oxidases.[3] Herein, we present high-yielding syntheses of $\{CpFe(CO)(\mu\text{-}CN)_2Cu(CH_3CN)_2\}_2$, $\{CpFe(CO)(\mu\text{-}CN)_2Cu(PCy_3)\}_2$, $\{CpFe(CO)(\mu\text{-}CN)_2Cu(PCy_3)_2\}_2$, $\{CpFe(CO)(\mu\text{-}CN)_2Cu(P\text{-}(p\text{-}tolyl)_3)_2\}_2$, and $\{CpFe(CO)(\mu\text{-}CN)_2Cu(dcpp)\}_2$ [dcpp = 1,3-bis(dicyclohexylphosphino)propane] according to the reactions[4] shown in Scheme 1.

■ **Caution.** *The cyanide anion is a strong respiratory inhibitor. As such, it is a strong base, thus producing the weak, highly toxic, hydrocyanic acid on protonation. Hence, alkali metal cyanide salts must be handled with great care. Toxic CO is evolved in Section 37.A. All operations should be conducted in an efficient fume hood.*

General Comments

All manipulations are performed under argon or dry nitrogen using standard Schlenk and glovebox techniques. Phosphines (Strem), cyclopentadienylirondicarbonyl dimer, and tetrafluoroboric acid (Aldrich) are commercial samples used as received.

A. POTASSIUM CARBONYLDICYANO(CYCLOPENTADIENYL) FERRATE(II), K[CpFe(CO)(CN)₂]

$$\{CpFe(CO)_2\}_2 + Br_2 \rightarrow 2CpFe(CO)_2Br$$

$$CpFe(CO)_2Br + 2KCN \rightarrow K[CpFe(CO)(CN)_2] + KBr + CO$$

*Department of Chemistry, Texas A&M University, College Station, TX 77843.
†Departamento de Química Inorgánica, Instituto de Ciencia de Materiales de Aragón, Universidad de Zaragoza CSIC, E-50009 Zaragoza, Spain.

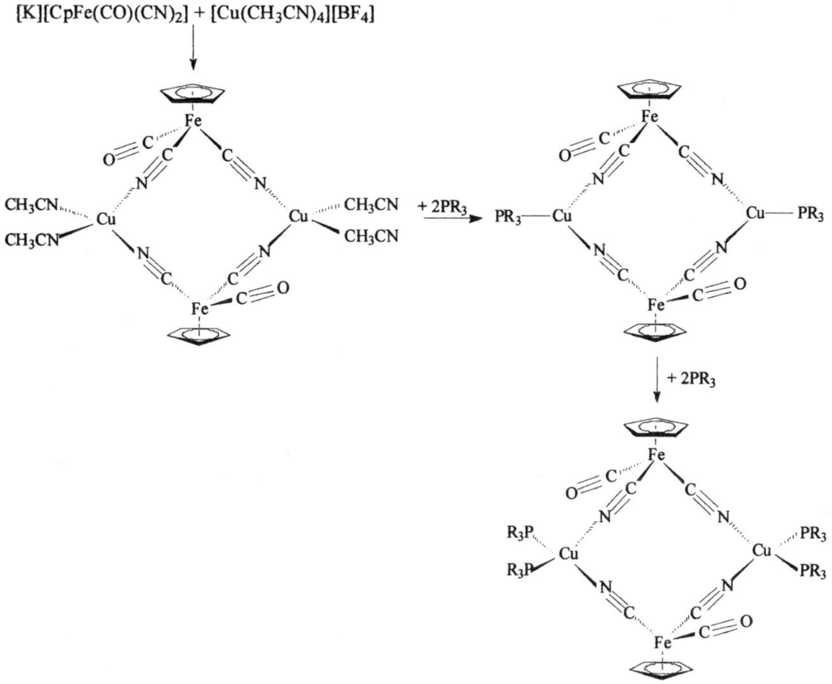

$[K][CpFe(CO)(CN)_2] + [Cu(CH_3CN)_4][BF_4]$

Scheme 1. Reactions involving cyanide-containing Fe(II)-Cu(I) complexes.

An ice-cooled mixture of 3.50 g (10 mmol) of $\{CpFe(CO)_2\}_2$ in 25 mL of methanol is treated dropwise with 1.60 g (10 mmol) of bromine. To this solution of $CpFe(CO)_2Br$ is added 5.0 g (77 mmol) of KCN in 25 mL of water, and the solution is heated under reflux for 30 min. The reaction solution is cooled to ambient temperature and then is evaporated to dryness under reduced pressure to provide a yellow-brown powder. The residue is extracted with three portions (15 mL) of hot ethanol, and the solvent is reduced by half-volume under vacuum. The solid product is isolated by filtration and dried under vacuum. Yield: 4.0g (83%).

Anal. Calcd for $C_8H_5N_2OFeK$: C, 40.0; H, 2.10; N, 11.68. Found: C, 40.3; H, 2.09; N, 11.76.

Properties

The complex $K[CpFe(CO)(CN)_2]$[5] is stable in air as a solid but decomposes in solution in the presence of oxygen. The complex is soluble in water, alcohol, or acetonitrile but insoluble in nonpolar solvents. The infrared spectrum of $K[CpFe(CO)(CN)_2]$ in acetonitrile exhibits two ν_{CN} vibrations at 2088 and 2094 cm^{-1} and a strong ν_{CO} vibration at 1949 cm^{-1}.

B. BIS{CARBONYL(CYCLOPENTADIENYL)IRONDI(μ-CYANO)-DIACETONITRILECOPPER}, $\{CpFe(CO)(\mu\text{-}CN)_2Cu(CH_3CN)_2\}_2$

$$2K[CpFe(CO)(CN)_2] + 2[Cu(CH_3CN)_4][BF_4] \rightarrow \{CpFe(CO)(\mu\text{-}CN)_2Cu(CH_3CN)_2\}_2 + 2K[BF_4]$$

Procedure

By use of a cannula or syringe, 20 mL of acetonitrile (distilled from CaH_2 and from P_2O_5 and freshly distilled from CaH_2 immediately prior to use) is added to a mixture of $K[CpFe(CO)(CN)_2]$ (0.4 mmol, 0.096 g) and $[Cu(CH_3CN)_4][BF_4]$[6] (0.4 mmol, 0.126 g) contained in a 50-mL Schlenk flask equipped with a stirring bar. The mixture is stirred for 30 min at ambient temperature, and a pale gray solid (KBF_4) precipitates from the yellow solution. The KBF_4 salt is separated by filtration, and the acetonitrile solution of $\{CpFe(CO)(\mu\text{-}CN)_2Cu(CH_3CN)_2\}_2$ is used immediately in the following syntheses.

Properties

$\{CpFe(CO)(\mu\text{-}CN)_2Cu(CH_3CN)_2\}_2$ is stable in acetonitrile solution, but gradually (after 4 h) decomposes in dichloromethane (dried and distilled over P_2O_5) to form a green solution containing Cu(II). The compound can also be obtained as a light yellow solid on removal of solvent under vacuum. IR (CH_3CN): 2125 (sh), 2118 (s) cm^{-1} (ν_{CN}); 1974 (vs) cm^{-1} (ν_{CO}). IR (CH_2Cl_2): 2137 (s), 2123(s) cm^{-1} (ν_{CN}), 1990 (vs) cm^{-1} (ν_{CO}). IR (KBr): 2131 (sh), 2118 (s) cm^{-1} (ν_{CN}); 1981 (vs) cm^{-1} (ν_{CO}).

C. BIS{CARBONYL(CYCLOPENTADIENYL)IRONDI(μ-CYANO)(TRICYCLOHEXYLPHOSPHINE)COPPER}, $\{CpFe(CO)(\mu\text{-}CN)_2Cu(PCY_3)\}_2$

$$\{CpFe(CO)(\mu\text{-}CN)_2Cu(CH_3CN)_2\}_2 + 2\ PCy_3 \rightarrow \{CpFe(CO)(\mu\text{-}CN)_2Cu(PCy_3)\}_2$$

Procedure

An acetonitrile solution of $\{CpFe(CO)(\mu\text{-}CN)_2Cu(CH_3CN)_2\}_2$, prepared as described in Section 37.B, is added to a 50-mL Schlenk flask containing 0.4 mmol of tricyclohexylphosphine (PCy_3, 0.112 g). A yellow solid precipitates from solution. On removal of the solvent by cannula filtration and drying the residue under vacuum, the crude product is isolated in ~91% yield. X-ray quality crystals are obtained on slow diffusion of acetonitrile into a dichloromethane solution of $\{CpFe(CO)(\mu\text{-}CN)_2Cu(PCy_3)\}_2$ maintained at 10°C for several days.

Anal. Calcd. for $[CpFe(CO)(CN)_2Cu(PCy_3)]_2 \cdot 1.5CH_2Cl_2$: C, 52.79; H, 6.54; N, 4.60. Found: C, 52.90; H, 7.21; N, 4.36.

Properties

$\{CpFe(CO)(\mu\text{-}CN)_2Cu(PCy_3)\}_2$ is formed as a yellow powder that is soluble in THF and dichloromethane and insoluble in acetonitrile. A dichloromethane solution of the complex is stable in an inert atmosphere for weeks. The infrared spectrum of the compound in dichloromethane shows ν_{CN} and ν_{CO} bands at 2115 (s) and 1974 (vs) cm^{-1}, respectively. The ^{31}P NMR spectrum in dichloromethane at room temperature has a broad peak at 15.16 ppm. The peak sharpens and shifts slightly upfield to 14.84 ppm as the temperature is lowered to −80°C.

D. BIS{CARBONYL(CYCLOPENTADIENYL)IRONDI (μ-CYANO)BIS(TRICYCLOHEXYLPHOSPHINE) COPPER}, $\{CpFe(CO)(\mu\text{-}CN)_2Cu(PCy_3)_2\}_2$

$$\{CpFe(CO)(\mu\text{-}CN)_2Cu(CH_3CN)_2\}_2 + 4\ PCy_3 \rightarrow \{CpFe(CO)(\mu\text{-}CN)_2Cu(PCy_3)_2\}_2$$

$\{CpFe(CO)\ \mu\text{-}CN)_2Cu(PCy_3)_2\}_2$ is synthesized directly by following the procedure described in Section 37.C and employing 0.8 mmol of PCy_3 (0.224 g). A clear yellow solution of $\{CpFe(CO)(\mu\text{-}CN)_2Cu(PCy_3)_2\}_2$ is obtained in acetonitrile. On removal of the solvent under vacuum, a crude product of $\{CpFe(CO)(\mu\text{-}CN)_2Cu(PCy_3)_2\}_2$ is obtained as a yellow solid (yield 92%). Yellow orange crystals are obtained by slow diffusion of hexane into a THF solution of $\{CpFe(CO)(\mu\text{-}CN)_2Cu(PCy_3)_2\}_2$. The compound can also be synthesized from the product obtained in Section 37.C by adding a stoichiometric amount of PCy_3 in dichloromethane.

Anal. Calcd. for $[CpFe(CO)(CN)_2Cu(PCy_3)_2]_2 \cdot 2THF$: C, 64.24; H, 8.87; N, 3.12. Found: C, 63.60; H, 10.13; N, 2.43.

Properties

$\{CpFe(CO)(\mu\text{-}CN)_2Cu(PCy_3)_2\}_2$ is a yellow solid that is soluble in THF, acetonitrile, and dichloromethane. The complex is sensitive to air and must be handled in an inert atmosphere. However, under these conditions a dichloromethane solution of $[CpFe(CO)(\mu\text{-}CN)_2Cu(PCy_3)_2]_2$ is stable for a week. IR (CH_3CN): 2105 (s), 2095 (s) cm^{-1} (ν_{CN}); 1959 (vs) cm^{-1} (ν_{CO}). IR (CH_2Cl_2): 2105 (s), 2093 (s) cm^{-1} (ν_{CN}), 1963 (vs) cm^{-1} (ν_{CO}). ^{31}P NMR (CH_2Cl_2): 11.7 (b) ppm (23°C); 10.9 (s) ppm (−80°C).

E. BIS{CARBONYL(CYCLOPENTADIENYL)IRONDI (μ-CYANO)BIS(TRI(*para*-TOLYL)PHOSPHINE)COPPER}, $\{CpFe(CO)(\mu\text{-}CN)_2Cu(p\text{-}tolyl)_3)_2\}_2$

$$\{CpFe(CO)(\mu\text{-}CN)_2Cu(CH_3CN)_2\}_2 + 4\ P(p\text{-}tolyl)_3 \rightarrow \{CpFe(CO)(\mu\text{-}CN)_2Cu(P(p\text{-}tolyl)_3)_2\}_2$$

Procedure

The acetonitrile solution of $\{CpFe(CO)(\mu\text{-}CN)_2Cu(CH_3CN)_2\}_2$ synthesized as in Section 37.B, is added to a 50-mL Schlenk flask containing 0.8 mmol of tri-*p*-tolylphosphine (0.244 g). The reaction solution is stirred for 30 min, and $\{CpFe(CO)(\mu\text{-}CN)_2Cu(P(p\text{-}tolyl)_3)_2\}_2$ is formed as a yellow precipitate. The solvent is removed by cannula filtration, and the yellow solid residue is dried under vacuum (90% yield). Yellow crystals are obtained on slow diffusion of hexane into a THF solution of $\{CpFe(CO)(\mu\text{-}CN)_2Cu(P(p\text{-}tolyl)_3)_2\}_2$ at 10°C for several days.

Anal. Calcd. for $[CpFe(CO)(CN)_2Cu(P(p\text{-}tolyl)_3)_2]_2$: C, 68.77; H, 5.42; N, 3.21. Found: C, 68.23; H, 5.50; N, 3.13.

Properties

$\{CpFe(CO)(\mu\text{-}CN)_2Cu(P(p\text{-}tolyl)_3)_2\}_2$ is a yellow solid that is soluble in THF and dichloromethane. A dichloromethane solution of the complex under an inert atmosphere is less stable than that of $\{CpFe(CO)(\mu\text{-}CN)_2Cu(PCy_3)_2\}_2$. IR (THF): 2120 (s) cm^{-1} (ν_{CN}); 1968 (vs) cm^{-1} (ν_{CO}). IR (CH_2Cl_2): 2120 (sh), 2112 (s) cm^{-1} (ν_{CN}), 1971 (vs) cm^{-1} (ν_{CO}).

F. BIS{CARBONYL(CYCLOPENTADIENYL)IRONDI(μ-CYANO)-1,3-BIS(DICYCLOHEXYLPHOSPHINO) PROPANECOPPER}, {CpFe(CO)(μ-CN)$_2$Cu(dcpp)}$_2$

$$\{CpFe(CO)(\mu\text{-}CN)_2Cu(CH_3CN)_2\} + 2\ dcpp \rightarrow \{CpFe(CO)(\mu\text{-}CN)_2Cu(dcpp)\}_2$$

Procedure

The reaction between K[CpFe(CO)(CN)$_2$] (0.4 mmol, 0.096 g) and [Cu(CH$_3$CN)$_3$][BF$_4$] (0.4 mmol, 0.126 g) is carried out directly in dichloromethane for 30 min, then the solution is filtered to remove the KBF$_4$ precipitate. A solution of dcpp (0.4 mmol, 0.174 g) in 10 mL of dichloromethane is added dropwise to the yellow filtrate, and the resulting yellow solution is stirred at room temperature for 30 min. Evaporation of the solvent under vacuum provides the product as a yellow powder in 91% yield. Yellow crystals are obtained on slow diffusion of hexanes into a THF solution of {CpFe(CO)(μ-CN)$_2$Cu(dcpp)}$_2$ maintained at 10°C for several days.

Anal. Calcd. for [CpFe(CO)(CN)$_2$Cu(dcpp)]$_2$: C, 59.43; H, 7.77; N, 4.08. Found: C, 59.10; H, 8.00; N, 3.98.

Properties

{CpFe(CO)(μ-CN)$_2$Cu(dcpp)}$_2$ is a yellow solid that is soluble in THF, acetonitrile, and dichloromethane. The complex is sensitive to air but stable in an inert atmosphere for weeks. IR (CH$_2$Cl$_2$): 2111 (s), 2093 (s) cm^{-1} (ν_{CN}), 1964 (vs) cm^{-1} (ν_{CO}). The phosphorus NMR resonance of the complex (−6.71 ppm) at ambient temperature is shifted slightly upfield from the position of the free phosphine (−5.55 ppm).

References

1. W. P. Fehlhammer and M. Fritz, *Chem. Rev.* **93**, 1243 (1993); S. M. Contakes, K. K. Klausmeyer, and T. B. Rauchfuss, *Inorg. Chem.* **39**, 2069 (2000); K. K. Klausmeyer, T. B. Rauchfuss, and S. R. Wilson, *Angew, Chem. Int. Ed.* **37**, 1694 (1998).
2. H. Vahrenkamp, A. Geiss, and G. N. Richardson, *J. Chem. Soc., Dalton Trans.* 3643 (1997).
3. B. S. Lim and R. H. Holm, *Inorg. Chem.* **37**, 4898 (1998).
4. D. J. Darensbourg, W.-Z. Lee, M. J. Adams, D. L. Larkins, and J. H. Reibenspies, *Inorg. Chem.* **38**, 1378 (1999); D. J. Darensbourg, W.-Z. Lee, M. J. Adams, and J. C. Yarbrough, *Eur. J. Inorg. Chem.* 2811 (2001).
5. C. E. Coffey, *Inorg. Nucl. Chem.* **25**, 179 (1965); C.-H. Lai, W.-Z. Lee, M. L. Miller, J. H. Reibenspies, D. J. Darensbourg, and M. Y. Darensbourg, *J. Am. Chem. Soc.* **120**, 10103 (1998).
6. G. J. Kubas, *Inorg. Synth.* **19**, 90 (1979).

38. CYANIDE-BRIDGED IRON(II)–M(II) MOLECULAR SQUARES, M(II) = Fe(II),Co(II),Cu(II)

Submitted by HIROKI OSHIO[*] and TASUKU ITO[†]
Checked by CAMERON SPAHN[‡] and THOMAS B. RAUCHFUSS[‡]

Cyanide ions bridge a variety of metal ions to form assembled systems that have been the focus of intense research interest for the purpose of developing new magnetic and magnetooptical materials. Prussian blue ($Fe^{III}_4[Fe^{II}(CN)_6]_3 \cdot 15H_2O$), a cyanide-bridged three-dimensional system, shows ferromagnetic ordering at 5.6 K.[1] A variety of other systems formulated as $M_x[M(CN)_{6 \text{ or } 7}]$ (M and M′ = transition metal ions) have been shown to have bulk magnetism with Curie temperatures ranging from 5 K to room temperature.[2] However, these three-dimensional systems are amorphous, and the lack of the structural information prevents further understanding of the magnetic and electronic interactions between metal centers. Therefore, the development of versatile synthetic routes to prepare cyanide-bridged mixed-metal clusters provides an opportunity for improved control of the interactions between metal ions. We present here the syntheses of a series of cyanide-bridged cluster molecules, $[Fe_2M_2(\mu\text{-CN})_4(bpy)_{6 \text{ or } 8}]^{4+}$ [M = Fe(II), Co(II), or Cu(II) ions and bpy = 2,2′-bipyridine], in which cyanide ions bridge four metal ions to form a cyclic tetranuclear core having a nearly square geometry. In the square, the initial two iron(II) ions coordinated by cyanide carbon atoms are in a low-spin state and the other transition metal ions introduced are coordinated by cyanide nitrogen atoms (see Fig. 1).

A. CYCLO{OCTAKIS(2,2′-BIPYRIDINE)TETRA(μ-CYANO) TETRAIRON(II)} TETRAKIS(HEXAFLUOROPHOSPHATE), $[Fe_4(bpy)_8(\mu\text{-CN})_4][PF_6]_4$

$$2Fe(bpy)_2(CN)_2 + 2FeCl_2 \cdot 4H_2O + 4bpy + 4[NH_4][PF_6] \rightarrow$$
$$[Fe_4(bpy)_8Cu\text{-}CN)_4][PF_6]_4 + 4NH_4Cl + 8H_2O$$

Procedure

All operations are carried out under a nitrogen atmosphere. A solution of $FeCl_2 \cdot 4H_2O$ (79 mg, 0.4 mmol) in methanol (10 mL) is added to a solution of

[*]Department of Chemistry, University of Tsukuba, Tsukuba 305-8571, Japan.
[†]Department of Chemistry, Graduate School of Science, Tohoku University, Sendai 980-8578, Japan.
[‡]School of Chemical Sciences, University of Illinois at Urbana—Champaign, Urbana, IL 61801.

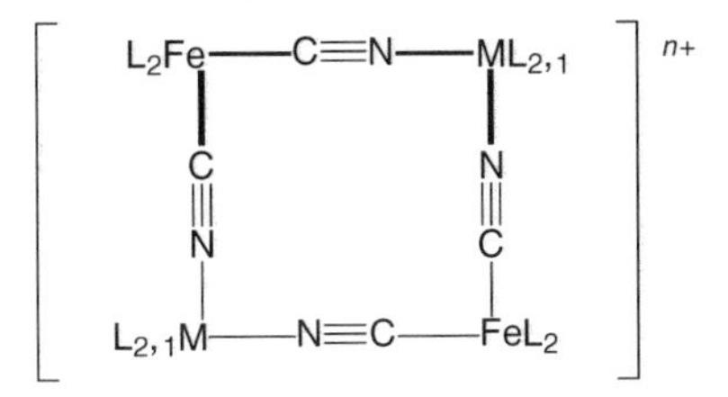

Figure 1. Structure of a cyanide-bridged Fe(II)–M(II) molecular square (L = bpy; M = Fe,Co,Cu).

Fe $(bpy)_2(CN)_2$[3] (170 mg, 0.4 mmol) in methanol (10 mL). To the resulting dark red solution, 2,2′-bipyridine (124 mg, 0.8 mmol) and $[NH_4][PF_6]$ (260 mg, 1.6 mmol) are added. The dark violet precipitate is isolated by suction filtration. Recrystallization by slow diffusion of chloroform into an acetonitrile solution gives dark violet microcrystals. Yield: 327 mg (0.15 mmol, 76%).

Anal. Calcd. for $C_{84}H_{64}F_{24}Fe_4N_{20}P_4$ C, 46.78; H, 2.99; N, 12.99. Found: C, 46.78; H, 3.20; N, 12.89.

Properties

The solid complex is diamagnetic and not air-sensitive. The cyclic voltammogram in acetonitrile shows two quasireversible waves at 0.67 V ($\Delta E_p = 70$ mV) and 0.86 V ($\Delta E_p = 80$ mV) versus SSCE, and controlled potential coulometry at 1.0 V shows that each quasireversible wave corresponds to an overall two-electron process for the Fe^{II}/Fe^{III} process.[4] The IR spectrum (KBr disk) has three ν_{CN} bands at 2083, 2112, and 2129 cm^{-1}.

B. CYCLO{OCTAKIS(2,2′-BIPYRIDINE)TETRA(μ-CYANO)-2,4-DICOBALT(II)-1,3-DIRON(II)} TETRAKIS (HEXAFLUOROPHOSPHATE), $[Fe_2Co_2(bpy)_8(\mu\text{-}CN)_4][PF_6]_4$

$$2Fe(bpy)_2(CN)_2 + 2CoCl_2 \cdot 6H_2O + 4bpy + 4[NH]_4[PF]_6 \rightarrow [Fe_2Co_2(bpy)_8(\mu\text{-}CN)_4][PF_6]_4 + 4NH_4Cl + 12H_2O$$

Procedure

$Fe^{II}(bpy)_2(CN)_2$ (55 mg, 0.13 mmol) in MeOH (25 mL) is added to a solution of $CoCl_2 \cdot 6H_2O$ (31 mg, 0.13 mmol) and bpy (41 mg, 0.26 mmol) in methanol

(25 mL). After stirring for a day, addition of $[NH_4][PF_6]$ (85 mg, 0.52 mmol) in methanol (25 mL) produces a red powder precipitate, which is isolated by suction filtration. Recrystallization by slow diffusion of chloroform into an acetonitrile solution yields red microcrystals. Yield: 90 mg (0.04 mmol, 64%).

Anal. Calcd. for $C_{84}H_{64}Co_2F_{24}Fe_2N_{20}P_4$: C, 46.65; H, 2.98; N, 12.95. Found: C, 46.90; H, 2.83; N, 12.71.

Properties

The solid complex is paramagnetic and not air-sensitive. The cyclic voltammogram in acetonitrile shows three oxidation peaks at 0.9, 1.1, and 1.51 V (vs. SCE) on scanning to positive potentials, whereas two reduction peaks at 1.39 and 0.23 V are observed in the subsequent negative scan. The waves at 0.9 and 1.1 V are assigned to Co^{II}/Co^{III}, and the quasireversible wave at 1.45 V [=(1.51 + 1.39)/2] is assigned to two single–electron processes for the two Fe ions (Fe^{II}/Fe^{III}). The IR spectrum has a strong ν_{CN} band at 2095 cm^{-1}.

C. CYCLO{HEXAKIS(2,2′-BIPYRIDINE)TETRA(μ-CYANO)-2,4-DICOPPER(II)-1,3-DIRON(II)} TETRAKIS (HEXAFLUOROPHOSPHATE), $[Fe_2Cu_2(bpy)_6(\mu\text{-}CN)_4][PF_6]_4$

$$2Fe(bpy)_2(CN)_2 + 2[Cu^{II}(bpy)(CH_3OH)_2](NO_3)_2 \cdot 3H_2O + 4[NH_4][PF_6] \rightarrow$$
$$[Fe_2Cu_2(bpy)_6(\mu\text{-}CN)_4][PF_6]_4 + 4[NH_4]NO_3 + 6H_2O + 4CH_3OH$$

Procedure

$Fe^{II}(bpy)_2(CN)_2$ (55 mg, 0.13 mmol) in MeOH (25 mL) is added to a solution of $[Cu^{II}(bpy)(CH_3OH)_2](NO_3)_2 \cdot 3H_2O$[5] (50 mg, 0.13 mmol) in methanol (25 mL). After stirring for 3 days, addition of $[NH_4][PF]_6$ (62 mg, 0.38 mmol) in methanol (25 mL) precipitates a deep red powder. Recrystallization by slow diffusion of chloroform into an acetonitrile solution yields dark red microcrystals. Yield: 110 mg (0.05 mmol, 78%).

Anal. Calcd. for $C_{64}H_{48}Cu_2F_{24}Fe_2N_{16}P_4$: C, 41.33; H, 2.60; N, 12.05%. Found: C, 41.20; H, 2.40; N, 11.86%.

Properties

The solid complex is paramagnetic and not air-sensitive. The iron(II) and copper(II) ions are alternately bridged by four cyanide ions, forming a cyclic structure. In the square, the paramagnetic copper(II) ions are magnetically isolated by the low-spin iron(II) ions. In the cycic voltammogram in acetonitrile, two quasireversible two-electron transfer processes (0.01 and 0.96 V vs. SSCE) are clearly observed with peak separations of 120 and 150 mV, respectively. The redox waves at 0.01 and 0.96 V are attributable to the Cu^{I}/Cu^{II} and Fe^{II}/Fe^{III} oxidations, respectively. The IR spectrum has a strong ν_{CN} band at 2113 cm^{-1}.

References

1. A. N. Holden, B. T. Matthias, P. W. Anderson, and H. W. Lewis, *Phys. Rev.* **102**, 1463 (1956).
2. A. Ito, M. Suenaga, and K. Ono, *J. Chem. Phys.* **48**, 3597 (1968). T. Mallah,.S. Thiebaut, M. Verdaguer, and P. Veillet, *Science* **262**, 1554 (1993); W, R. Entley and G. S. Girolami, *Science* **268**, 397 (1995); V. Gadet, T. Mallah, I. Castro, P. Veillet, and M. Verdaguer, *J. Am. Chem. Soc.* **114**, 9213 (1992); S. Ferlay, T. Mallah, R. Quaès, P. Veillet, and M. Verdaguer, *Inorg. Chem.* **38**, 229 (1999); S. Ferlay, T. Mallah, R. Ouahès, P. Veillet, and M. Verdaguer, *Nature* **378**, 701 (1995); O. Sato, T. Iyoda, A. Fujishima, and K. Hashimoto, *Nature* **271**, 49 (1996); W. R. Entley and G. S. Girolami, *Inorg. Chem.* **33**, 5165 (1994); J. Larionova, R. Clèrac, J. Sanchiz, O. Kahn, S. Golhen, and L. Ouahab, *J. Am. Chem. Soc.* **120**, 13088 (1998).
3. A. A. Shilt, *Inorg. Synth.* **12**, 249 (1970).
4. H. Oshio, H. Onodera, O. Tamada, H. Mizutani, T. Hikichi, and T. Ito, *Chem. Eur. J.* **6**, 2523 (2000).
5. F. M. Jaeger and J. A. Dijk, *Z. Anorg. Chem.* **227**, 273 (1936).

39. POLY{(2,2′-(BIPYRIDINE)(TRIPHENYLPHOSPHINE) BIS(DICYANOCOPPER(I)}, $\{(CuCN)_2(bpy)(PPh_3)\}_n$

Submitted by MAOCHUN HONG*
Checked by DOUGLAS J. CHESNUT† and JON ZUBIETA†

Copper(I) cyanide has shown promise in the construction of self-assembled zeolitic frameworks[1] and as a precursor in the synthesis of $YBa_2Cu_3O_{7-x}$ superconductors.[2] A number of complexes of copper(I) cyanide have been reported over the years,[3] but studies since the late 1990s have shown that there is still much to be learned about such systems.[4–6] Laser ablation of CuCN yields positive and negative ions of composition $[Cu_n(CN)_{n+1}]^-$ ($n = 1$–5) and $\{Cu_n(CN)_{n-1}]^+$ ($n = 1$–6).[7] Calculations showed that the possible structural identities of these species are chain, cyclic, or metal-bridged arrangements. Thus, an interesting challenge is generation, stabilization, isolation, and structural characterization of such structures from solution.[8] Herein, we report the synthesis of the polymeric complex $\{(CuCN)_2(2,2\text{-bipyridine})(PPh_3)\}_n$.[9]

*Fujian Institute of Research on the Structure of Matter, Chinese Academy of Sciences, Fuzhou, Fujian, 350002 China.

†Department of Chemistry, Syracuse University, Syracuse, NY 13244.

At room temperature CuCN is not soluble in most solvents. Therefore, the direct reaction of CuCN with PPh_3 in DMF results in the formation of an amorphous precipitate that is not soluble in any solvent. But when 2,2′-bipyridine is introduced into the reaction system of CuCN and PPh_3 in DMSO, a red solution is obtained. By slowly diffusing diethyl ether into this solution, red block crystals of the complex $\{(CuCN)_2(2,2\text{-bipyridine})(PPh_3)\}_n$ are obtained. In contrast, when pyridine-2-thiolate replaces 2,2′-bipyridine, the complex $\{CuCN(PPh_3)_2\}_6$ is formed instead.[9]

■ **Caution.** *CuCN is corrosive and toxic, so protective clothing should be worn when handling it. The procedure should be performed using standard inert-atmosphere techniques and degassed solvents.*

Procedure

A mixture of copper(I) cyanide (0.45 g, 5 mmol) and 2,2′-bipyridine (0.80 g, 5 mmol) is added under a nitrogen atmosphere into a 100-mL Schlenk flask containing 40 mL deoxygenated DMSO. Stirring the mixture at room temperature for 2 h generates a yellow solution with some undissolved solid. Then PPh_3 (2.62 g, 10 mmol) is added under nitrogen to the solution with vigorous stirring. After 10 min, all yellow precipitate is dissolved and a clear yellow solution is formed. Stirring is continued for 2 h, then the solution is filtered, and the filtrate is transferred to one side of a U-shaped tube. The other side, separated by a glass frit at the bottom of the U-tube, is filled with diethyl ether. Slow diffusion of diethyl enter into the DMSO solution for 10 days provides red block crystals of the product. The overall yield is 1.17 g (78%).

Anal. Calcd. for $(CuCN)_2(2,2'\text{-bipyridine})(PPh_3)$: C, 60.20; H, 3.85; N, 9.36. Found: C, 60.92; H, 3.27; N, 9.13.

Properties

The $[(CuCN)_2(2,2'\text{-bipyridine}(PPh_3)]_n$ crystals are red blocks. The space group is $P2(1)/c$, where $a = 15.983$ (2), $b = 21.968$ (3), $c = 8.8715$ (11) Å, $\beta = 94.066$ (3)°. The calculated density is 1.44 g/cm. The compound is not soluble in CH_2Cl_2, THF, or DMF.

References

1. B. F. Hoskins and R. Robson, *J. Am. Chem Soc.* **112**, 1546 (1990); A. K. Brimah, E. Siebel, R. D. Fischer, N. A. Davis, D. C. Apperley, and R. K. Harris, *J. Organometal Chem.* **475**, 85 (1994); H. Yuge and T. Iwamoto, *J. Inclusion Phenom.* **26**, 119 (1996); L. C. Brousseau, D. Williams, J. Kouvetakis, and M. O'Keefe, *J. Am. Chem. Soc.* **119**, 6292 (1997).

2. N. A. Khan, N. Baber, M. Z. Iqbal, and M. Mazhar, *Chem. Mater.* **5**, 1283 (1993).
3. B. J. Hathaway, in *Comprehensive Coordination Chemistry*, G. Wilkinson, ed., Pergamon, Oxford, (1987), Vol. 5, p. 533; G. A. Bowmaker, H. Hartl, and V. Urban, *Inorg. Chem.* **39**, 4548 (2000).
4. F. B. Stocker, M. A. Troester, and D. Britton, *Inorg. Chem.* **35**, 3145 (1966); F. B. Stocker, T. P. Staeva, C. M. Rienstra, and D. Britton, *Inorg. Chem.* **38**, 984 (1999).
5. A. J. Blake, J. P. Danks, V. Lippolis, S. Parsons, and M. Schröder, *New J. Chem.* **22**, 1301 (1998).
6. D. J. Chesnut, A. Kusnetzow, R. Birge, and J. Zubieta, *Inorg. Chem.* **38**, 5484 (1999).
7. I. G. Dance, P. A. W. Dean, and K. J. Fisher, *Inorg. Chem.* **33**, 6261 (1994).
8. S. Kroeker, R. E. Wasylishen, and J. V. Hanna, *J. Am. Chem. Soc.* **121**, 1582 (1998).
9. Y. J. Zhao, M. C. Hong, W. P. Su, R. Cao, Z. Y. Zhou, and A. S. C. Chan, *J. Chem. Soc., Dalton Trans.* 1685 (2000).

Chapter Five

CLUSTER AND POLYNUCLEAR COMPOUNDS

40. A TETRANUCLEAR LANTHANIDE–HYDROXO COMPLEX FEATURING THE CUBANE-LIKE $[Ln_4(\mu_3\text{-}OH)_4]^{8+}$ CLUSTER CORE, $[Nd_4(\mu_3\text{-}OH)_4(H_2O)_{10}(alanine)_6][ClO_4]_8$

Submitted by RUIYAO WANG* and ZHIPING ZHENG*
Checked by QIN-DE LIU† and SUNING WANG†

$$4\,Nd(ClO_4)_3 + 6\text{ alanine} + 4\,NaOH + 10\,H_2O \longrightarrow [Nd_4(\mu_3\text{-}OH)_4(H_2O)_{10}(alanine)_6](ClO_4)_8 + 4\,NaClO_4$$

Polynuclear lanthanide–hydroxo complexes are of interest as chemically modified and controllable precursors to sol-gel materials,[1] synthetic hydrolases capable of catalyzing the hydrolytic cleavage of DNA and RNA,[2] and new paradigms for radiographic contrast agents.[3] However, primarily because of the lack of a general synthetic methodology to such materials, systematic exploration of their chemistry has been hampered, as has their implementation in practical roles. In fact, the formation of such substances is frequently characterized by random organization,[4] and many literature examples have been the result of serendipitous discovery.[5–8]

Adventitious hydrolysis is believed to be responsible for the formation of these high-nuclearity hydroxo species.[5,6] The hydrolytic process is presumably

*Department of Chemistry, University of Arizona, Tucson, AZ 85721.
†Department of Chemistry, Queens University, Kingston, Ontario, Canada K7L 2N6.

Inorganic Syntheses, Volume 34, edited by John R. Shapley
ISBN 0-471-64750-0

limited by various multidentate ligands present in these complexes so that discrete cluster-type polynuclear species, instead of intractable polymeric lanthanide hydroxides and/or oxohydroxides, can be isolated and structurally characterized.[9] Stimulated by such observations and with the hope of developing a general approach to these useful precursors, we have systematically studied the hydrolysis of the lanthanide ions utilizing α-amino acids as auxiliary ligands. This methodology is generally applicable to all the lanthanide ions, but the nature of the final product is influenced significantly by the supporting amino acids. Specifically, tetranuclear hydroxo complexes of the general formula $[Ln_4(\mu_3\text{-}OH)_4(AA)_6]^{8+}$ (Ln = trivalent lanthanide ion; AA = α-amino acid) are obtained when glycine, phenylalanine, proline, serine, and valine are utilized.[12–14] The core component of these complexes features a common cluster motif whereby the constituent lanthanide ions and triply bridging hydroxo groups occupy the alternate vertices of a distorted cube. When an α-amino acid bearing a carboxylate group in the sidechain (e.g., glutamic or aspartic acid) is utilized, an extended network structure supported by the $[Ln_4(\mu_3\text{-}OH)_4]^{8+}$ clusters and cluster-linking carboxylate groups is obtained.[12,13] Interestingly, when tyrosine is employed, a drastically different type of clusters composed of five vertex-sharing $[Ln_4(\mu_3\text{-}OH)_4]^{8+}$ units centered on a templating Cl^- are isolated.[15,16] Detailed structural descriptions of both single- and multicubane complexes can be found in the literature.[12,15]

We report below the straightforward synthesis of $[Nd_4(\mu_3\text{-}OH)_4(\text{alanine})_6][ClO_4]_8$, representative of all complexes featuring a single central cubane-like cluster core.[12] Analogous complexes with other lanthanide ions and/or α-amino acids can be prepared in exactly the same fashion, with the use of appropriate starting lanthanide materials and supporting ligands.

■ **Caution.** *Metal perchlorates are potentially explosive! Only a small amount of materials should be prepared and handled with great care. The digestion of Nd_2O_3* should be carried out in a well-ventilated fume hood with drop-wise addition of concentrated perchloric acid to a slurry of the oxide in deionized water.*

Procedure

An aqueous solution of $Nd(ClO_4)_3$ is prepared by digesting Nd_2O_3 in concentrated perchloric acid. A suitable concentration is achieved by dilution with deionized water of the concentrated solution, and the exact concentration is determined by complexometric titration with EDTA, using xylenol orange as

*Available from Strem Chemicals, Inc., 7 Mulliken Way, Newburyport, MA 01950. All other chemicals including solvents can be purchased from Aldrich.

the indicator. A 10-mL disposable vial, equipped with a magnetic stirring bar, is charged with 4.0 mL of aqueous $Nd[ClO_4]_3$ ($\sim$1.0 M). D,L-Alanine (89 mg, 1.0 mmol) is added as a solid to this solution, and the mixture is stirred at about 80°C on a heating plate. Aqueous NaOH ($\sim$0.5 M) is added dropwise to the abovementioned mixture until an incipient but permanent precipitate, presumably of lanthanide hydroxides and/or oxohydroxides, is observed. The amount of NaOH necessary to reach this endpoint (pH$\sim$6) depends on the pH of the solution mixture prior to the addition of base, but does not affect the reaction outcome. While hot, the mixture is filtered using a small funnel equipped with a piece of fine-sized filter paper. The filtrate is collected in a second 10-mL disposable vial, and its volume is reduced to approximately 4 mL by slow evaporation (with stirring) on the heating plate. The stirring bar is then removed, and the vial is tightly covered with Parafilm and allowed to stand on the gradually cooling heating plate. Pink-colored parallelepipeds appear in about one week. The solid is collected by filtration, washed with an ice-cold mixture of diethyl ether and tetrahydrofuran (1 : 1, v/v), and dried under dynamic vacuum in a desiccator charged with Silica gel until constant weight. More product can be obtained by concentrating the filtrate. A typical yield of 50–60% (based on alanine) is obtained. [*Note*: The actual lanthanide/alanine ratio (4 : 1) used is empirically chosen on the basis of solubility consideration instead of reaction stoichiometry. When a stoichiometric amount of alanine (Nd : alanine = 2 : 3) is used, precipitation of the free ligand results; the putative cluster complex remains in solution and resists crystallization.]

Anal. Calcd. For $[Nd_4(OH)_4(H_2O)_{10}(Ala)_6][(Cl)_4]4NaCl_4 \cdot 23H_2O$: C, 7.06; H, 3.70; N, 2.75; Nd, 18.89. Found: C, 7.04; H, 3.22; N, 2.69; Nd, 18.94.

Properties

The compound is very soluble in water, and an aqueous solution of the complex turns turbid after stirring overnight at room temperature, presumably because of further hydrolysis of the complex. The compound is also readily soluble in common organic solvents, such as tetrahydrofuran, dichloromethane, acetone, and acetonitrile. Its infrared spectrum (in a Nujol mull) shows absorptions at 3419 (s), 1620 (s), 1480 (m), 1428 (m), 1380 (w), 1347 (w), 1305 (w), 1142 (s), 1108 (s), 636 (s), 627 (s), 548 (m) cm^{-1}. Its MALDI-TOF (matrix-assisted laser desorption/ionization–time of flight) mass spectrum shows a peak envelope centered at 2020 amu, corresponding to the molecular ion minus one ClO_4^- and two aqua ligands. Its magnetic moment at room temperature, determined by Evans' method,[17] is 6.78 μ_B, which is in good agreement with the value (7.36 μ_B) calculated for four magnetically noninteractive Nd(III) ions using the Van Vleck equation.

References

1. L. G. Hubert-Pfalzgraf, *New J. Chem.* **19**, 727 (1995).
2. S. J. Franklin, *Curr. Opin. Chem. Biol.* **5**, 201 (2001).
3. S. Yu and A. Watson, *Chem. Rev.* **99**, 2353 (1999).
4. R. Anwander, *Angew. Chem., Int. Ed.* **37**, 599 (1998).
5. M. R. Bürgstein and P. W. Roesky, *Angew. Chem., Int. Ed.* **39**, 549 (2000).
6. T. Dubé, S. Gambarotta, and G. Yap, *Organometallics* **17**, 3967 (1998).
7. X. M. Chen, Y. L. Wu, Y. X. Tong, Z. Sun, and D. N. Hendrickson, *Polyhedron* **16**, 4265 (1997).
8. J. C. Plakatouras, I. Baxter, M. B. Hursthouse, K. M. Abdul Malik, J. McAleese, and S. R. Drake, *J. Chem. Soc., Chem. Commun.* 2455 (1994).
9. G. L. Abbati, A. Cornia, A. C. Fabretti, A. Caneschi, and D. Gatteschi, *Inorg. Chem.* **37**, 3759 (1998).
10. Z. Zheng, *Chem. Commun.* 2521 (2001).
11. R. Wang and Z. Zheng, *Comments Inorg. Chem.* **22**, 1 (2000).
12. R. Wang, H. Liu, M. D. Carducci, T. Jin, C. Zheng, and Z. Zheng, *Inorg. Chem.* **40**, 2743 (2001).
13. B. Ma, D. Zhang, S. Gao, T. Jin, C. Yan, and G. X. Xu, *Angew. Chem., Int. Ed.* **39**, 2743 (2001).
14. B. Ma, D. Zhang, S. Gao, T. Jin, and C. Yan, *New J. Chem.* **24**, 251 (2000).
15. R. Wang, T. Jin, Z. Zheng, and R. J. Staples, *Angew. Chem., Int. Ed.* **38**, 1813 (1999).
16. R. Wang, H. D. Selby, H. Liu, Z. Zheng, M. D. Carducci, T. Jin, J. W. Anthis, and R. J. Staples, *Inorg. Chem.* **41**, 278 (2002).
17. D. F. Evans, *J. Chem. Soc.* 2003 (1959).
18. J. H. Van Vleck, *Prog. Sci. Technol. Rare Earths* **2**, 1 (1966).

41. TETRADECACHLOROHEXATANTALUM OCTAHYDRATE, $Ta_6Cl_{14} \cdot 8H_2O$

Submitted by FRIEDRICH W. KOKNAT[*] and DAVID J. MARKO[*]
Checked by DANIEL N. T. HAY[†] and LOUIS MESSERLE[†]

$$20\,NaCl + 14\,TaCl_5 + 16\,Ta \longrightarrow 5\,Na_4Ta_6Cl_{18}$$

$$Na_4Ta_6Cl_{18} + 8\,H_2O \longrightarrow Ta_6Cl_{14} \cdot 8\,H_2O + 4\,NaCl$$

Coordination complexes that have metal halide clusters as their central cations represent a unique branch of coordination chemistry. The $[M_6X_{12}]^{n+}$ cluster

[*]Department of Chemistry, Youngstown State University, Youngstown, OH 44555.
[†]Department of Chemistry, University of Iowa, Iowa City, IA 52242.

cations of niobium and tantalum consist of metal–metal-bonded M_6 octahedra with 12 halogens bridging over the octahedral edges.[1] Each metal atom has one site open for attachment of a terminal ligand, and negative,[2,3] neutral,[4] and positive[5,6] complexes have been prepared with the $[M_6X_{12}]^{n+}$ central cation in various oxidation states. Syntheses of complexes containing $[M_6X_{12}]^{n+}$ central cations require as starting materials hydrates of the formula $M_6X_{14} \cdot 8H_2O$. These hydrates represent neutral complexes $[(M_6X_{12})X_2(H_2O)_4] \cdot 4H_2O$ with two halide ions and four water molecules ligated to the $[M_6X_{12}]^{2+}$ core.[7]

Preparations of niobium and tantalum cluster halides $M_6X_{14} \cdot 8H_2O$ employ a high temperature reaction that supplies a product containing $[M_6X_{12}]^{2+}$ cations. This material is extracted, and from the filtered aqueous solution, the hydrate is precipitated with concentrated aqueous HX. Early preparations[8–10] used sodium amalgam to reduce the metal pentahalide. Harned's method,[11] which has been widely used, involves the reduction of the pentahalide with cadmium metal, possibly with the formation of $Cd_2[(M_6X_{12})X_6]$. A variation of Harned's method, developed in McCarley's group,[12] uses a lithium halide/potassium halide melt as a solvent for the reduction of the pentahalides with cadmium or aluminum metal. Another method, also due to McCarley and co-workers,[13] involves the high-temperature disproportionation of Nb_3X_8 in the presence of an alkali halide AX into A_2NbX_6 and $A_4Nb_6X_{18}$. When performed in niobium tubes or in the presence of niobium metal, the disproportionation goes over into an outright reduction of Nb_3X_8 to $A_4Nb_6X_{18}$. Broll et al.[14] extended this procedure to the preparation of $[Ta_6X_{12}]^{2+}$-containing material.

The synthesis described here employs a high-temperature conproportionation in which tantalum metal and its pentachloride in the presence of stoichiometric amounts of sodium chloride react to form $Na_4[(Ta_6Cl_{12})Cl_6]$. This complex is readily hydrolyzed in water, and addition of hydrochloric acid yields the hydrate $Ta_6Cl_{14} \cdot 8H_2O$. The total procedure takes 2–3 days. This method supplies the hydrated cluster chloride in high yield from commercial available starting materials. Analogous procedures to prepare the tantalum bromide cluster as well as both niobium chloride and bromide clusters have been described.[15] Important for the success of the high-temperature syntheses are the choice of the alkali halide and the use of a stoichiometric amount rather than an excess of the alkali halide. Excessively high reaction temperatures and prolonged reaction times lower the yield substantially. These aspects have been discussed in detail in the original paper.[15]

General Comments

Commercially available starting materials are employed for the preparations. Tantalum metal may be used in granular form (with 1 mm diameter) or as a powder. Tantalum pentachloride is extremely moisture-sensitive. It should be handled under strictly anhydrous conditions in a drybox or drybag.

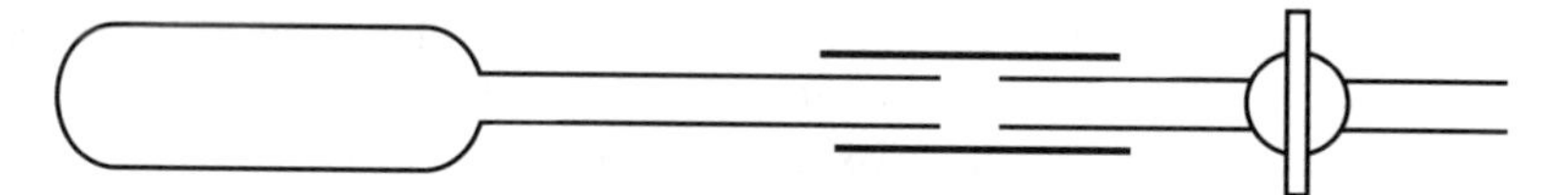

Figure 1. Reaction tube with vacuum hose and stopcock.

The simple Vycor reaction tube used for the high-temperature reaction is shown in Fig. 1. It consists of a wide body of 30 mm outside diameter and ~100 mm length and of a narrow neck of 9 mm outer diameter and 180 mm length. For handling outside of the drybox, it is equipped with a vacuum hose and a stopcock. Prior to use, it should be flamed out under a dynamic vacuum. To keep the neck of the reaction tube clean, the tube should be filled using a funnel with a narrow neck of 180 mm length.

Procedure

In a drybox or drybag, stoichiometric amounts of tantalum pentachloride (25.1 g, 70 mmol) and sodium chloride (5.84 g, 100 mmol) are intimately mixed by grinding in a mortar. The mixture is placed in the Vycor reaction tube (see Fig. 1), and excess tantalum powder (58 g, 320 mmol) is added. Using a gas/oxygen torch, the reaction vessel is sealed under a dynamic vacuum near the midpoint of its neck.

■ **Caution.** *Welder's glasses should be worn for eye protection.*

After the tube has cooled, it is shaken for several minutes to mix the contents thoroughly and then placed in a tube furnace. Over a 10-h period it is gradually heated to 680–700°C and is kept at this temperature overnight.

■ **Caution.** *During the heatup period the reaction tube may explode in the furnace if the temperature is raised too fast, if the tube has not been filled under strictly anhydrous conditions, or if the metal pentahalide is partially hydrolyzed.*

Then the cooled tube is wrapped in aluminum foil except for an area near the bottom of its neck, which is scratched with a glass cutter. The neck is snapped to open the tube, and the solid reaction product is removed and ground in a mortar. The powder is transferred to a 3-L beaker or Erlenmeyer flask and is stirred for 5 h with 1500 mL of deionized water. The mixture is allowed to settle for up to 3 h, then it is filtered through fluted filter paper.* The solid residue

*The checkers found this step to be very slow, possibly due to excess tantalum metal powder.

including the filter paper is extracted 3 more times in the same way. A small amount (up to 1.2 g per extract) of $SnCl_2 \cdot 2H_2O$ may be added periodically during this operation to suppress air oxidation to brown $[Ta_6X_{12}]^{3+}$. The emerald green filtrate is treated with an equal volume of concentrated hydrochloric acid, and the solution is heated (not to boiling) with stirring until precipitation of $Ta_6Cl_{14} \cdot 8H_2O$ is complete (1–2 h). The supernatant solution should be nearly colorless. The dark green precipitate is collected on a fritted-glass filter, washed twice with 150-mL portions of concentrated hydrochloric acid and twice with 150-mL portions of ether, and dried in vacuo over phosphorus pentoxide. The yield ranges from 35 to 39 g (81–90%).

Anal. Calcd. for $Ta_6Cl_{14} \cdot 8H_2O$: Cl, 28.75; Ta, 62.90. Found: Cl, 28.94; Ta, 63.06; Cl : Ta, 14.05 : 6.00.

Properties

The compound $Ta_6Cl_{14} \cdot 8H_2O$ is insoluble in diethyl ether, chloroform, and carbon tetrachloride. It dissolves readily in methanol. Aqueous solutions are best prepared by adding water to concentrated methanolic solutions. Solutions of $Ta_6Cl_{14} \cdot 8H_2O$ show the characteristic intense emerald green color of the $[Ta_6Cl_{12}]^{2+}$ ion. Solutions of the compound are subject to air oxidation of the $[Ta_6Cl_{12}]^{2+}$ unit. This ease of oxidation extends even to the solid compound. Over a period of several weeks, solid green $Ta_6Cl_{14} \cdot 8H_2O$ in closed vials may change its color to shades of brown or reddish brown, indicating the formation of the $[Ta_6Cl_{12}]^{3+}$ or $[Ta_6Cl_{12}]^{4+}$ ions, respectively. Absorption spectra of the solid hydrated halides $M_6X_{14} \cdot 8H_2O$ have been reported by Spreckelmeyer[16] and infrared spectra, by Mattes.[17]

References

1. P. A. Vaughn, J. H. Sturdivant, and L. Pauling, *J. Am. Chem. Soc.* **72**, 5477 (1950).
2. P. B. Fleming, T. A. Dougherty, and R. E. McCarley, *J. Am. Chem. Soc.* **89**, 159 (1967).
3. (a) N. Prokopuk, C. S. Weinert, V. O. Kennedy, D. P. Siska, H.-J. Jeon, C. L. Stern, and D. F. Shriver, *Inorg. Chimica Acta* **300**, 951 (2000); (b) V. O. Kennedy, C. L. Stern, and D. F. Shriver, *Inorg. Chem.* **33**, 5967 (1994).
4. R. A. Field and D. L. Kepert, *J. Less-Common Metals* **13**, 378 (1967).
5. P. B. Fleming, J. L. Meyer, W. K. Grindstaff, and R. E. McCarley, *Inorg. Chem.* **9**, 1769 (1970).
6. N. Brnicevic, D. Nöthig-Hus, B. Kojic-Prodic, Z. Ruzic-Toros, Z. Danilovic, and R. E. McCarley, *Inorg. Chem.* **31**, 3924 (1992).
7. R. D. Burbank, *Inorg. Chem.* **5**, 1491 (1966).
8. P. C. Chabrie, *Compt. Rend.* **144**, 804 (1907).
9. W. H. Chapin, *J. Am. Chem. Soc.* **32**, 323 (1910).
10. H. S. Harned, *J. Am. Chem. Soc.* **35**, 1078 (1913).
11. H. S. Harned, C. Pauling, and R. B. Corey, *J. Am. Chem Soc.* **82**, 4815 (1960).

12. P. B. Fleming, Ph.D. thesis, Iowa State University, Ames, Iowa, 1968.
13. P. B. Fleming, L. A. Mueller, and R. E. McCarley, *Inorg. Chem.* **6**, 1 (1967).
14. A. Broll, D. Juza, and H. Schäfer, *Z. Anorg. Allg. Chem.* **382**, 69 (1969).
15. F. W. Koknat, J. A. Parsons, and A. Vongvusharintra, *Inorg. Chem.* **13**, 1699 (1974).
16. B. Spreckelmeyer, *Z. Anorg. Allg. Chem.* **365**, 225 (1969).
17. R. Mattes, *Z. Anorg. Allg. Chem.* **364**, 279 (1969).

42. POLYOXOMOLYBDATE CLUSTERS: GIANT WHEELS AND BALLS

Submitted by ACHIM MÜLLER,* SAMAR K. DAS,* ERICH KRICKEMEYER,* and CHRISTOPH KUHLMANN*
Checked by MASAHIRO SADAKANE,† MICHAEL H. DICKMAN,† and MICHAEL T. POPE†

The syntheses of aesthetically beautiful nanoscaled polyoxomolybdates of the type $\{Mo_{11}\}_n$, which have spherical-/ball- ($n = 12$) or wheel-shaped ($n = 14$) anions (see Figs. 1–3) and are interesting not only for inorganic chemistry, are reported. In spite of their completely different overall structure, they have similar building blocks[1,2] (see also Refs. 3, 4) (Fig. 1), resulting in the remarkable equivalent stoichiometry.[1] Important in this context is that the species can probably be considered as belonging to the class of the most complex discrete inorganic species and consequently, their characterization was correspondingly complex.

A formulation of the discrete wheel-type species (Fig. 2), which is useful for reactivity studies, corresponds to $[\{Mo_2^{VI}O_4\ (\mu_2\text{-}O)\ L_2\}_{n-x}^{2+}\ \{Mo_8^{VI/V}O_{26}\ (\mu_3\text{-}O)_2\ \ H(H_2O)_3\ \ Mo^{VI/V}\}_n^{3-}]^{(n+2x)-} \equiv [\{Mo_2\}_{n-x}\ \ \{Mo_8\}_n\ \ \{Mo_1\}_n]^{(n+2x)-}$ ($L = H_2O$ or other ligands), where x refers to the number of defects or missing $\{Mo_2\}$ groups introduced into the system, $n = 14$ corresponds to a tetradecameric, and $n = 16$ corresponds to a hexadecameric species.[1,2,5] Here we report the synthesis of a relevant mixed-crystal compound C, which contains the tetradecameric ring ($n = 14$) both as a complete entity and as one with a missing $\{Mo_2\}$ group in the ratio 1 : 1, as well as the compound D with the rings linked to layers. It should be mentioned that it is a difficult task to determine the formula of a compound belonging to class II or III of the Robin–Day classification with a

*Lehrstuhl für Anorganische Chemie I, Fakultät für Chemie, Universität Bielefeld, D-33501 Bielefeld, Germany.
†Department of Chemistry, Georgetown University, Washington, DC 20057.

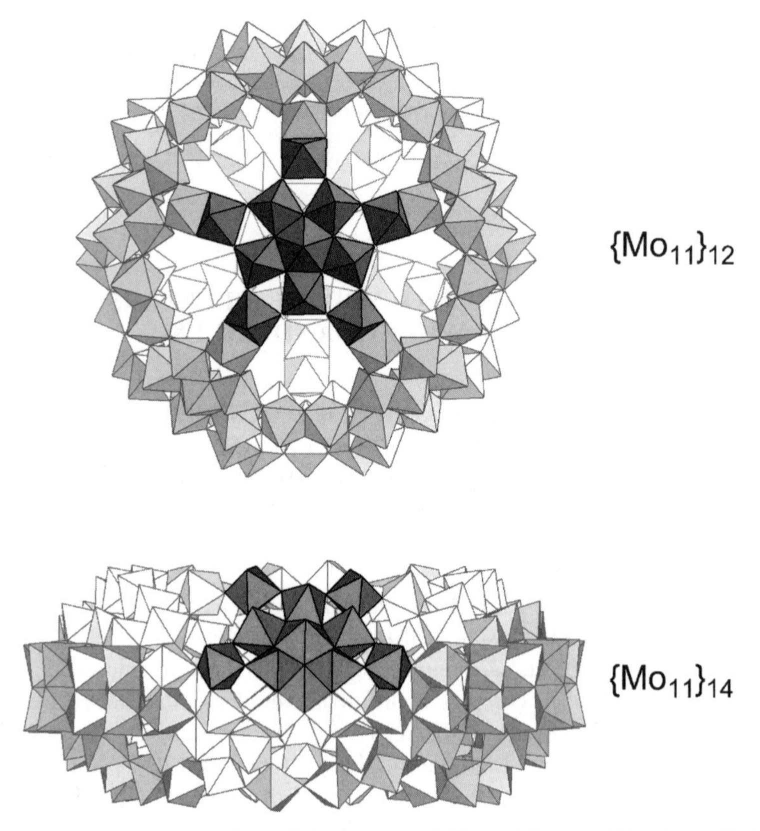

Figure 1. Schematic comparison of the pentagonal C_5- and C_s-type $\{Mo_{11}\}$ motifs (dark gray) as present, for example, in the ball-shaped $\{Mo_{11}\}_{12}$-(above; corresponding to the anion of compound A) and in the $\{Mo_{11}\}_{14}$-type cluster with view perpendicular to that of Fig. 2 (below; corresponding to an anion of compound C without defect).

protonated, mixed-valence anionic species with a very high relative molecular mass, especially if a very low concentration of disordered cations in the lattice complicates the determination of the anion charge.[5]

In the case of giant wheel (molybdenum blue) compounds, the general synthetic strategy involves the acidification (pH $\sim$1) and reduction of an aqueous molybdate(VI) solution [possible reducing agents: iron powder, tin(II) chloride, molybdenum(V) chloride, ascorbic acid, cysteine, hydroxylamine, hypophosphorous acid, sodium dithionite, or hydrazine sulfate].[2–5] On the other hand, an icosahedral ball-shaped cluster can be formed in an aqueous Mo(VI)

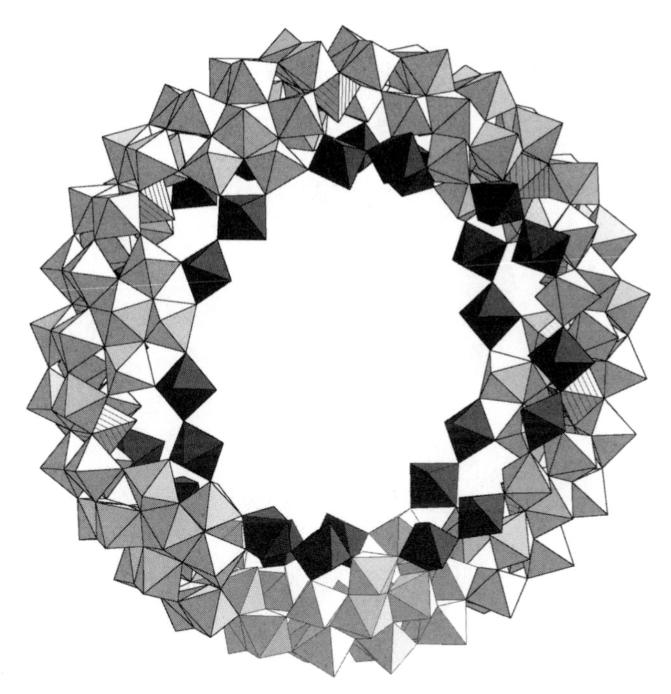

Figure 2. Polyhedral representation of the $\{Mo_{11}\}_{14}$-type cluster ($\equiv[\{Mo_2\}\ \{Mo_8\}\ \{Mo_1\}]_{14}{}^{14-}$) with one $\{Mo_8\}$ group more pale at the bottom ($\{Mo_1\}$ units: hatched; $\{Mo_2\}$ groups: dark gray).

containing solution at relatively higher pH values (ranging from 2 to 4) in the presence of an appropriate bidentate bridging ligand L, for instance, acetate, stabilizing the dinuclear $\{Mo_2^V\}$ units formed in the presence of a reducing agent.

The formula for the discrete ball-shaped species described here is $[\{Mo^{VI}(Mo_5^{VI}O_{21})(H_2O)_6\}_{12}\ \{Mo_2^VO_4L\}_{30}]^{42-}$ ($\equiv P_{12}S_{30}$) ($L = CH_3COO^-$, $PO_2H_2^-$). In agreement with the formula, 30 dinuclear spacer groups S for 12 pentagonal units P (spanning an icosahedron) are positioned on the corners of a (distorted) truncated icosahedron (Fig. 3).[1,2,6,7]

The compounds have been characterized by elemental analyses, spectroscopy [electronic absorption, IR, Raman (or resonance Raman for the blue compounds with $\lambda_e = 1064$ nm)], redox titration [for the determination of the (formal) number of Mo^V centers], thermal gravimetric analyses (TGA), and single-crystal X-ray structure analyses (see Refs. 2, 5 and literature cited therein). Mainly the crystals of the compounds that contain discrete wheel-shaped anions lose lattice water rapidly on removal from the mother liquor even at room temperature,[2,5] a

Figure 3. Polyhedral representation of the ball-shaped cluster anion of compound A highlighting the binuclear spacer units (see Fig. 1 highlighting the pentagons).

fact that complicates the analyses. (In particular, the crystal water content* is difficult to determine, especially for species where the cluster units are loosely packed in the crystal lattice, e.g., in the case of C.) As the above mentioned spectra and most values of the elementary analyses of different molybdenum blue compounds (e.g., with and without defects or with linked and nonlinked units) are necessarily almost identical, it is practically impossible to distinguish the compounds alone on the basis of such data. Therefore, in order to ensure their purity, it is essential to determine in addition the characteristic unit cell dimensions of several crystals of each type of compound, and thus these values are included here. For the syntheses of related compounds other than those described here, see papers cited in Refs. 1, 2, and 5.

Materials and Hazards

All chemicals used are obtained commercially and are of analytical grade. No further purification is done prior to use. Concentrated acids, such as hydrochloric

*The theoretical number of crystal water molecules is calculated from the residual volume of the unit cell not occupied by the refined lattice components.

or acetic acid, are corrosive. Sodium molybdate is harmful if swallowed in small doses. Hydrazinium sulfate is toxic and corrosive and harmful on inhalation, contact with the skin, or if swallowed. Sodium dithionite is harmful if swallowed, and in contact with acids produces toxic SO_2. Both reducing agents used have to be fresh, stored cool, dry, and airtight.

A. $(NH_4)_{42}$ $[Mo^{VI}_{72}Mo^{V}_{60}O_{372}$ $(CH_3COO)_{30}$ $(H_2O)_{72}]$ · HYDRATE

$$132\,MoO_4^{2-} + 15\,N_2H_6^{2+} + 30\,CH_3COOH + 192\,H^+ \longrightarrow$$

$$[Mo^{VI}_{72}Mo^{V}_{60}O_{372}(CH_3COO)_{30}(H_2O)_{72}]^{42-} + 15\,N_2 + 84\,H_2O$$

Procedure

After adding $N_2H_6 \cdot SO_4$ (0.8 g, 6.1 mmol) to a solution of $(NH_4)_6\,Mo_7O_{24} \cdot 4H_2O$ (5.6 g, 4.5 mmol) and CH_3COONH_4 (12.5 g, 162.2 mmol) in H_2O (250 mL) and stirring for 10 min (color change to blue-green), 50% (v/v) CH_3COOH (83 mL) is added. The reaction solution, now green, is stored in an open 500-mL Erlenmeyer flask at 20°C without further stirring (slow color change to dark brown). After 4 days the precipitated red-brown crystals are filtered off through a glass frit (D2), washed with 90% ethanol and diethyl ether, and finally dried in air. Yield: 3.3 g (52% based on Mo). (*Note*: Compound A can in principle be recrystallized from a very concentrated aqueous solution by adding additional NH_4Cl.)

Anal. Calcd. for $(NH_4)_{42}$ $[Mo^{VI}_{72}Mo^{V}_{60}O_{372}$ $(CH_3COO)_{30}$ $(H_2O)_{72}]$ · ~300 H_2O · ~10 CH_3COONH_4: C, 3.4; H, 3.8; N, 2.6; Mo, 44.3 (Mo^V, 20.1); crystal H_2O, 18.9. Found: C, 3.4; H, 3.2; N, 2.7; Mo, 45.8 (Mo^V, 20.2; cerimetric titration); crystal H_2O, 18.0 (TGA value).

Properties

The water-soluble red-brown compound crystallizes in a cubic space group [$a = 46.0576\ (14)$ Å][6] and forms octahedral crystals (including related truncated ones). The IR spectrum (Fig. 4) shows main peaks at 1626 (m) (δ_{H_2O}), 1546 (m) ($\nu_{as,COO}$), 1440 (sh), 1407 (m) (δ_{CH_3}, $\nu_{s,COO}$, δ_{as,NH_4^+}) 969 (m), 936 (w-m) ($\nu_{Mo=O}$), 853 (m), 792 (s), 723 (s), 567 (s) cm^{-1}, while the electronic absorption spectrum (H_2O/CH_3COOH, pH = 4) shows an intense band at 450 nm ($\varepsilon = 3.5 \cdot 10^5\ M^{-1}cm^{-1}$). Because of the high symmetry, relatively

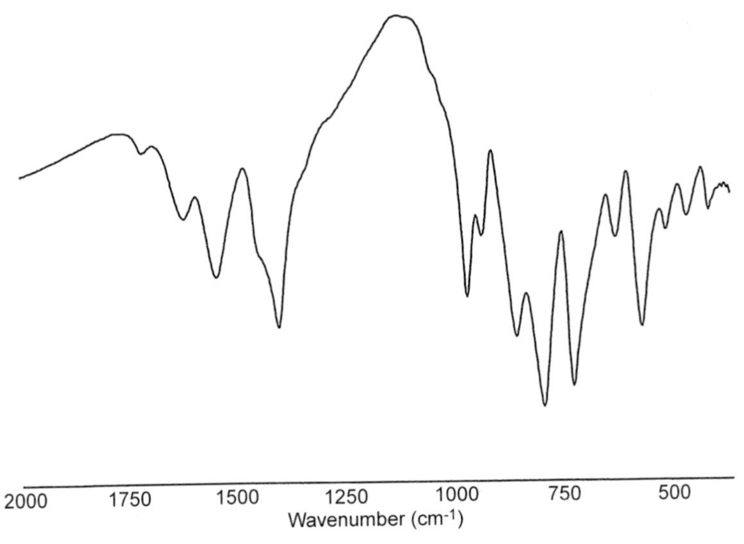

Figure 4. IR spectrum of $(NH_4)_{42}[Mo^{VI}_{72}Mo^{V}_{60}O_{372}(CH_3COO)_{30}(H_2O)_{72}]\cdot\sim300\ H_2O\cdot\sim10\ CH_3COONH_4$ (compound A) (KBr disk).

few characteristic Raman bands [953 (m), 935 (m), 875 (s) ($\nu_{Mo=O}$), ~845 (sh), 374 (m-s), 314 (m), 212 (w) cm^{-1}] are observed that are assignable to the irreducible representations H_g and A_{1g} (bands with highest intensity).

B. $(NH_4)_{42}$ $[Mo^{VI}_{72}Mo^{V}_{60}O_{372}$ $(H_2PO_2)_{30}$ $(H_2O)_{72}]$ HYDRATE

$$[Mo^{VI}_{72}Mo^{V}_{60}O_{372}(CH_3COO)_{30}(H_2O)_{72}]^{42-} + 30\,H_3PO_2 \longrightarrow$$
$$[Mo^{VI}_{72}Mo^{V}_{60}O_{372}(H_2PO_2)_{30}(H_2O)_{72}]^{42-} + 30\,CH_3COOH$$

Procedure

A stirred solution of compound A (1.0 g, 0.04 mmol) in H_2O (50 mL) is treated with $NaH_2PO_2\cdot H_2O$ (1.0 g, 9.4 mmol) and subsequently 3 mL of 1 M HCl is added dropwise, giving a pH value of ~2. After stirring for 24 h in a closed flask at 20°C and addition of NH_4Cl (2.0 g, 37.4 mmol), the mixture is kept at 15°C without further stirring. The dark brown crystals of $(NH_4)_{42}$ $[Mo^{VI}_{72}Mo^{V}_{60}O_{372}$ $(H_2PO_2)_{30}$ $(H_2O)_{72}]\cdot\sim300\ H_2O$, which slowly precipitate from the dark brown solution over a period of 4 days, are filtered through a glass frit (D2), washed with 10 mL of ice-cooled 98% 2-propanol, and dried in air. Yield: 0.8 g (81% based on compound A).

Anal. Calcd. for $(NH_4)_{42}$ $[Mo^{VI}_{72}Mo^{V}_{60}O_{372}$ $(H_2PO_2)_{30}$ $(H_2O)_{72}]\cdot{\sim}300$ H_2O: H, 3.5; N, 2.1; P, 3.3; Mo, 45.2 (Mo^{V}, 20.5); crystal H_2O, 19.3. Found: H, 3.2; N, 2.1; P, 3.4; Mo, 45.5 (Mo^{V}, 20.8; cerimetric titration); crystal H_2O, 18.5 (TGA value).

Properties

The compound [space group $R\bar{3}$, $a = 32.719$ (1), $c = 73.567$ (2) Å][7] is obtained by a ligand exchange (30 CH_3COO^- by 30 $H_2PO_2^-$) reaction. The IR spectrum is practically identical to that of compound A except for the appearance of $H_2PO_2^-$ bands [1118 (m), 1075 (w), 1033 (m) cm^{-1}] instead of CH_3COO^- bands. Also the Raman and electronic absorption spectra are almost identical to those of compound A.

C. $Na_{15}[Mo^{VI}_{126}Mo^{V}_{28}O_{462}H_{14}(H_2O)_{70}]_{0.5}[Mo^{VI}_{124}Mo^{V}_{28}O_{457}H_{14}(H_2O)_{68}]_{0.5}$ HYDRATE

$$154\,MoO_4^{2-} + 14\,S_2O_4^{2-} + 322\,H^+ \longrightarrow$$

$$[Mo^{VI}_{126}Mo^{V}_{28}O_{462}H_{14}(H_2O)_{70}]^{14-} + 28\,SO_2 + 84\,H_2O$$

■ **Caution.** *This reaction evolves SO_2; hence, it should be carried out in a good fume hood.*

Procedure

To a solution of $Na_2MoO_4\cdot 2H_2O$ (3.0 g, 12.4 mmol) in 10 mL of water, freshly powdered* $Na_2S_2O_4$ (0.2 g, 1.15 mmol) is added (light yellow coloration). Immediately afterward, under continuous stirring, 30 mL of hydrochloric acid (1 M) is rushed into the solution (color changes to deep blue). The solution is stirred in an open 100-mL Erlenmeyer flask for a further 10 min and then stored undisturbed in a closed flask at 20°C (not higher!) for 3 days.† The precipitated blue crystals (plate crystal aggregation may occur because of the high concentrations used) are removed by filtration, washed quickly with a small amount of cold water (note the rather high solubility!), and dried at room temperature over $CaCl_2$. Yield: 0.7 g (28% based on Mo).

*Sodium dithionite should be fresh. The stability of $S_2O_4^{2-}$ ions in acidic media has not been considered for the formulation of the equation (formulated for the non-defect cluster). It is used as it gives the best results as reducing agent.

†Crystals filtered off after less than 3 days might be too small to be identified by X-ray diffraction. On standing much longer than 3 days (e.g., 2 weeks), other types of crystals (Mo_{176}) are found. (The checkers found these after 2 weeks, also.)

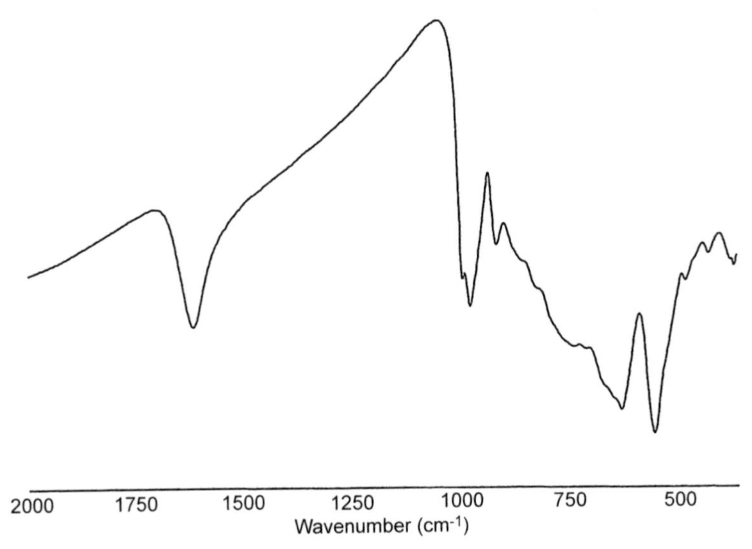

Figure 5. IR spectrum of $Na_{15}[Mo^{VI}_{126}Mo^{V}_{28}O_{462}H_{14}(H_2O)_{70}]_{0.5}[Mo^{VI}_{124}Mo^{V}_{28}O_{457}H_{14}(H_2O)_{68}]_{0.5} \cdot \sim 400\ H_2O$ (compound C) (KBr disk).

Anal. Calcd. for $Na_{15}[Mo^{VI}_{126}Mo^{V}_{28}O_{462}H_{14}(H_2O)_{70}]_{0.5}[Mo^{VI}_{124}Mo^{V}_{28}O_{457}H_{14}(H_2O)_{68}]_{0.5} \cdot \sim 400\ H_2O$: Na, 1.1; Mo, 47.6 ($Mo^V$, 8.7); crystal H_2O, 23.4. Found: Na, 1.2; Mo, 49.0 (Mo^V, 8.9; cerimetric titration); crystal H_2O, 23.0 (TGA value).

Properties

The molybdenum blue compound C crystallizes in the form of deep blue plates, which are extremely soluble at room temperature in water and even in low-molecular-weight alcohols. When removed from the mother liquor, a type of fast weathering process is observed associated with the loss of crystal water and crystallinity* [space group $P\bar{1}$ with $a = 30.785$ (2), $b = 32.958$ (2), $c = 47.318$ (3) Å; $\alpha = 90.53$ (1), $\beta = 89.86$ (1), $\gamma = 96.85$ (1)°].[8] The IR spectrum (Fig. 5) shows peaks at 1616 (m) (δ_{H_2O}), 975 (m), 913 (w-m) ($\nu_{Mo=O}$), 820 (sh), 750 (s), 630 (s), 555 (s) cm^{-1}. The electronic absorption spectrum is dominated by two bands characteristic for all molybdenum blue species [H_2O/HCl, pH = 1: λmax, nm (ε, $M^{-1}cm^{-1}$) = 745 (1.8×10^5), 1070 (1.4×10^5)] that have to be assigned to intervalence (Mo^V/Mo^{VI}) charge transfer transitions

*For the same reason the single crystals taken from the mother liquor were immediately cooled to liquid nitrogen temperature for data collection within the process of single-crystal X-ray structure determination.

(IVCT). The second ε value corresponds to the abundance of 28 Mo^V centers (see Ref. 5). Using an excitation line within the contour of the 1070-nm band gives rise to a resonance–Raman spectrum showing five bands in the region between 900 and 200 cm^{-1} very characteristic for all molybdenum blue species [802 (s), 535 (m), 462 (s), 326 (s), 215 (s) cm^{-1}].

D. Na_{21} [$Mo^{VI}_{126}Mo^{V}_{28}O_{462}H_{14}$ $(H_2O)_{54}$ $(H_2PO_2)_7$] HYDRATE

$$154\,MoO_4^{2-} + 21\,H_2PO_2^- + 294\,H^+ \longrightarrow$$

$$[Mo^{VI}_{126}Mo^{V}_{28}O_{462}H_{14}(H_2O)_{54}(H_2PO_2)_7]^{21-} + 14\,H_2PO_3^- + 86\,H_2O^\dagger$$

Procedure

To a solution of $Na_2MoO_4 \cdot 2H_2O$ (3.0 g, 12.4 mmol) and NaCl (1.0 g, 17.1 mmol) in 25 mL hydrochloric acid (1.1 M), $NaH_2PO_2 \cdot H_2O$* (0.212 g, 2.0 mmol) is added. After stirring for 15 min under bubbling nitrogen gas, the resulting solution is kept undisturbed in a closed flask at room temperature. After 3 days‡ the precipitated, blue bipyramid-shaped crystals are filtered, washed quickly with a small amount of cold water, and dried at room temperature under argon. Yield: 0.43 g (18% based on Mo).

Anal. Calcd. for Na_{21} [$Mo^{VI}_{126}Mo^{V}_{28}O_{462}H_{14}$ $(H_2O)_{54}$ $(H_2PO_2)_7$]$\cdot \sim$300 H_2O: Na, 1.6; Mo, 50.1 (Mo^V, 9.1); P, 0.7; crystal H_2O, 18.3. Found: Na, 1.7; Mo, 50.9 (Mo^V, 9.5; cerimetric titration); P, 0.7; crystal H_2O, 18.5 (TGA value).

Properties

The compound crystallizes mainly in the form of blue square bipyramidal crystals. The IR spectrum is practically identical to that of C except for the additional IR bands at 1124 (w)/1076 (vw)/1039 (vw) (τ_{PH_2}, γ_{PH_2}, δ_{PO_2}) showing nicely the presence of the $H_2PO_2^-$ ligand. Also, the resonance Raman and electronic absorption spectra are almost identical to those of compound C. The crystal structure[9] [space group *Cmca*, $a = 50.075$ (3), $b = 56.049$ (4), $c = 30.302$ (2) Å] shows the abundance of nanosized ring-shaped units (crystallographic site symmetry 2/*m*), which are linked through covalent Mo–O–Mo bonds: each cluster ring is surrounded by four rings, resulting in a layer structure with condensed

*The quality of the NaH_2PO_2 used strongly affects this reaction. The checkers found material from Aldrich to be best.

†Equation formulated by the checkers.

‡Disturbance of the reaction mixture might result in the precipitation of a less crystalline material. The crystals should be separated out from the reaction mixture not later than 3 days, to avoid the coprecipitation of amorphous materials as aftereffect.

ring-shaped units parallel to the *ac* plane. The packing of the layers gives rise to the formation of a type of nanosized channel with the "encapsulated" $H_2PO_2^-$ ligands, which are coordinated at the well-defined sites of $\{Mo_2\}$ groups, replacing the H_2O ligands.

References

1. A. Müller, P. Kögerler, and H. Bögge, *Struct. Bond.* **96**, 203 (2000).
2. A. Müller, P. Kögerler, and C. Kuhlmann, *J. Chem. Soc., Chem. Commun.* (feature article) 1347 (1999).
3. M. T. Pope, *Heteropoly and Isopoly Oxometalates*, Springer, Berlin, 1983.
4. (a) N. V. Sidgwick, *The Chemical Elements and Their Compounds*, Clarendon, London, 1962, Vol. II, p. 1046; (b) *Gmelins Handbuch der anorganischen Chemie*, Verlag Chemie, Berlin, 1935, Vol. 53 (Mo), pp. 134–147; (c) *Gmelin Handbook of Inorganic Chemistry*, Springer, Berlin, 1987, Mo Suppl., Vol. B3a, pp. 63–65; 1989, Mo Suppl., Vol. B3b, pp. 15–16; (d) J. W. Mellor, *A Comprehensive Treatise on Inorganic and Theoretical Chemistry*, Longman, London, 1931, Vol. XI, pp. 526–531.
5. A. Müller and C. Serain, *Acc. Chem. Res.* **33**, 2 (2000).
6. A. Müller, E. Krickemeyer, H. Bögge, M. Schmidtmann, and F. Peters, *Angew. Chem., Int. Ed. Engl.* **37**, 3360 (1998).
7. A. Müller, S. Polarz, S. K. Das, E. Krickemeyer, H. Bögge, M. Schmidtmann, and B. Hauptfleisch, *Angew. Chem., Int. Ed. Engl.*, **38**, 3241 (1999).
8. A. Müller, S. K. Das, V. P. Fedin, E. Krickemeyer, C. Beugholt, H. Bögge, M. Schmidtmann, and B. Hauptfleisch, *Z. Anorg. Allg. Chem.* **625**, 1187 (1999).
9. A. Müller, S. K. Das, H. Bögge, C. Beugholt, and M. Schmidtmann, *J. Chem Soc., Chem. Commun.* 1035 (1999).

43. TETRAKIS-{(η^6-1-ISOPROPYL-4-METHYLBENZENE) RUTHENIUM(II)TETRAOXOMOLYBDATE(VI)}

Submitted by BRUNO THERRIEN,[*] LAURENT PLASSERAUD,[*] and GEORG SÜSS-FINK[*]
Checked by DANIELLE LAURENCIN[†] AND ANNA PROUST[†]

$$2(\eta^6\text{-}p\text{-MeC}_6\text{H}_4{}^i\text{Pr})_2\text{Ru}_2\text{Cl}_4 + 4\text{Na}_2\text{MoO}_4 \longrightarrow$$

$$(\eta^6\text{-}p\text{-MeC}_6\text{H}_4{}^i\text{Pr})_4\text{Ru}_4\text{Mo}_4\text{O}_{16} + 8\text{NaCl}$$

Since the early 1980s there has been a steadily growing interest in molecules containing both organometallic groups and oxometallic entities,[1] particularly since they provide molecular models for heterogeneous catalysts derived from

[*]Institut de Chimie, Université de Neuchâtel, CH-2000, Neuchâtel, Switzerland.
[†]Laboratoire de Chimie Inorganique et Matériaux Moleculaires, CNRS 7071, Université Pierre & Marie Curie, 75252 Paris, France.

organometallic complexes adsorbed at metal oxide surfaces.[2] Organometallic metaloxo clusters contain soft as well as hydrophilic ligands. Since the discovery of the first species of this type, $[(C_5H_5)TiPW_{11}O_{39}]^{4-}$ in 1978,[3] this field has been pioneered mainly by the groups of Klemperer,[4] Isobe,[5] Finke,[6] and Proust and Gouzerh.[7] The combination of low- and high-valence transition metals and their amphiphilic character predispose these molecules also as homogeneous catalysts for oxidation reactions;[8] the catalytic potential of these compounds has been reviewed.[9] Herein, we describe the synthesis of $(\eta^6\text{-}p\text{-MeC}_6\text{H}_{4i}\text{Pr})_4$ $Ru_4Mo_4O_{16}$, a neutral organoruthenium oxomolybdenum cluster that exists as two structural isomers in solution, one of which can be crystallized from dichloromethane/toluene. In the crystal line state, the molecule contains an unprecedented $Ru_4Mo_4O_{12}$ framework, which can be described as a central Mo_4O_4 cube with four folded ORuO flaps resembling the sails of a windmill.[10] In solution, the windmill structure (Fig. 1a) was found to isomerize to a triple-cubane structure (Fig. 1b); the equilibrium was dependent on the nature of the solvent. In chloroform, both isomers are present in a nearly 1 : 1 ratio, while in dichloromethane the triple-cubane isomer is predominant.[11]

Procedure

To a suspension of $(\eta^6\text{-}p\text{-MeC}_6\text{H}_4{}^i\text{Pr})_2\text{Ru}_2\text{Cl}_4$[12] (500 mg, 0.82 mmol) in 50 mL of distilled water is added dropwise an aqueous solution (25 mL) of $Na_2MoO_4 \cdot 2$ H_2O (Fluka; 2.0 g, 8.3 mmol). The mixture is stirred at room temperature for 4 h. After evaporation of the dark orange solution, the precipitate is extracted with CH_2Cl_2 and chromatographed on silicagel (column 50 × 80 mm, 1.5 L of a 5% MeOH in dichloromethane) to remove a brown residue. The solvent is removed on a rotary evaporator, and the solid is washed with ether and then dried to give 422 mg (65% yield) of product.

Anal. Calcd. for $C_{40}H_{56}Mo_4O_{16}Ru_4$: C, 30.39; H, 3.57 Found: C, 30.33; H, 3.51.

Properties

The product is an orange solid, sparingly soluble in water and aromatic hydrocarbons, and amply soluble in polar organic solvents. It is stable in air at room temperature for several months, and decomposition occurred when this product was heated over 230°C. The compound is readily identified by its IR spectrum (KBr), which contains the following Mo=O absorptions 921 (m), 874 (w), and Mo–O 785 (s), 739 (s), 642 (w), and 602 (m) cm^{-1} The FAB mass spectrum shows the molecular ion at m/z 1581. Characteristic ^1H NMR (200 MHz, $CDCl_3$)*: δ 1.39 (d, 6H), 2.27 (s, 3H), 2.97 (sp, 1H), 5.30 (d, 2H), 5.37 (d, 2H).

*Only the predominant set of signals is given, corresponding to the triple-cubane isomer.

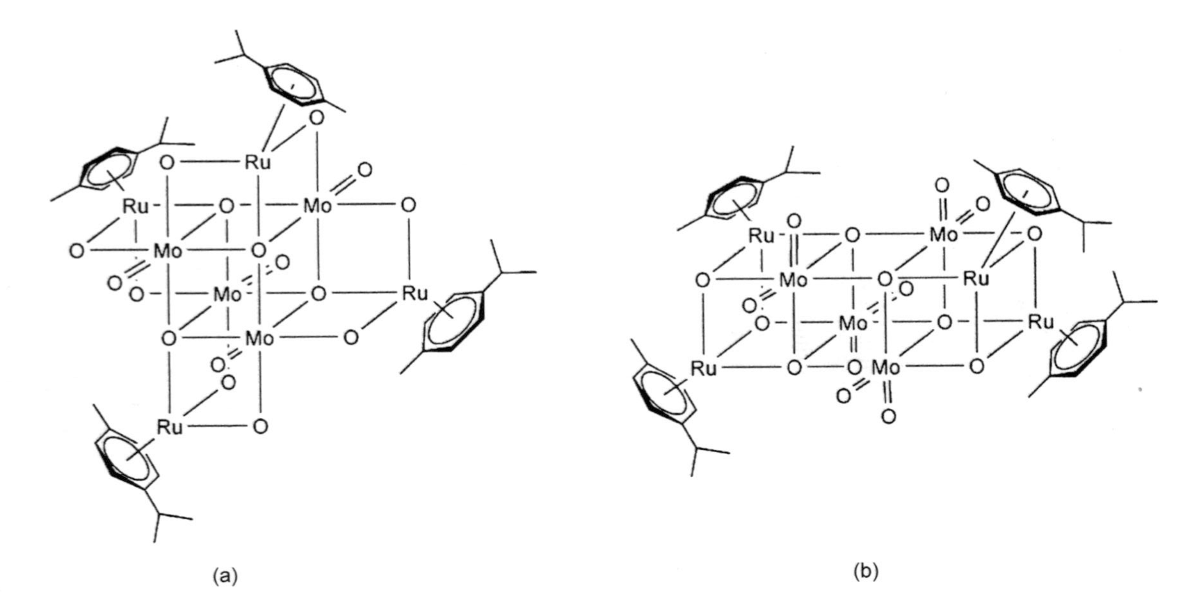

Figure 1. Structural isomers of an organoruthenium oxomolybdenum cluster: (a) windmill-like and (b) triple-cubane configurations.

References

1. V. W. Day and W. G. Klemperer, *Science* **228**, 533 (1985).
2. K. Isobe and A. Yagasaki, *Acc. Chem. Res.* **26**, 524 (1993).
3. R. K. C. Ho and W. G. Klemperer, *J. Am. Chem. Soc.* **100**, 6772 (1978).
4. H. K. Chae, W. G. Klemperer, and V. W. Day, *Inorg. Chem.* **28**, 1424 (1989); H. K. Chae, W. G. Klemperer, D. E. Paez Loyo, V. W. Day, and T. A. Ebersbacher, *Inorg. Chem.* **31**, 3187 (1991); H. K. Chae, W. G. Klemperer, and T. A. Marquart, *Coord. Chem. Rev.* **128**, 209 (1993).
5. Y. Hayashi, K. Toriumi, and K. Isobe, *J. Am. Chem. Soc.* **110**, 3666 (1988); Y. Do, X.-Z. You, C. Zhang, Y. Ozawa, and K. Isobe, J. Am. Chem. Soc. **113**, 5892 (1991); Y. Hayashi, Y. Ozawa, and K. Isobe, *Inorg. Chem.* **30**, 1025 (1991); J. T. Park, T. Nishioka, T. Suzuki, and K. Isobe, *Bull. Chem. Soc. Jpn.* **67**, 1968 (1994).
6. Y. Lin, K. Nomiya, and R. G. Finke, *Inorg. Chem.* **32**, 6040 (1993); A. Trovarelli and R. G. Finke, *Inorg. Chem.* **32**, 6034 (1993); M. Pohl and R. G. Finke, *Organometallics* **12**, 1453 (1993); B. M. Rapko, M. Pohl, and R. G. Finke, *Inorg. Chem.* **33**, 3625 (1994); M. Pohl, D. K. Lyon, N. Mizuno, K. Nomiya, and R. G. Finke, *Inorg. Chem.* 34, 1413 (1995); K. Nomiya, Ch. Nozaki, M. Kaneko, R. G. Finke, and M. Pohl, *J. Organomet. Chem.* **505**, 23 (1995); M. Pohl, Y. Lin, T. J. R. Weakley, K. Nomiya, M. Kaneko, H. Weiner, and R. G. Finke, *Inorg. Chem.* **34**, 767 (1995).
7. R. Villanneau, R. Delmont, A. Proust, and P. Gouzerh, *Chem. Eur. J.* **6**, 1184 (2000).
8. P. Gouzerh and A. Proust, *Chem. Rev.* **98**, 77 (1998).
9. C. L. Hill and C. M. Prosser-McCartha, *Coord. Chem. Rev.* **143**, 407 (1995).
10. G. Süss-Fink, L. Plasseraud, V. Ferrand, and H. Stoeckli-Evans, *Chem. Commun.* 1657 (1997); G. Süss-Fink, L. Plasseraud, V. Ferrand, S. Stanislas, A. Neels, H. Stoeckli-Evans, M. Henry, G. Laurenczy, and R. Roulet, *Polyhedron* **17**, 2817 (1998).
11. V. Artero, A. Proust, P. Herson, R. Thouvenot, and P. Gouzerh, *Chem. Commun.* **883** (2000).
12. M. A. Bennett, T.-N. Huang, T. W. Matheson, and A. K. Smith, *Inorg. Synth.* **21**, 74 (1982).

44. DIRUTHENAPENTA- AND DIRUTHENAHEXABORANES

Submitted by ANTONIO DI PASQUALE,[*] XINJIAN LEI,[*] and THOMAS P. FEHLNER[*]
Checked by MITSUHIRO HATA[†] and LAWRENCE BARTON[†]

Compounds containing maingroup transition element covalent bonds, other than carbon, present the interesting possibilities of transition-element-promoted main-group chemistry and maingroup-element-promoted transition element chemistry totally analogous to organometallic chemistry.[1] Development depends on the progress of synthesis, and for boron the preparation of metallaboranes containing polyborane fragments has been hindered historically by the necessity of obtaining and handling the higher boranes.[2–4] The procedures described here

[*]Department of Chemistry and Biochemistry, University of Notre Dame, Notre Dame, IN 46556.
[†]Department of Chemistry, University of Missouri—St. Louis, St. Louis, MO 63121.

demonstrate practical preparations of metallaboranes containing both tri- and tetraborane fragments utilizing monoboranes.[5] This method has the advantages of (1) readily available starting materials, (2) high yield of the principal product, and (3) the use of common Schlenk apparatus.

General Comments

All of the following manipulations are performed under argon using standard Schlenk techniques. Solvents are degassed and distilled under nitrogen immediately before use unless otherwise noted. $(Cp^*RuCl_2)_2$ is available from Strem Chemicals, Inc., and is used as received. Lithium borohydride, $LiBH_4$ (2.0 M in tetrahydrofuran), and borane tetrahydrofuran, $BH_3 \cdot THF$ (1.0 M in tetrahydrofuran), are available from Aldrich Chemical Co. and are used as received. Note that these solutions decrease in molarity depending on time and use, and, if solutions are not titrated before use, appropriate increases in the volumes of solution used may be required.

■ **Caution.** *Care should be used in handling lithium borohydride as it is toxic, and corrosive and ignites in the presence of water. Lithium borohydride should be stored under nitrogen or argon to maintain its purity and reactivity.*

A. NIDO-1,2-$\{(\eta^5\text{-}C_5Me_5)Ru\}_2B_3H_9$

$$[Cp^*RuCl_2]_2 + 4LiBH_4 \longrightarrow (Cp^*Ru)_2B_3H_9 + 4LiCl + 2H_2 + BH_3 \cdot THF$$

Procedure

A 0.500-g (0.81-mmol) sample of $(Cp^*RuCl_2)_2$ is transferred in air into a 250-mL Schlenk reaction flask containing a magnetic stirring bar and fitted with a rubber septum. The flask is evacuated, slowly at first to avoid loss of $(Cp^*RuCl_2)_2$, and then charged with argon. Then 35 mL of tetrahydrofuran (THF) is added to the flask by syringe to generate a red suspension, and the flask is placed in a −40°C cold bath. With stirring, 2.2 mL (4.4 mmol) of 2.0 M $LiBH_4$ is added dropwise over a 5-min period by syringe. Then the flask is allowed to warm to room temperature with continued stirring. After one hour, no more H_2 gas evolves from the solution, the shiny crystals of $(Cp^*RuCl_2)_2$ have disappeared, and a dull precipitate is present. The THF is removed under vacuum, and 40 mL of hexane is added by syringe to the flask. After sonication, the hexane and remaining traces of THF are removed under vacuum. Then a second 40-mL portion of hexane is added to the flask and the mixture is sonicated. The solution is filtered through a fine glass frit, and its volume is reduced by half under vacuum.

■ **Caution.** *Any unreacted $LiBH_4$ will collect on the glass frit. This must be deactivated using an alcohol, such as 2-propanol or t-butanol, to prevent fire.*

Chromatography is performed on a short (2.5 × 8-cm) silicagel (60/200-mesh silica gel, from J. T. Baker Inc., dried in a 140°C oven overnight) column. First, a yellow band elutes with hexane and contains 0.330 g (0.64 mmol) of $(Cp^*Ru)_2B_3H_9$ (79% yield). Second, a red-orange band elutes with diethyl ether and contains ~0.067 g (0.089 mmol) of $(Cp^*Ru)_3B_3H_8$ (5–10% yield depending on conditions.) (*Note*: The chromatography is sensitive to the nature of the silicagel and column length. The checkers suggest 20 : 1 hexane/Et_2O for convenient elution.) After removal of the solvent under vacuum, the solution yields pure microcrystals. Recrystallization from a saturated hexane solution at 4°C for several days yields X-ray-quality crystals. Since $(Cp^*Ru)_2B_3H_9$ reacts with $BH_3 \cdot THF$ to yield $(Cp^*Ru)_2 B_4H_{10}$, excessive reaction times or elevated temperatures can lead to the formation of $(Cp^*Ru)_2 B_4H_{10}$ as an impurity in this procedure.

Anal. Calcd. for $C_{20}H_{39}B_3Ru_2$: C, 46.73; H, 7.65. Found: C, 46.65; H, 7.79.

Properties

$(Cp^*Ru)_2B_3H_9$ is a modestly stable solid at room temperature. As a solid, it can be manipulated in the open air, but should be stored for any extended period of time in an inert atmosphere at −40°C. In solution, it is air-sensitive and should be manipulated in an inert atmosphere. Heating leads to decomposition at temperatures of ⩾80°C. MS (FAB), $P^+ = 516 = {}^{101}Ru_2{}^{11}B_3{}^{12}C_{20}{}^{1}H_{39}{}^+$, 516.1418 calculated, 516.1404 observed. ^{11}B NMR (hexane, 22°C): δ −0.5 (d, $J_{B-H} = 120$ Hz, $\{^1H\}$, s, 1B), δ −2.5 (dd, $J_{B-H} = 100$, 60 Hz, $\{^1H\}$, s, 2B) ^1H NMR(C_6D_6, 22°C): δ 3.28 (partially collapsed quartet, pcq, 1H, B–H_t), δ 2.66 (pcq, 2H, B-H_t), δ 1.93 (s, 15H, C_5Me_5), δ 1.79 (s, 15H, C_5Me_5), δ −4.05 (s, br, 2H, B–H–B), δ −11.25 (pcq, 2H, B–H–Ru), δ −13.55 (s, 2H, Ru–H–Ru). IR (hexane): 2498 (m), 2450 (m), 2426 (w) cm^{-1}(B–H_t).

B. NIDO-2,3-$\{(\eta^5\text{-}C_5Me_5)Ru\}_2B_4H_{10}$

This compound can be prepared directly from $(Cp^*RuCl_2)_2$ and $BH_3 \cdot THF$ according to the following reaction:

$$(Cp^*RuCl_2)_2 + 8BH_3 \cdot THF \longrightarrow 4BH_2Cl + 3\,H_2 + (Cp^*Ru)_2B_4H_{10} + 8THF$$

However, in terms of purification, it is more efficient to prepare it by cage expansion according to

$$(Cp^*Ru)_2B_3H_9 + BH_3 \cdot THF \longrightarrow H_2 + (Cp^*Ru)_2B_4H_{10} + THF$$

Procedure

Under an inert atmosphere, 2 equiv of $BH_3 \cdot THF$ (0.19 mL, 0.19 mmol) are added to $(Cp^*Ru)_2B_3H_9$ (0.10 g, 0.097 mmol) dissolved in 10 mL THF. The solution is stirred at 60°C for 10 h. Filtration and removal of the THF affords 0.096 g of orange crystals of **2** (95% based on the Ru).

Anal. Calcd. for $C_{20}H_{40}B_4Ru_2$: C, 45.68; H, 7.67. Found: C, 45.84, H, 7.44.

Properties

MS (FAB), $P^+ = 528$, 4 B, 2 Ru atoms, calcd. for weighted average of isotopomers lying within the instrument resolution, 528.1589, observed, 528.1602. ^{11}B NMR (hexane, 22°C): δ 17.1 (dd, $J_{B\text{–}H} = 110$ Hz, 56 Hz,$\{^1H\}$, s), 3.9 (d, $J_{B\text{–}H} = 120$ Hz, $\{^1H\}$, s) ^1H NMR (C_6D_6, 22°C): 3.36 (pcq, 2H, B–H_t), 2.70 (pcq, 2H, B–H_t), 1.85 (s, 15H, C_5Me_5), 1.81 (s, 15H, C_5Me_5), −3.62 (s, br, 1H, B–H–B), −3.93 (s, br, 2H, B–H–B), −13.39 (b, 2H, B–H–Ru), −13.50 (s, 1H, Ru–H–Ru). IR (hexane): 2500 (m), 2434 (m) cm^{-1}.

References

1. T. P. Fehlner, ed., *Inorganometallic Chemistry*, Plenum, New York, 1992.
2. R. N. Grimes, in *Metal Interactions with Boron Clusters*; R. N. Grimes, ed., Plenum, New York, 1982, p. 269.
3. J. D. Kennedy, *Prog. Inorg. Chem.* **32**, 519 (1984).
4. J. D. Kennedy, *Prog. Inorg. Chem.* **34**, 211 (1986).
5. X Lei, M. Shang, and T. P. Fehlner, *J. Am. Chem. Soc.* **121**, 1275 (1999).

45. MIXED-METAL CLUSTERS OF IRIDIUM WITH RUTHENIUM AND OSMIUM

Submitted by ENRIQUE LOZANO DIZ,* SUSANNE HAAK,* and GEORG SÜSS-FINK*
Checked by ZOLTAN BENI† and GABOR LAURENCZY†

The chemistry of mixed-metal clusters[1] has received much attention because of the catalytic potential of these complexes[2]. Whereas the cluster anions

*Institut de Chimie, Université de Neuchâtel, CH -2000, Neuchâtel, Switzerland.
†Institut de Chimie Moleculaire et Biologigue, École Polytechnique Fédérale de Lausanne CH-1015 Lausanne, Switzerland.

$[Ru_3Co(CO)_{13}]^{-3}$ and $[Os_3Co(CO)_{13}]^{-4}$ have been known since 1980 from the reaction of $[Co(CO)_4]^-$ with $M_3(CO)_{12}$ (M = Ru or Os), the analogous reaction of $[Rh(CO)_4]^-$ with $Ru_3(CO)_{12}$ gave instead $[Ru_2Rh_2(CO)_{12}]$.[2–5] Here we report the synthesis of the compounds $[PPN][Ru_3Ir(CO)_{13}]$ and $[PPN][Os_3Ir(CO)_{13}]$ from the reaction of $[PPN][Ir(CO)_4]$ with $M_3(CO)_{12}$ (M = Ru or Os).[6]

■ **Caution:** *Toxic carbon monoxide is evolved during these reactions. Proper ventilation with a good fume hood is required.*

General Remarks

The metal carbonyls $Ru_3(CO)_{12}$[7] and $Os_3(CO)_{12}$[8] as well as $[PPN][Ir(CO)_4]$[9] are available according to published methods.

A. BIS(TRIPHENYLPHOSPHORANYLIDENE)AMMONIUM TRIDECACARBONYLTRIRUTHENIUMIRIDATE, $[PPN][IrRu_3(CO)_{13}]$

$$[PPN][Ir(CO)_4] + Ru_3(CO)_{12} \longrightarrow [PPN][IrRu_3(CO)_{13}] + 3\,CO$$

Procedure

In a 250-mL Schlenk tube under N_2 atmosphere, a mixture of $[PPN][Ir(CO)_4]$ (0.26 g, 0.31 mmol) and $Ru_3(CO)_{12}$ (0.2 g, 0.31 mmol) dissolved in tetrahydrofuran (50 mL) is refluxed for 1 h with vigorous stirring under N_2 atmosphere. After removal of the solvent, the brown residue is dissolved in diethyl ether (20 mL), and the resulting solution is filtered, then concentrated ~50% in volume. Addition of hexane (60 mL) to the stirred ether solution causes the precipitation of $[PPN][Ru_3Ir(CO)_{13}]$; it is isolated in pure form by decantation as a red-brown powder, which is washed with hexane (3 × 10 mL) and dried in vacuo (0.36 g, 84%).

Anal. Calcd. for $C_{49}H_{30}IrNO_{13}P_2Ru_3$: C, 42.09; H, 2.16; N. 1.00. Found: C, 42.07; H, 2.35; N, 0.92.

Properties

The product is a red-brown microcrystalline powder that appears to be stable toward air and moisture but should be stored under N_2. The compound is soluble in polar organic solvents such as diethyl ether, tetrahydrofuran, dichloromethane, or methanol. The IR spectrum presents four vibrations in the region of terminal carbonyl ligands and two absorptions that are attributed to bridging carbonyl groups [ether, ν_{CO} (cm^{-1}): 2068 (w), 2017 (vs), 1970 (s), 1924 (w), 1821 (m), 1803 (m)]. The ^{1}H NMR spectrum (200 MHz, $CDCl_3$) displays only the signals of the counterion $[PPN]^+$: δ 7.40–7.70 (30H, m). The 13 carbonyl ligands in the $[Ru_3Ir(CO)_{13}]^-$ anion are highly fluxional in solution. The ^{13}C NMR spectrum of $[PPN][Ru_3Ir(CO)_{13}]$ displays only one ^{13}C resonance in the carbonyl region at δ 201.6 ($CDCl_3$) down to -60°C. In the solid state, two isomers of $[Ru_3Ir(CO)_{13}]^-$ crystallize together. The single-crystal structure analysis reveals in the same crystal the presence of $[Ru_3Ir(\mu\text{-}CO)_2(CO)_{11}]^-$, in which 2 of the 13 carbonyls are bridging, and of $[Ru_3Ir(\mu\text{-}CO)_4(CO)_9]^-$, in which 4 of the 13 carbonyls are bridging.

B. BIS(TRIPHENYLPHOSPHORANYLIDENE)AMMONIUM TRIDECACARBONYLTRIOSMIUMIRIDATE, $[PPN][IrOs_3(CO)_{13}]$

$$[PPN][Ir(CO)_4] + Os_3(CO)_{12} \longrightarrow [PPN][IrOs_3(CO)_{13}] + 3CO$$

The compound $[PPN][Os_3Ir(CO)_{13}]$ is prepared in a 250 mL pressure Schlenk tube (Fig. 1) under N_2 atmosphere.

■ **Caution.** *An appropriate explosion shield should be used.*

A mixture of $[PPN][Ir(CO)_4]$ (0.40 g, 0.47 mmol) and $Os_3(CO)_{12}$ (0.43 g, 0.47 mmol) dissolved in tetrahydrofuran (100 mL) is stirred and heated at 100°C for 4 h; the carbon monoxide formed is released every 30 min. After removal of the solvent, the brown residue is dissolved in diethyl ether (40 mL), and the resulting solution is filtered and concentrated in volume by 50%. Addition of hexane (80 mL) to the stirred ether solution causes the precipitation of $[PPN][Os_3Ir(CO)_{13}]$; it is isolated in pure form by decantation as an orange-red powder, which is washed with hexane (3×15 mL) and dried in vacuo (0.42 g, 54%).

Anal. Calcd. for $C_{49}H_{30}IrNO_{13}P_2Os_3$: C, 35.32; H, 1.93; N, 0.85. Found: C, 35.34; H, 1.82; N, 0.84.

Properties

The compound is an orange-red microcrystalline powder, which appears to be stable toward air and moisture, but should be stored under N_2. The compound is soluble in polar organic solvents such as diethyl ether, tetrahydrofuran, dichloromethane or methanol. The IR spectrum presents three vibrations in the

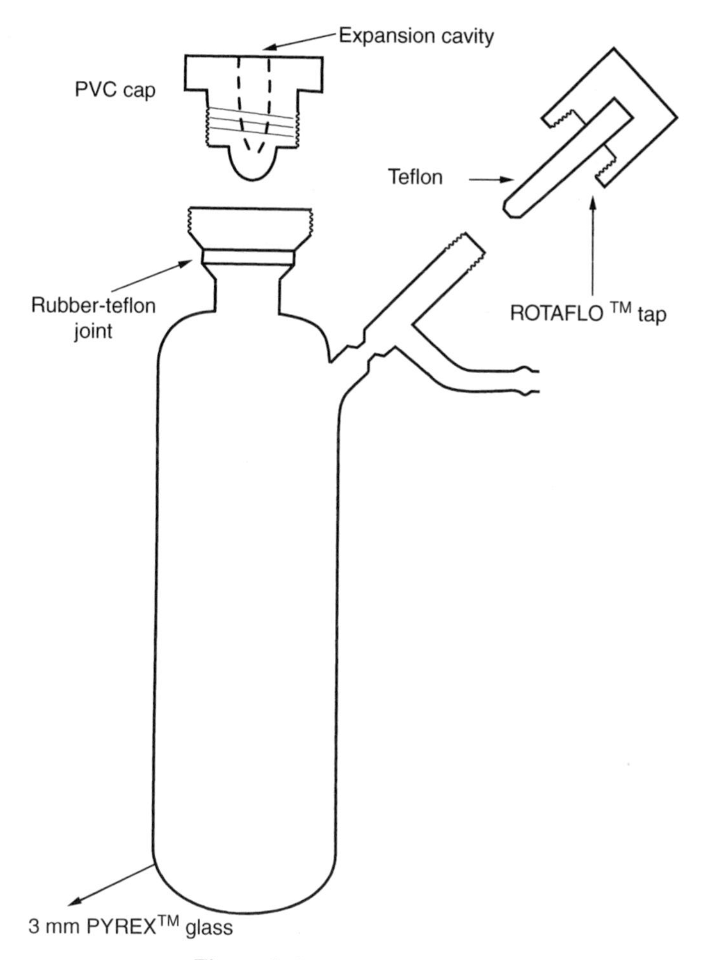

Figure 1. Pressure Schlenk tube.

region of the terminal carbonyl ligands and one in the bridging carbonyl region [KBr, ν_{CO} (cm^{-1}): 2068 (w), 2011 (vs), 1969 (s), 1790 (m)]. The ^{1}H NMR spectrum (200 MHz, $CDCl_3$) displays only the signals of the counter ion $[PPN]^+$: δ 7.40–7.70 (30H, m).

References

1. D. A. Roberts and G. L. Geoffrey, in *Comprehensive Organometallic Chemistry,* G. Wilkinson, F. G. A. Stone, and E. W. Abel, eds. Pergamon, Oxford, 1982, Vol. 2, p. 763; R. D. Adams, in *Comprehensive Organometallic Chemistry,* 2nd ed., E. W. Abel, F. G. A. Stone, E. W. Abel, and F. G. Wilkinson, eds., Elsevier, Oxford, 1995, Vol. 10, p. 1.

2. P. Braunstein and J. Rose, in *Comprehensive Organometallic Chemistry*, 2nd ed., E. W. Abel, and F. G. A. Stone, G. Wilinson, eds., Elsevier, Oxford, 1995, Vol. 10, p. 351.
3. P. C. Steinhardt, W. L. Gladfelter, A. D. Harley, J. R. Fox, and G. L. Geoffrey, *Inorg. Chem.* **19**, 332 (1980).
4. E. W. Burkhart and W. L. Geoffrey, *J. Organometal. Chem.* **198**, 179 (1989).
5. A. Fumagelli, D. Italia, M. C. Malatesta, G. Ciani, M. Moret, and A. Sironi, *Inorg. Chem.* **35**, 1765 (1996).
6. G. Süss-Fink, S. Haak, V. Ferrand, and H. Stoeckli-Evans, *J. Chem. Soc., Dalton Trans.* 3681 (1997).
7. M. I. Bruce, C. M. Jensen, and N. L. Jones, *Inorg. Synth.* **28**, 216 (1990).
8. S. R. Drake and P. A. Loveday, *Inorg. Synth.* **28**, 230 (1990).
9. L. Garlaschelli, R. della Pergola, and S. Martinengo, *Inorg. Synth.* **28**, 211 (1990).

46. TRI- AND HEXARUTHENIUM CARBONYL CLUSTERS

Submitted by ELENA CARIATI,[*] CLAUDIA DRAGONETTI,[*] ELENA LUCENTI,[†] and DOMINIQUE ROBERTO[*]
Checked by AURORA CASTRO[‡] and PETER M. MAITLIS[‡]

Inspired by the development of silica-mediated syntheses of various neutral [$Ru_3(CO)_{12}$, $H_4Ru_4(CO)_{12}$, and $Ru_3(CO)_{10}Cl_2$] or anionic [$[Ru_6C(CO)_{16}]^{2-}$, $[H_3Ru_4(CO)_{12}]^-$, $[HRu_3(CO)_{11}]^-$, and $[HRu_6(CO)_{18}]^-$] high-nuclearity ruthenium carbonyl clusters,[1] some of us have found that the majority of these ruthenium carbonyl clusters can be prepared efficiently by reductive carbonylation at atmospheric pressure, starting from $RuCl_3 \cdot nH_2O$ or $\{Ru(CO)_3Cl_2\}_2$ dissolved in ethylene glycol, a high-boiling solvent carrying nonacidic OH groups that could mimic the role and polarity of the OH groups of the silica surface.[2] In particular, easily reproducible yields are obtained by a two-step methodology, analogous to the two-step route to convert silica-supported $RuCl_3 \cdot nH_2O$ into Ru carbonyl clusters via [$Ru(CO)_3Cl_2(HOSi{\equiv})$].[1] This involves (1) first carbonylation of $RuCl_3 \cdot nH_2O$ to give species of the kind $Ru(CO)_xCl_2L$ ($x = 2, 3$; L = ethylene glycol) and (2) addition of specific amounts of alkali carbonates and further reductive carbonylation to give the desired cluster.[3] Neutral clusters are formed by working in the presence of the stoichiometric amount of alkali carbonate necessary to remove the chloro ligands from the coordination sphere of ruthenium, whereas anionic clusters are obtained by working with an excess of alkali carbonate.[2,3] Here details are given about the one-pot two-step synthesis in

*Dipartimento di Chimica Inorganica, Metallorganica e Analitica dell'Università di Milano, Unità di Ricerca dell'INSTM di Milano and Centro CIMAINA, 20133 Milano, Italy.
†Istituto di Scienze e Tecnologie Molecolari del CNR (ISTM), 20133, Milano, Italy.
‡Department of Chemistry, University of Sheffield, Sheffield S3 7HF, England.

ethylene glycol solution of neutral $Ru_3(CO)_{12}$ and anionic $[Ru_6C(CO)_{16}]^{2-}$ starting from $RuCl_3 \cdot nH_2O$.[2–3]

■ **Caution.** *Because of the toxicity of carbon monoxide and ruthenium carbonyl compounds, all manipulations should be carried out in an efficient fume hood, wearing gloves and eye protection.*

Reagents

$RuCl_3 \cdot nH_2O$ (41.20–42.68 wt% Ru) is available commercially from Engelhard. The exact content of water may be determined before each reaction by thermogravimetric analysis.

A. DODECACARBONYLTRIRUTHENIUM, $Ru_3(CO)_{12}$

$$RuCl_3 \cdot nH_2O + CO \xrightarrow{\text{ethylene glycol}} Ru(CO)_xCl_2(\text{ethylene glycol}) \xrightarrow[\text{CO}]{Na_2CO_3} Ru_3(CO)_{12}$$
$$(x = 2, 3)$$

Several methods have been reported in the literature for the synthesis of $Ru_3(CO)_{12}$.[5–8] Some of us have reported that $Ru_3(CO)_{12}$ is obtained in high yields by bubbling CO (1 atm) at 95°C through an ethylene glycol solution of $\{Ru(CO)_3Cl_2\}_2$ (93% yield) or $RuCl_3 \cdot nH_2O$ (70% yield) in the presence of Na_2CO_3 (molar ratio Na : Cl = 1 : 1).[2] Here we report details on an improved, easily reproducible, high-scale, one-pot, two-step synthesis involving species such as $Ru(CO)_xCl_2L$ ($x = 2$ or 3; L = ethylene glycol) as intermediates.[3]

Procedure

$RuCl_3 \cdot nH_2O$ (40.88 wt% Ru; 2.173 g, 8.790 mmol) is dissolved in ethylene glycol (600 mL) in a three-necked flask containing a magnetic stirring bar and equipped with a condenser and a thermometer. CO (1 atm) is bubbled through the resulting dark brown solution for 5 min and then is allowed to flow only on the top of the condenser. The reaction mixture is heated under CO at 110°C for 5 h. At this stage the solution is pale green, and its infrared spectrum shows carbonyl stretching bands at ν_{CO} = 2138 (sh), 2129 (m, br), 2066 (vs), 1998 (s) cm^{-1} due to the formation of species such as $Ru(CO)_xCl_2L$ ($x = 2,3$; L = ethylene glycol). After cooling the reaction mixture to 25°C, Na_2CO_3 (1.385 g, 13.086 mmol; molar ratio Na_2CO_3 : Ru = 3 : 2) is added, and CO (1 atm) is allowed to flow on the top of the condenser. Thermal treatment at 80°C for 15 h affords an orange material ($Ru_3(CO)_{12}$) that precipitates in the reaction flask, whereas, in parallel, some $Ru_3(CO)_{12}$ sublimes on the cold walls of the condenser.* Extraction of the sublimate and reaction mixture with dichloromethane [$\sim 4 \times 150$ mL, until no more $Ru_3(CO)_{12}$ is present in the CH_2Cl_2 phase] at room temperature, followed by evaporation to dryness of the dichloromethane phase, gives an orange solid residue, which is washed with water (3×10 mL) and then pentane (3×10 mL), in order to eliminate traces of ethylene glycol, affording spectroscopically pure $Ru_3(CO)_{12}$ (1.145 g; 1.790 mmol; 61 % yield). Yields vary slightly from preparation to preparation and are typically in the range 60–70%.

Anal. Calcd. for $C_{12}O_{12}Ru_3$: C, 22.53. Found: C, 22.44.

[*Note*: The reduction of $RuCl_3 \cdot nH_2O$ must be carried out with the stream of CO at the top of the condenser in order to obtain reproducible high yields of $Ru_3(CO)_{12}$. In fact, if the reaction is carried out by bubbling CO in the flask, a nonreproducible quantity of HCl is lost during the first step working at 110°C; consequently, during the second step, a considerable amount of anionic ruthenium carbonyl complexes can be formed as byproducts even using a molar ratio $Na_2CO_3 : Ru = 3:2$. In contrast the synthesis of $[Ru_6C(CO)_{16}]^{2-}$ can be carried out under CO bubbling, since it occurs in the presence of a large excess of alkali carbonate and therefore it is not sensitive to the exact quantity of base.]

Properties

$Ru_3(CO)_{12}$ is an orange air-stable crystalline solid, soluble in a wide variety of organic solvents, insoluble in water, and volatile.[5–8] Its IR spectrum, in hexane, shows ν_{CO} bands at 2061 (vs), 2031 (s), and 2011 (m) cm^{-1}. No band assignable to a bridging carbonyl ligand is observed.[5–8]

B. BIS(TETRABUTYLAMMONIUM) CARBIDOHEXADECACARBONYLHEXARUTHENATE, $[NBu_4]_2[Ru_6C(CO)_{16}]$

$$RuCl_3 \cdot nH_2O + CO \xrightarrow{\text{ethylene glycol}} Ru(CO)_xCl_2(\text{ethylene glycol}) \xrightarrow[CO]{K_2CO_3} K_2[Ru_6C(CO)_{16}]$$

$$(x = 2, 3)$$

$$2[NBu_4]I + K_2[Ru_6(CO)_{16}] \longrightarrow [NBu_4]_2[Ru_6(CO)_{16}] + 2KI$$

*Some brown unidentified complex (characterized by ν_{CO} at 1998 (vs) and 1975 (sh) cm^{-1} in Nujol mull) may precipitate along with $Ru_3(CO)_{12}$ in the reaction flask but, in contrast to the latter cluster, it is insoluble in CH_2Cl_2 and therefore readily eliminated.

The synthesis of $[Ru_6C(CO)_{16}]^{2-}$ usually requires $Ru_3(CO)_{12}$ as starting material (yields 60–90%).[9] It was found that treatment of $Ru(CO)_3Cl_2(HOSi\equiv)$, prepared in situ by carbonylation (1 atm of CO, 100°C) of silica-supported $RuCl_3 \cdot nH_2O$,[4] with 1 atm of CO at 150°C for 10 h in the presence of K_2CO_3 (molar ratio $K_2CO_3 : Ru = 10:1$) affords $K_2[Ru_6C(CO)_{16}]$ in very high yield (95%).[1] This silica-mediated synthesis led to new syntheses by bubbling CO through an ethylene glycol solution of $\{Ru(CO)_3Cl_2\}_2$ or $RuCl_3 \cdot nH_2O$ in the presence of K_2CO_3 at 160°C,[2] but it turned out later that the synthesis starting from $RuCl_3 \cdot nH_2O$ is not always reproducible. However, a modified one-pot, two-step synthesis involving species such as $Ru(CO)_xCl_2L$ ($x = 2, 3$; L = ethylene glycol) as intermediates gives reproducible, high yields of $K_2[Ru_6C(CO)_{16}]$, as detailed here.[3]

Procedure

$RuCl_3 \cdot nH_2O$ (41.2 wt% Ru; 2.01 g, 8.17 mmol) is dissolved at room temperature in ethylene glycol (550 mL) in a three-necked flask equipped with a condenser and a thermometer. CO (1 atm) is bubbled through the resulting dark brown solution, and the reaction mixture is heated under CO at 110°C. After 3–4 h the solution is pale green, and its infrared spectrum shows carbonyl stretching bands at ν_{CO} = 2138 (sh), 2129 (m, br), 2066 (vs), 1998 (s) cm^{-1} due to the formation of species such as $Ru(CO)_xCl_2L$ ($x = 2, 3$; L = ethylene glycol). The reaction mixture is cooled to room temperature, K_2CO_3 (11.29 g, 81.73 mmol; molar ratio $K_2CO_3 : Ru = 10:1$) is added and CO (1 atm) is bubbled through the resulting mixture. The solution is heated under CO at 160°C for 8 h, then cooled to room temperature. The infrared spectrum of the resulting dark red solution shows carbonyl stretching bands at ν_{CO} = 2034 (vw), 1977 (vs), 1953 (w, sh), 1917 (w) cm^{-1}, typical for $[Ru_6C(CO)_{16}]^{2-}$. Repeated extraction of this anion under N_2 with a solution (3×200 mL) of NBu_4I (2.22 g, 6.03 mmol) in CH_2Cl_2 (600 mL) gives a dark red solution.* A further extraction is carried out until the CH_2Cl_2 phase is no longer colored, with 2×50 mL of NBu_4I (0.952 g, 2.58 mmol) in CH_2Cl_2 (100 mL). Evaporation of the combined CH_2Cl_2 solutions affords an oily slurry that is washed with H_2O (3×50 mL) and pentane (2×20 mL) in order to obtain spectroscopically pure $[NBu_4]_2[Ru_6C(CO)_{16}]$ as a red powder (1.73 g, 1.11 mmol, 82% yield). Lower yields are obtained by working with higher concentrations.

*CH_2Cl_2 is partially miscible with ethylene glycol. For this reason, and because of the dark color of both phases, in the first extractions it is quite difficult to see a phase separation; at this stage the difference in the viscosity of the solvents is quite useful for their separation.

Anal. Calcd. for $C_{49}H_{72}N_2O_{16}Ru_6$: C, 37.92; H, 4.64; N, 1.81. Found : C, 37.83; H, 4.67; N, 1.81.

Properties

$[NBu_4]_2[Ru_6C(CO)_{16}]$ is obtained as a red, slightly air-sensitive powder, soluble in a wide variety of organic solvents and insoluble in water. Its IR spectrum, in CH_2Cl_2, shows ν_{CO} bands at 2030 (w), 1978 (s), 1952 (sh, m), 1918 (m), 1820 (sh, m), 1780 (m) cm^{-1}.[9c] The ^{13}C NMR shows ^{13}C resonances at δ 213 (carbonyls) and 459 (interstitial carbon atom).[9b]

References

1. D. Roberto, E. Cariati, E. Lucenti, M. Respini, and R. Ugo, *Organometallics* **16**, 4531 (1997).
2. C. Roveda, E. Cariati, E. Lucenti, and D. Roberto, *J. Organomet. Chem.* **580**, 117 (1999).
3. E. Lucenti, E. Cariati, C. Dragonetti, and D. Roberto, *J. Organomet. Chem.* **669**, 44 (2003).
4. D. Roberto, R. Psaro, and R. Ugo, *Organometallics* **12**, 2292 (1993).
5. (a) J. L. Dawes and I. D. Holmes, *Inorg. Nucl. Chem. Lett.* **7**, 847 (1971); (b) A. Mantovani and S. Cenini, *Inorg. Synth.* **16**, 47 (1976).
6. (a) B. R. James and G. L. Rempel, *Chem. Indust.* 1036 (1971); (b) B. R. James, G. L. Rempel, and W. K. Teo, *Inorg. Synth.* **16**, 45 (1976).
7. (a) M. I. Bruce and F. G. A. Stone, *Chem. Commun.* 684 (1966); (b) M. I. Bruce and F. G. A. Stone, *J. Chem. Soc. A* 1238 (1967); (c) C. R. Eady, P. F. Jackson, B. F. G. Johnson, J. Lewis, M. C. Malatesta, M. McPartlin, and W. J. H. Nelson, *J. Chem. Soc., Dalton Trans.* 383 (1980); (d) M. I. Bruce, J. G. Matisons, R. C. Wallis, J. M. Patrick, B. W. Skelton, and A. H. White, *J. Chem. Soc., Dalton Trans.* 2365 (1983); (e) M. I. Bruce, C. M. Jensen, and N. L. Jones, *Inorg. Synth.* **28**, 216 (1987).
8. B. F. G. Johnson and J. Lewis, *Inorg. Synth.* **13**, 92 (1972).
9. (a) B. F. G. Johnson, J. Lewis, S. W. Sankey, K. Wong, M. McPartlin, and W. J. H. Nelson, *J. Organomet. Chem.* **191**, C3 (1980); (b) J. S. Bradley, G. B. Ansell, and E. W. Hill, *J. Organomet. Chem.* **184**, C33 (1980); (c) C. M. T. Hayward and J. R. Shapley, *Inorg. Chem.* **21**, 3816 (1982); (d) S. H. Han, G. L. Geoffroy, B. D. Dombek, and A. L. Rheingold, *Inorg. Chem.* **27**, 4355 (1988).

47. SILICA-MEDIATED SYNTHESIS OF $Os_3(CO)_{10}(\mu\text{-}H)(\mu\text{-}OH)$

Submitted by CLAUDIA DRAGONETTI,[*] ELENA LUCENTI,[†] and DOMINIQUE ROBERTO[*]
Checked by WENG KEE LEONG[‡] and QI LIN[‡]

$$Os_3(CO)_{12} + HOSi{\equiv} \longrightarrow Os_3(CO)_{10}(\mu\text{-H})(\mu\text{-OSi}{\equiv}) + 2\,CO$$

$$Os_3(CO)_{10}(\mu\text{-H})(\mu\text{-OSi}{\equiv}) + H_2O \longrightarrow Os_3(CO)_{10}(\mu\text{-}H)(\mu\text{-OH}) + HOSi{\equiv}$$

The synthesis of $Os_3(CO)_{10}(\mu\text{-H})(\mu\text{-OH})$ was reported to occur in solution by reaction of $Os_3(CO)_{12}$ with $NaBH_4$ (27% yield)[1] or by hydrolysis of the reactive intermediates $Os_3(CO)_{10}(\mu\text{-H})(\mu\text{-OCH=CH}_2)$ [36% yield; total yield starting from $Os_3(CO)_{12}$, 20%],[2] $Os_3(CO)_{10}$(cyclohexa-1,3-diene) [20% yield; total yield starting from $Os_3(CO)_{12}$, 9%],[3] or $Os_3(CO)_{10}(\mu\text{-H})(\mu\text{-NCHNMe}_2)$ [60% yield; total yield starting from $Os_3(CO)_{12}$, 33%].[4] The facile activation of $Os_3(CO)_{12}$ by the surface of silica, via reaction with surface silanol groups to give silica-anchored $Os_3(CO)_{10}(\mu\text{-H})(\mu\text{-OSi}{\equiv})$ in nearly quantitative yield,[5] provided subsequent alternative routes to the synthesis of $Os_3(CO)_{10}(\mu\text{-H})(\mu\text{-OH})$. In fact, $Os_3(CO)_{10}(\mu\text{-H})(\mu\text{-OH})$ was obtained in fair yields [56% yield starting from $Os_3(CO)_{12}$] by treatment of $Os_3(CO)_{10}(\mu\text{-H})(\mu\text{-OSi}{\equiv})$ with aqueous HF, which dissolves silica.[6] Excellent and much higher yields [total yields starting from $Os_3(CO)_{12}$, 91%] were however achieved by mild hydrolysis of $Os_3(CO)_{10}(\mu\text{-H})(\mu\text{-OSi}{\equiv})$ at 95°C in a biphasic water/toluene system.[7] This latter synthetic method is described here. The compound $Os_3(CO)_{10}(\mu\text{-H})(\mu\text{-OH})$ is a convenient starting material for the synthesis of a large variety of clusters of the type $Os_3(CO)_{10}(\mu\text{-H})(\mu\text{-Y})$ (Y = a three-electron donor).[7–9]

■ **Caution.** *Because of the toxicity of osmium carbonyl compounds, all manipulations should be carried out in an efficient fume hood, wearing gloves and eye protection.*

Reagents

A nonporous SiO_2 (Aerosil 200 Degussa, with a nominal surface area of 200 m^2/g)[#] is used for the synthesis of the intermediate $Os_3(CO)_{10}(\mu\text{-H})(\mu\text{-OSi}{\equiv})$. $Os_3(CO)_{12}$

[*]Dipartimento di Chimica Inorganica, Metallorganica e Analitica dell'Università di Milano, Unità di Ricerca dell' INSTM di Milano and Centro CIMAINA, 20133 Milano, Italy.
[†]Istituto di Scienze e Tecnologie Molecolari del CNR (ISTM), 20133 Milano, Italy.
[‡]Department of Chemistry, National University of Singapore, 119260 Singapore.
[#]The checkers employed conventional chromatography grade silicagel (Merck, silicagel 60, 230–400 mesh) with comparable results.

is available commercially from Aldrich or Strem; it may also be prepared according to the literature.[10] All solvents are deoxygenated by bubbling N_2 for 15–30 min.

A. DECACARBONYLTRIOSMIUM ON SILICA, $Os_3(CO)_{10}(\mu\text{-}H)(\mu\text{-}OSi\equiv)$

Procedure

8.14 g of SiO_2 are added to a 1-L three-necked, round-bottomed flask, equipped with a magnetic stirring bar, and kept under vacuum (10^{-2} torr) at 25°C for 3 h. The flask is refilled with N_2 and, under a gentle stream of N_2, 0.270 g (0.298 mmol; 2.08 wt% Os with respect to SiO_2) of $Os_3(CO)_{12}$ and 600 mL of deoxygenated *n*-octane are added. The flask is then equipped with a condenser, and the resulting slurry is stirred *vigorously* under N_2 at reflux temperature until the solvent is decolorized (10–12 h) because of the anchoring of the yellow $Os_3(CO)_{12}$ cluster on the silica surface to produce $Os_3(CO)_{10}(\mu\text{-}H)(\mu\text{-}OSi\equiv)$. The reaction mixture is cooled to room temperature, filtered under N_2 in a 1-L Pyrex Büchner filter funnel with fritted disk, and washed with 3 × 40 mL of deoxygenated dichloromethane to remove traces of unreacted $Os_3(CO)_{12}$. The silica is dried under vacuum (10^{-2} torr) to give a pale yellow powder, which is kept under N_2. The osmium loading on the silica surface (2.01 wt% Os with respect to SiO_2) is indirectly established by evaporation of the combined filtrates, affording unreacted $Os_3(CO)_{12}$. The osmium loading varies slightly from preparation to preparation and is typically in the range 1.90–2.04 wt% Os with respect to SiO_2 [yields of $Os_3(CO)_{10}(\mu\text{-}H)(\mu\text{-}OSi\equiv)$ are in the range 91–98%].

B. DECACARBONYL-μ-HYDRIDO-μ-HYDROXOTRIOSMIUM, $Os_3(CO)_{10}(\mu\text{-}H)(\mu\text{-}OH)$

In this procedure, 7.56 g (2.04 wt% Os with respect to SiO_2; 0.783 mmol of Os) of silica-anchored $Os_3(CO)_{10}(\mu\text{-}H)(\mu\text{-}OSi\equiv)$, 200 mL of doubly distilled deoxygenated water, and 300 mL of deoxygenated toluene are added under N_2 to a 1-L two-necked, round-bottomed flask equipped with a condenser, a thermometer, and a magnetic stirring bar. The mixture is stirred *vigorously* at 95°C under N_2 for 5 h. The silica powder and the organic phase become white and yellow, respectively, due to the formation of $Os_3(CO)_{10}(\mu\text{-}H)(\mu\text{-}OH)$, which goes into the organic phase. After cooling to room temperature, the toluene phase is separated and the suspension of silica in water extracted with 2 × 70 mL of dichloromethane to recover the last traces of $Os_3(CO)_{10}(\mu\text{-}H)(\mu\text{-}OH)$ absorbed on silica. Evaporation of the combined organic extracts affords $Os_3(CO)_{10}(\mu$-H)(μ-OH) contaminated by traces of silica and, in some cases, of high-boiling organic compounds. Addition of ~10–15 mL of dichloromethane to this residue, followed by filtration (to remove silica) through a 25-mL Pyrex Büchner filter funnel with fritted disk, and evaporation of the solution to dryness, affords crude $Os_3(CO)_{10}(\mu$-H)(μ-OH). Washing the residue with 1 mL of pentane (to remove high-boiling organic compounds) gives the spectroscopically pure cluster (0.213 g; 0.245 mmol; 94% yield; total yield starting from $Os_3(CO)_{12}$, 91%). Alternatively, crude $Os_3(CO)_{10}(\mu$-H)(μ-OH) can be purified by column chromatography on silicagel (column diameter 20 mm, length 250 mm) using CH_2Cl_2 as eluant. Total yields starting from $Os_3(CO)_{12}$ vary slightly from preparation to preparation and are typically in the range 85–91%.

Properties

The complex $Os_3(CO)_{10}(\mu$-H)(μ-OH) is obtained as an orange air-stable powder, which is soluble in most organic solvents, but insoluble in water. The IR spectrum, in CH_2Cl_2, shows ν_{CO} bands at 2112 (w), 2070 (vs), 2060 (s), 2021 (vs), 2000 (m), and 1983 (m) cm^{-1}, whereas the ^1H NMR spectrum, in $CDCl_3$ solution, has resonances at δ −12.64 (H–Os) and 0.14 (OH) ppm. In the electrospray ionization mass spectrum, there is the molecular ion peak at $m/z = 870\ [M]^+$.

References

1. B. F. G. Johnson, J. Lewis, and P. A. Kilty, *J. Chem. Soc. A* 2859 (1968).
2. A. J. Arce, A. J. Deeming, S. Donovan-Mtunzi, and S. E. Kabir, *J. Chem. Soc., Dalton Trans.* 2479 (1985).
3. E. G. Bryan, B. F. G. Johnson, and J. Lewis, *J. Chem. Soc., Dalton Trans.* 1328 (1977).
4. J. Banford, M. J. Mays, and P. R. Raithby, *J. Chem. Soc., Dalton Trans.* 1355 (1985).
5. R. Barth, B. C. Gates, H. Knözinger, and J. Hulse, *J. Catal.* **8**, 147 (1983).
6. C. Dossi, A. Fusi, M. Pizzotti, and R. Psaro, *Organometallics* **9**, 1994 (1990).
7. D. Roberto, E. Lucenti, C. Roveda, and R. Ugo, *Organometallics* **16**, 5974 (1997).
8. E. Lucenti, D. Roberto, C. Roveda, R. Ugo, and A. Sironi, *Organometallics* **19**, 1051 (2000).
9. M. W. Lum and W. K. Leong, *J. Chem. Soc., Dalton Trans.* 2476 (2001).
10. C. Roveda, E. Cariati, E. Lucenti, and D. Roberto, *J. Organomet. Chem.* **580**, 117 (1999) and refs. therein.

48. EFFICIENT BASE- AND SILICA-MEDIATED SYNTHESES OF OSMIUM CLUSTER ANIONS FROM α-{$Os(CO)_3Cl_2$}$_2$

**Submitted by ELENA CARIATI,[*] CLAUDIA DRAGONETTI,[*]
ELENA LUCENTI,[†] and DOMINIQUE ROBERTO[*]
Checked by BRIAN F. G. JOHNSON[‡]**

Although the reported syntheses in solution of high-nuclearity osmium carbonyl clusters usually require $Os_3(CO)_{12}$ as starting material, it has been reported[1,2] that various neutral or anionic osmium carbonyl clusters can be synthesized in high yields and rather easily by controlled reduction at atmospheric pressure of silica-supported α-{$Os(CO)_3Cl_2$}$_2$ or silica-bound $Os(CO)_3Cl_2(HOSi\equiv)$ (generated in situ by controlled reductive carbonylation of silica-supported $OsCl_3 \cdot 3H_2O$)[3] in the presence of specific amounts of alkali carbonates. These silica-mediated syntheses were the springboard to new convenient syntheses of $Os_3(CO)_{12}$, $H_4Os_4(CO)_{12}$, $[H_3Os_4(CO)_{12}]^-$, and $[H_4Os_{10}(CO)_{24}]^{2-}$ starting from $OsCl_3 \cdot 3H_2O$ or α-{$Os(CO)_3Cl_2$}$_2$ and working in ethylene glycol solution, a high-boiling solvent carrying nonacidic OH groups that could mimic the role and polarity of the OH groups of the silica surface.[4-5] However, because addition of bases to glycols working at high temperatures (higher than $\sim$200°C) leads to degradation of the solvent with exothermic reactions proceeding rapidly and uncontrollably,[6,7] glycols cannot be used as a safe reaction medium for the synthesis of high-nuclearity anionic osmium carbonyl clusters that require both high temperatures and strong basic conditions. Obviously this inconvenience does not exist with the silica surface as a reaction medium. Therefore, when high temperatures and basic conditions are required, as in the case of the synthesis of $[Os_{10}C(CO)_{24}]^{2-}$ and $[H_2Os_4(CO)_{12}]^{2-}$ by reductive carbonylation of α-{$Os(CO)_3Cl_2$}$_2$ in the presence of alkali carbonates (200°C; molar ratio Na_2CO_3 or $K_2CO_3 : Os = 10–20 : 1$),[2] the use of the silica surface as a convenient and safe reaction medium is recommended. Here we give details on the silica-mediated synthesis of α-{$Os(CO)_3Cl_2$}$_2$,[3] $[Os_{10}C(CO)_{24}]^{2-}$ and $[H_2Os_4(CO)_{12}]^{2-}$,[2] and on the synthesis in ethylene glycol solution of $[H_4Os_{10}(CO)_{24}]^{2-}$.[5]

■ **Caution.** *Because of the toxicity of carbon monoxide and of osmium carbonyl compounds, and the flamability of hydrogen, all manipulations should be carried out in an efficient fume hood, wearing gloves and eye protection.*

[*]Dipartimento di Chimica Inorganica, Metallorganica e Analitica dell'Università di Milano, Unità di Ricerca dell' INSTM di Milano and Centro CIMAINA, 20133 Milano, Italy.
[†]Istituto di Scienze e Tecnologie Molecolari del CNR (ISTM), 20133 Milano, Italy.
[‡]Chemistry Laboratories, University of Cambridge, Cambridge, CB2 IEW England.

Reagents

Nonporous SiO_2 (Aerosil 200 Degussa, with a nominal surface area of 200 m^2/g) is available commercially. $OsCl_3 \cdot 3H_2O$ can be purchased from Strem Chemicals or Sigma–Aldrich.

A. HEXACARBONYL(DI-μ-CHLORO)(DICHLORO)DIOSMIUM, α-$\{Os(CO)_3Cl_2\}_2$

$$OsCl_3 \cdot nH_2O + CO \xrightarrow{SiO_2} \alpha\text{-}\{Os(CO)_3Cl_2\}_2$$

Several routes for preparation of α-$\{Os(CO)_3Cl_2\}_2$ have been reported.[8] An efficient way to prepare the compound is by heating anhydrous or hydrated $OsCl_3$ physisorbed on silica at 180°C under static CO (1 atm). During the reaction chemisorbed $Os(CO)_3Cl_2(HOSi\equiv)$ is formed, whereas some α-$\{Os(CO)_3Cl_2\}_2$ and *cis*-$Os(CO)_4Cl_2$ sublime onto the cold walls of the reactor. However, extraction of the sublimate and of the silica powder with hot $CHCl_3$ affords only α-$\{Os(CO)_3Cl_2\}_2$ (80–90% yield), because any *cis*-$Os(CO)_4Cl_2$ is thermally converted to α-$\{Os(CO)_3Cl_2\}_2$.[3] The details of this silica-mediated synthesis are reported as follows.

Procedure

First, 4.43 g of SiO_2 is added to a 500-mL two-necked round-bottomed flask, equipped with a magnetic stirring bar, and treated under vacuum (10^{-2} torr) at 25°C for 3 h. The flask is refilled with N_2 and, under a gentle stream of N_2, 1.13 mmol (5.0 wt% Os with respect to SiO_2) of anhydrous $OsCl_3$ (0.335 g) or $OsCl_3 \cdot 3H_2O$ (0.396 g) and 150 mL of deoxygenated water are added. The resulting slurry is stirred overnight at room temperature under N_2 and then dried under vacuum (10^{-2} torr), using a water bath at 80°C, to give a gray powder. The powder is placed into a cylindrical Pyrex vessel (diameter 60 mm, length 350 mm), originally described for the reductive carbonylation of silica-supported metal chlorides at atmospheric pressure,[3] evacuated (10^{-2} torr) at room temperature and then exposed to CO at atmospheric pressure. The bottom of the vessel (about half of the cylinder) is put into an oven and heated at 180°C until the silica powder becomes completely white (about 2 days). During the reaction some α-$\{Os(CO)_3Cl_2\}_2$ and *cis*-$Os(CO)_4Cl_2$ sublime on the cold walls of the vessel, whereas $Os(CO)_3Cl_2(HOSi\equiv)$ is formed on the silica surface. The sublimate and the silica powder are combined and extracted with $CHCl_3$ (150 mL) in a Soxhlet apparatus affording pure α-$\{Os(CO)_3Cl_2\}_2$ (0.352 g; 0.510 mmol;

90% yield). A similar yield is obtained on a three-fold scale. Alternatively, the sublimate and the silica powder can be extracted with acetone (150 mL). Evaporation of the solvent followed by a 2-h reflux in chloroform (to convert any remaining *cis*-$Os(CO)_4Cl_2$ into α-$\{Os(CO)_3Cl_2\}_2$) and addition of pentane at room temperature affords α-$\{Os(CO)_3Cl_2\}_2$ as a white crystalline powder.[3]

Anal. Calcd. for $C_6O_6Cl_4Os_2$: C, 10.43. Found : C, 10.39.

Properties

α-$\{Os(CO)_3Cl_2\}_2$ is a white air-stable crystalline solid, soluble in a wide variety of organic solvents but insoluble in water. Its IR spectrum, in chloroform, shows ν_{CO} bands at 2137 (m) and 2064 (s) cm^{-1}.[1,8]

B. POTASSIUM DODECACARBONYLDI(μ-HYDRIDO) TETRAOSMATE, $K_2[H_2Os_4(CO)_{12}]$

$$\alpha\text{-}\{Os(CO)_3Cl_2\}_2 + CO \xrightarrow{SiO_2, K_2CO_3} K_2[H_2Os_4(CO)_{12}]$$

$[H_2Os_4(CO)_{12}]^{2-}$ was prepared for the first time in 39% yield by reduction of $Os_3(CO)_{12}$ with sodium borohydride in refluxing dioxane for 4 h.[9] Although $H_4Os_4(CO)_{12}$ is readily deprotonated by KOH in methanol to form the soluble monoanion $[H_3Os_4(CO)_{12}]^-$,[10] removal of a second proton by a base is difficult, and there is no report of the preparation in solution of $[H_2Os_4(CO)_{12}]^{2-}$ by this route.[11] However, Gates et al.[11] reported that deprotonation of $H_4Os_4(CO)_{12}$ at room temperature on highly dehydroxylated MgO gives a mixture of the supported anions $[H_3Os_4(CO)_{12}]^-$ and $[H_2Os_4(CO)_{12}]^{2-}$. More recently some of us found a convenient high-yield route to $[H_2Os_4(CO)_{12}]^{2-}$ by heating at 200°C α-$\{Os(CO)_3Cl_2\}_2$ supported on silica doped with excess K_2CO_3 (molar ratio $K_2CO_3 : Os = 20 : 1$) under 1 atm of CO.[2] Here are reported details of this silica-mediated synthesis.

Procedure

1. Support of α-$[Os(CO)_3Cl_2]_2$ and K_2CO_3 on silica: 7.65 g of SiO_2 is added to a 500-mL two-necked round-bottomed flask, equipped with a magnetic stirring bar, and treated under vacuum (10^{-2} torr) at 25°C for 3 h. The flask is refilled with N_2 and, under a gentle stream of N_2, 0.278 g (0.403 mmol; 2.0 wt% Os with respect to SiO_2) of α-$\{Os(CO)_3Cl_2\}_2$, 2.55 g (18.45 mmol; molar ratio $K_2CO_3 : Os = 23 : 1$) of powdered K_2CO_3 and 250 mL of deoxygenated (by

bubbling N_2 for 15 min) dichloromethane are added. The resulting slurry is stirred vigorously under N_2 at room temperature for 2 days. The solvent is then evaporated at room temperature under vacuum (10^{-2} torr), affording a white powder, which is stored under N_2.

2. Preparation of $[H_2Os_4(CO)_{12}]^{2-}$: 9.01 g (2.0 wt% Os with respect to SiO_2; 0.693 mmol of Os) of α-$\{Os(CO)_3Cl_2\}_2$ supported on silica doped with K_2CO_3 is placed into a cylindrical Pyrex vessel (diameter 60 mm, length 350 mm), originally described for the reductive carbonylation of silica-supported metal chlorides at atmospheric pressure,[3] evacuated (10^{-2} torr) at room temperature and then exposed to static CO at atmospheric pressure. The bottom of the vessel (about half of the cylinder) is put into an oven and heated at 200°C for 2 days. After cooling to room temperature, 200 mL of acetonitrile (deoxygenated and dried over 4-Å molecular sieves activated under vacuum at 300°C) is added under N_2 in the cylindrical vessel, and the resulting slurry is filtered under N_2 through a Pyrex Büchner filter funnel with fritted disk. Evaporation of the solvent under vacuum (10^{-2} torr) at room temperature gives spectroscopically pure $K_2[H_2Os_4(CO)_{12}]$ (0.177 g; 0.150 mmol; 87% yield), which must be kept under N_2. Yields vary slightly from preparation to preparation and are typically in the range 87–92%.

Anal. Calcd. for $C_{12}H_2O_{12}K_2Os_4$: C, 12.23; H, 0.17. Found: C, 12.28; H, 0.18.

Properties

The cluster $K_2[H_2Os_4(CO)_{12}]$ is obtained as an orange air-sensitive powder, soluble in polar solvents such as acetone and acetonitrile. The IR spectrum, in CH_3CN, shows ν_{CO} bands at 2042 (vw), 1984 (vs), 1954 (s), 1933 (s), and 1881 (m) cm^{-1} whereas the ^1H NMR spectrum, in CD_3CN solution, has a signal at δ −20.9 (H–Os). As the $[(Ph_3P)_2N]^+$ salt, the IR spectrum in CH_2Cl_2 shows ν_{CO} bands at 2012 (vw), 1978 (vs), 1948 (s), 1928 (s), and 1871 (m) cm^{-1}, whereas the ^1H NMR spectrum, in CD_3COCD_3 solution, has a signal at δ −20.58 ppm (H–Os).[9] Acidification of an acetonitrile solution of $K_2[H_2Os_4(CO)_{12}]$ with a few drops of concentrated H_2SO_4 affords quantitatively $H_4Os_4(CO)_{12}$.[2]

C. SODIUM CARBIDOTETRACOSACARBONYLDECAOSMATE, $Na_2[Os_{10}C(CO)_{24}]$

$$\alpha\text{-}\{Os(CO)_3Cl_2\}_2 + H_2 \xrightarrow{SiO_2, Na_2CO_3} Na_2[Os_{10}C(CO)_{24}]$$

By treatment of H_2OsCl_6 adsorbed on MgO with 1 atm of CO + H_2 at 275°C for 5 h, Gates et al.[12] reported the synthesis of the cluster anion $[Os_{10}C(CO)_{24}]^{2-}$ in a yield (65%) equivalent to that obtained in conventional syntheses in solution by reaction of $Os_3(CO)_{12}$ with Na (in tetraglyme, 230°C, 70 h)[13] or by pyrolysis of $Os_3(CO)_{11}(C_5H_5N)$ (vacuum, sealed tube, 250°C, 64 h)[14]. More recently some of us found that a convenient high-yield route to $[Os_{10}C(CO)_{24}]^{2-}$ is by heating at 200°C α-$\{Os(CO)_3Cl_2\}_2$ supported on silica doped with excess Na_2CO_3 under 1 atm of H_2 for 24 h.[2] The details of this silica-mediated synthesis are as follows.

Procedure

1. Support of α-$\{Os(CO)_3Cl_2\}_2$ and Na_2CO_3 on silica: 8.74 g of SiO_2 is added to a 500-mL two-necked round-bottomed flask, equipped with a magnetic stirring bar, and treated under vacuum (10^{-2} torr) at 25°C for 3 h. The flask is refilled with N_2 and, under a gentle stream of N_2, 0.318 g (0.461 mmol; 2.0 wt% Os with respect to SiO_2) of α-$[Os(CO)_3Cl_2]_2$, 0.975 g (9.20 mmol; molar ratio $Na_2CO_3:Os = 10:1$) of powdered Na_2CO_3 and 250 mL of deoxygenated (by bubbling N_2 for 15 min) dichloromethane are added. The resulting slurry is stirred vigorously under N_2 at room temperature for 2 days. The solvent is evaporated at room temperature under vacuum (10^{-2} torr), affording a white powder, which is stored under N_2.

2. Preparation of $[Os_{10}C(CO)_{24}]^{2-}$: 8.20 g (2.0 wt% Os with respect to SiO_2; 0.753 mmol of Os) of α-$\{Os(CO)_3Cl_2\}_2$ supported on silica + Na_2CO_3 is placed into the cylindrical Pyrex vessel (diameter 60 mm, length 350 mm), originally described for the reductive carbonylation of silica-supported metal chlorides at atmospheric pressure,[3] evacuated (10^{-2} torr) at room temperature and then exposed to static H_2 at atmospheric pressure. The bottom of the vessel (about half of the cylinder) is put into an oven and heated at 200°C for 24 h. During the reaction some $H_4Os_4(CO)_{12}$ sublimes as a pale yellow powder onto the cold part of the reaction vessel, whereas the silica powder becomes pink because of the formation of $Na_2[Os_{10}C(CO)_{24}]$. After cooling to room temperature, the silica powder is transferred under N_2 to a Pyrex Büchner filter funnel with fritted disk and extracted under N_2 with 100 mL of acetonitrile (deoxygenated by bubbling N_2 for 15 min). Evaporation of the solvent under vacuum (10^{-2} torr) at room temperature gives $Na_2[Os_{10}C(CO)_{24}]$ contaminated with some $Na[H_3Os_4(CO)_{12}]$ and acetamide. Washing of this mixture with 5 mL of dichloromethane removes the acetamide and $Na[H_3Os_4(CO)_{12}]$, which appears to be more soluble in the presence of the acetamide. Spectroscopically pure $Na_2[Os_{10}C(CO)_{24}]$ (0.149 g; 0.057 mmol; 75% yield) is thus obtained.

Anal. Calcd. for $C_{25}O_{24}Na_2Os_{10}$: C, 11.40. Found: C, 11.35.

[*Note:* The selectivity to $Na_2[Os_{10}C(CO)_{24}]$ decreases when this synthesis is carried out working with high metal loadings, as the amount of $H_4Os_4(CO)_{12}$ is much higher. For example, using a sample with 15 wt% Os/SiO_2 $Na_2[Os_{10}C(CO)_{24}]$ and $H_4Os_4(CO)_{12}$ are obtained in 39% and 33% yields, respectively.]

Properties

The cluster $Na_2[Os_{10}C(CO)_{24}]$ is obtained as a dark red-brown air-stable powder, soluble in polar organic solvents such as acetonitrile and acetone. In the fast-atom-bombardment (FAB) mass spectrum, using nitrobenzyl alcohol as the matrix, there is the molecular ion peak of $[Os_{10}C(CO)_{24}]^{2-}$ at m/z = 2586.[2] The IR spectrum in CH_3CN, as the Na^+ salt, shows ν_{CO} bands at 2038 (s) and 1990 (s) cm^{-1}, whereas that in acetone, as the $[Et_4N]^+$ salt,[13] shows ν_{CO} bands at 2034 (s) and 1992 (s) cm^{-1}. The ^{13}C NMR spectrum in CD_2Cl_2, as the $[(Ph_3P)_2N]^+$ salt, shows two signals of equal intensity due to two different sets of CO ligands at 189.9 and 178.2 ppm (these signals remain unchanged on cooling the solution from 30°C to −60°C, indicating that the molecule is not fluxional in solution), in agreement with the X-ray structure of $[Os_{10}C(CO)_{24}]^{2-}$, whereas no resonance due to the encapsulated carbon atom could be detected.[14b]

D. SODIUM TETRACOSACARBONYLTETRA(μ-HYDRIDO) DECAOSMATE, $Na_2[H_4Os_{10}(CO)_{24}]$

$$\alpha\text{-}\{Os(CO)_3Cl_2\}_2 + H_2 \xrightarrow{\text{ethylene glycol, } Na_2CO_3} Na_2[H_4Os_{10}(CO)_{24}]$$

$[H_4Os_{10}(CO)_{24}]^{2-}$ was first obtained in very low yields, along with other high-nuclearity clusters, by heating under reflux a solution of $Os_3(CO)_{12}$ in isobutanol.[15a] More efficient routes (yields ~20%) to this anionic cluster involving the thermolysis of derivatives of $Os_3(CO)_{12}$ in alcohols were reported only as unpublished results in a reference.[15b] More recently much better yields (81%) have been obtained by treatment of an ethylene glycol solution of α-$\{Os(CO)_3Cl_2\}_2$ and Na_2CO_3 (molar ratio Na_2CO_3 : Os = 1 : 1) at 160°C under 1 atm of H_2 for 6 h, details for which are described here.[5]

Procedure

First, α-$\{Os(CO)_3Cl_2\}_2$ (0.188 g; 0.272 mmol) is dissolved in ethylene glycol (40 mL) in a three-necked flask containing a magnetic stirring bar and equipped with a condenser and a thermometer. Na_2CO_3 (0.058 g; 0.543 mmol; molar ratio $Na_2CO_3 : Os = 1 : 1$) is added, and H_2 (1 atm) is bubbled through the resulting solution. After heating at 160°C under H_2 bubbling for ~6 h, the reaction is complete, as evidenced by the infrared spectrum of the solution, which shows the carbonyl bands characteristic of $Na_2[H_4Os_{10}(CO)_{24}]$. In some cases, $H_4Os_4(CO)_{12}$ is formed as byproduct; this neutral cluster is present as a precipitate and as a sublimate on the cold walls of the condenser. Extraction of the precipitate and sublimate with dichloromethane (~15–20 mL), followed by evaporation to dryness, gives a yellow solid residue, which is washed with water (2 × 2 mL) in order to eliminate traces of ethylene glycol to afford $H_4Os_4(CO)_{12}$ (0.009 g; 0.008 mmol; 6% yield). On the other hand, the ethylene glycol solution is evaporated to dryness (10^{-2} torr), and the resulting residue is extracted with acetone (~15–20 mL), cooled to 0°C, and filtered, in order to remove NaCl and in some cases traces of $H_4Os_4(CO)_{12}$. Evaporation to dryness of the acetone solution, followed by washing with water (2 × 2 mL) to eliminate traces of ethylene glycol, affords spectroscopically pure $Na_2[H_4Os_{10}(CO)_{24}]$ (0.115 g; 0.044 mmol; 81% yield) as a dark brown powder.*

Anal. Calcd. for $C_{24}H_4O_{24}Na_2Os_{10}$: C, 10.98; H, 0.15. Found: C, 11.02; H, 0.16.

Properties

$Na_2[H_4Os_{10}(CO)_{24}]$ is a dark brown air-stable powder soluble in polar solvents such as acetone and acetonitrile, but insoluble in dichloromethane, while $[(Ph_3P)_2N]_2[H_4Os_{10}(CO)_{24}]$[5] is soluble in this latter solvent. The ^1H NMR spectrum, at room temperature in acetone solution as the Na^+ salt, has a resonance at δ −16.46 (H–Os).[5] Variable-temperature ^1H and ^{13}C NMR studies of $PPN_2[H_4Os_{10}(CO)_{24}]$ show a high degree of ligand fluxionality in CD_2Cl_2 solution at room temperature. On cooling to −75°C, the single broad hydride resonance at $\delta = -16.48$ ppm splits into two signals of equal intensity at −14.70 and −19.08 ppm,[15] whereas the ^{13}C NMR spectrum at 22°C consists of two (exchange-broadened) carbonyl resonance signals at δ 175.8 and 192.9 ppm that coalesce at −43°C.[15b] The IR spectrum, in CH_2Cl_2, shows ν_{CO} bands at 2076 (w), 2035 (s), 1996 (sh, s), and 1986 (s) cm^{-1}.[15a] Addition of concentrated H_2SO_4 or CF_3CO_2H (one drop) to a solution (~15 mL) of $[H_4Os_{10}(CO)_{24}]^{2-}$ (0.011 mmol; as Na^+ salt, dissolved in acetonitrile; as $[NBu_4]^+$ or $[(Ph_3P)_2N]^+$ salt, dissolved in dichloromethane), followed by filtration through Celite and eventually silica, gives $[H_5Os_{10}(CO)_{24}]^-$ in quantitative yield.[5,15a,16]

*In some preparations $Na[H_3Os_4(CO)_{12}]$ is present as byproduct; it can be removed by column chromatography on silicagel (column diameter 20 mm, length 200 mm) using acetone : hexane (4:1) as eluant; in these cases, yields of $Na_2[H_4Os_{10}(CO)_{24}]$ are approximately 50%.

References

1. D. Roberto, E. Cariati, R. Psaro, and R. Ugo, *Organometallics* **13**, 734 (1994).
2. D. Roberto, E. Cariati, R. Ugo, and R. Psaro, *Inorg. Chem.* **35**, 2311 (1996).
3. D. Roberto, R. Psaro, and R. Ugo, *Organometallics* **12**, 2292 (1993).
4. C. Roveda, E. Cariati, E. Lucenti, and D. Roberto, *J. Organomet. Chem.* **580**, 117 (1999).
5. E. Lucenti, D. Roberto, C. Roveda, R. Ugo, and E. Cariati, *J. Cluster Sci.* **12**, 113 (2001).
6. (a) A. Wurtz, *Ann. Chim. Phys.* **55**, 417 (1859); (b) J. U. Nef, *Justus Liebigs Ann. Chem.* **335**, 310 (1904); (c) H. S. Fry and E. L. Schulze, *J. Am. Chem. Soc.* **50**, 1131 (1928); (d) M. H. Milnes, *Nature* **232**, 395 (1971).
7. L. Bretherick, *Handbook of Reactive Chemical Hazards*, 2nd ed., Butterworths, London, 1979.
8. (a) W. A. Hermann, E. Herdtweck, and A. Schäfer, *Chem. Ber.* **121**, 1907 (1988); (b) M. I. Bruce, M. Cooke, M. Green, and D. J. Westlake, *J. Chem. Soc. A* 987 (1969); (c) R. Psaro and C. Dossi, *Inorg. Chim. Acta* **77**, L255 (1983).
9. B. F. G. Johnson, J. Lewis, P. R. Raithby, G. M. Sheldrick, and G. Süss-Fink, *J. Organomet. Chem.* **162**, 179 (1978).
10. B. F. G. Johnson, J. Lewis, P. R. Raithby, G. M. Sheldrick, K. Wong, and M. McPartlin, *J. Chem. Soc., Dalton Trans.* 673 (1978).
11. H. H. Lamb, L. C. Hasselbring, C. Dybowski, and B. C. Gates, *J. Mol. Catal.* **56**, 36 (1989).
12. H. H. Lamb, A. S. Fung, P. A. Tooley, J. Puga, R. Krause, M. J. Kelley, and B. C. Gates, *J. Am. Chem. Soc.* **111**, 8367 (1989).
13. C. M. T. Hayward and J. R. Shapley, *Inorg. Chem.* **21**, 3816 (1982).
14. (a) P. F. Jackson, B. F. G. Johnson, J. Lewis, M. McPartlin, and W. J. H. Nelson, *J. Chem. Soc., Chem. Commun.* 224 (1980); (b) P. F. Jackson, B. F. G. Johnson, J. Lewis, W. J. H. Nelson, and M. McPartlin, *J. Chem. Soc., Dalton Trans.* 2099 (1982).
15. (a) D. Braga, B. F. G. Johnson, J. Lewis, M. McPartlin, W. J. H. Nelson, and M. D. Vargas, *J. Chem. Soc., Chem. Commun.* 241 (1983); (b) A. Bashall, L. H. Gade, J. Lewis, B. F. G. Johnson, G. J. McIntyre, and M. McPartlin, *Angew. Chem., Int. Ed. Engl.* **30**, 1164 (1991).
16. S. R. Drake, B. F. G. Johnson, and J. Lewis, *J. Chem. Soc., Dalton Trans.* 1517 (1988).

49. TRIOSMIUM COMPLEXES OF FULLERENE-60

Submitted by CHANG YEON LEE,* HYUNJOON SONG,* KWANGYEOL LEE,* BO KEUN PARK,* and JOON T. PARK*

Checked by THOMAS L. MALOSH† and JOHN R. SHAPLEY†

Exohedral metallofullerenes have attracted attention concerning the effect of metal coordination on the chemical and physical properties of C_{60}. It has been

*Department of Chemistry and School of Molecular Science, (BK21) Korea Advanced Institute of Science and Technology, Daejeon, 305-701, Korea.

†Department of Chemistry, University of Illinois at Champaign—Urbana, Urbana, IL 61801.

demonstrated that a variety of cluster frameworks [$Re_3(\mu\text{-}H)_3$,[1] Ru_3,[2] Os_3,[3] Ru_5C,[4] Os_5C,[5] $PtRu_5C$,[4] Rh_6[6]] can bind C_{60} via a face-capping cyclohexatriene-like bonding mode, $\mu_3\text{-}\eta^2:\eta^2:\eta^2\text{-}C_{60}$. In particular, the compound $Os_3(CO)_9(\mu_3\text{-}\eta^2:\eta^2:\eta^2\text{-}C_{60})$ and its derivatives have been extensively studied; interesting bonding mode conversions[7] of the C_{60} ligand and unusual electrochemical communication[8] between C_{60} and metal cluster centers were revealed. Herein we report the synthesis of two representative face-capping $\mu_3\text{-}\eta^2:\eta^2:\eta^2\text{-}C_{60}$-triosmium cluster complexes, $Os_3(CO)_9(\mu_3\text{-}\eta^2:\eta^2:\eta^2\text{-}C_{60})$ and $Os_3(CO)_8(PPh_3)(\mu_3\text{-}\eta^2:\eta^2:\eta^2\text{-}C_{60})$.

General Comments

All reactions are carried out under N_2 atmosphere with use of standard Schlenk techniques. Solvents are dried appropriately before use. C_{60} (99.5%, Southern Chemical Group, LLC) is used without further purification. Anhydrous trimethylamine *N*-oxide (mp 225–230°C) was obtained from $Me_3NO \cdot 2H_2O$ (98%, Aldrich) by sublimation (3 times) at 90–100°C under vacuum.

■ **Caution.** *These procedures must be performed in a good fume hood as toxic carbon monoxide is evolved.*

A. NONACARBONYL(μ_3-η^2:η^2:η^2-FULLERENE-60) TRIOSMIUM, $Os_3(CO)_9(\mu_3\text{-}\eta^2:\eta^2:\eta^2\text{-}C_{60})$

$$Os_3(CO)_{12} + 2\,Me_3NO/MeCN \longrightarrow Os_3(CO)_{10}(NCMe)_2 + 2CO_2 + 2Me_3N$$
$$Os_3(CO)_{10}(NCMe)_2 + C_{60} \longrightarrow Os_3(CO)_9(\mu_3\text{-}\eta^2:\eta^2:\eta^2\text{-}C_{60}) + CO$$

Procedure

To a dichloromethane/acetonitrile (200/15-mL) solution of $Os_3(CO)_{12}$ (300 mg, 0.331 mmol) is added dropwise an acetonitrile solution (10 mL) of anhydrous Me_3NO (55 mg, 0.732 mmol) at room temperature, and the reaction mixture is stirred for 1 h. After evaporation of the solvent in vacuo, the residue is dissolved in chlorobenzene (10 mL). The resulting yellow solution is added dropwise to a chlorobenzene (200 mL) solution of C_{60} (239 mg, 0.331 mmol), and the reaction mixture is stirred at 80°C for 30 min to give a blue-green solution. The solution is concentrated to ~50 mL under vacuum, and the resulting solution is heated to reflux for 90 min. After solvent removal in vacuo, the residue is purified by column chromatography* (silicagel, CS_2) to afford $Os_3(CO)_9(\mu_3\text{-}\eta^2{:}\eta^2{:}\eta^2\text{-}C_{60})$ as a reddish brown solid (89 mg, 0.058 mmol, 18%), along with recovered C_{60} (115 mg, 0.160 mmol, 48%).

Anal. Calcd. for $C_{69}O_9Os_3$: C, 53.70. Found : C, 53.94.

Properties

IR (CS_2) ν_{CO} 2081 (s), 2046 (vs), 2016 (m), 2002 (m), 1983 (sh) cm^{-1}; ^{13}C NMR (CS_2/ext. CD_2Cl_2, 298 K) δ 176.1 (s, 9CO); MS (FAB^+) m/z 1548 (M^+).

B. OCTACARBONYL(TRIPHENYLPHOSPHINE)(μ_3-η^2:η^2:η^2-FULLERENE-60)TRIOSMIUM, $Os_3(CO)_8(PPh_3)(\mu_3\text{-}\eta^2{:}\eta^2{:}\eta^2\text{-}C_{60})$

$$Os_3(CO)_9(\mu_3\text{-}\eta^2 : \eta^2 : \eta^2\text{-}C_{60}) + Me_3NO/MeCN + PPh_3 \longrightarrow$$
$$Os_3(CO)_8(PPh_3)(\mu_3\text{-}\eta^2 : \eta^2 : \eta^2\text{-}C_{60}) + CO_2 + Me_3N$$

$Os_3(CO)_9(\mu_3\text{-}\eta^2{:}\eta^2{:}\eta^2\text{-}C_{60})$ (20.0 mg, 0.0130 mmol) and an excess amount of PPh_3 (10.2 mg, 0.0389 mmol) are dissolved in chlorobenzene (20 mL). The solution is cooled to 0°C, and an acetonitrile (3 mL) solution of anhydrous Me_3NO (1.1 mg, 0.015 mmol) is added dropwise. The reaction mixture is allowed to warm to room temperature and is stirred for 30 min. After evaporation of the solvent in vacuo, the residue is dissolved in chlorobenzene (20 mL). The reaction mixture is stirred and heated at 100°C for 1 h. After solvent removal in vacuo, the residue is purified by preparative TLC (silicagel GF_{254}, CS_2) to afford $Os_3(CO)_8(PPh_3)(\mu_3\text{-}\eta^2{:}\eta^2{:}\eta^2\text{-}C_{60})$ (11.5 mg, 0.0065 mmol, 50%, $R_f = 0.4$) as a brown solid.

Anal. Calcd. for $C_{89}H_{15}O_8PS_6Os_3 \cdot 3CS_2$: C, 53.3; H, 0.75; S, 9.59. Found : C, 53.6; H, 0.40; S, 10.4.

Properties

IR (CS_2) ν_{CO} 2064 (vs), 2033 (s), 2013 (m), 1999 (m), 1983 (m), 1952 (w) cm^{-1}; ^1H NMR ($CDCl_3$, 298 K) δ 7.35 (m, 18H); $^{31}P\{^1H\}$ NMR ($CDCl_3$, 298 K) δ 9.64 (s); $^{13}C\{^1H\}$ NMR ($C_6H_4Cl_2$/ext. CD_2Cl_2, 298 K) δ 184.8 (d, $^2J_{PC} = 5.7$ Hz, 2CO), 177.6 (s, 6CO); MS (FAB^+) m/z 1778 (M^+).

*The Checkers preferred thin-layer chromatopgraphy for separation.

References

1. H. Song, Y. Lee, Z.-H Choi, K. Lee, J. T. Park, J. Kwak, and M.-G. Choi, *Organometallics* **20**, 3139 (2001).
2. H. F. Hsu and J. R. Shapley, *J. Am. Chem. Soc.* **118**, 9192 (1996).
3. J. T. Park, H. Song, J.-J. Cho, M. K. Chung, J.-H. Lee, and I. H. Suh, *Organometallics* **17**, 227 (1998).
4. K. Lee and J. R. Shapley, *Organometallics* **17**, 3020 (1998).
5. K. Lee, C. H. Lee, H. Song, J. T. Park, H. Y. Chang, and M.-G Choi, *Angew. Chem., Int. Ed.* **39**, 1801 (2000).
6. K. Lee, H. Song, B. Kim, J. T. Park, S. Park, and M.-G. Choi, *J. Am. Chem. Soc.* **124**, 2872 (2002).
7. H. Song, K. Lee, C. H. Lee, J. T. Park, H. Y. Chang, and M.-G. Choi, *Angew. Chem., Int. Ed.* **40**, 1500 (2001).
8. H. Song, K. Lee, J. T. Park, and M.-G. Choi, *Organometallics* **17**, 4477 (1998).

50. TRIPHENYLPHOSPHINE-STABILIZED GOLD NANOPARTICLES

Submitted by JAMES E. HUTCHISON,* EVAN W. FOSTER,* MARVIN G. WARNER,* SCOTT M. REED,* and WALTER W. WEARE*
Checked by WILLIAM BUHRO† and HENG YU†

$$H_2O + H\,AuCl_4 + 2\,PPh_3 \longrightarrow AuCl\,PPh_3 + O{=}PPh_3 + 3\,HCl$$

$$AuCl\,PPh_3 + Na\,BH_4 \longrightarrow Au_{101}(PPh_3)_{21}Cl_5{}^{\ddagger} + PPh_3 + Na\,Cl + 3\,HCl$$

Phosphine-stabilized gold nanoparticles possessing a core diameter of about 1.4 nm have been widely investigated since the early 1980s as models for metallic catalysts[1] and building blocks for nanoscale structures and electronic devices.[2–6] Although the chemical formula ($Au_{55}(PPh_3)_{12}Cl_6$) of the material first prepared by Schmid et al.[7] in 1981 has been disputed,[8–11] these particles have proved to be very useful because their small size and narrow dispersity (1.4 ± 0.4 nm)[10] largely eliminate the need for fractional crystallization or chromatographic size selection. Unfortunately, the synthesis[12] is inconvenient, requiring rigorously anaerobic conditions and diborane gas as a reducing agent. In addition, the product is thermally unstable in solution, decomposing via aggregation at room temperature.

*Department of Chemistry and Materials Science Institute, University of Oregon, Eugene, OR 97403.
†Department of Chemistry, Washington University, One Brookings Drive, St. Louis, MO 63130.
‡Average composition for 1.5 ± 0.4 nm particles prepared using this procedure.

More recently, we and others have shown that phosphine-stabilized gold nanoparticles are excellent precursors to other functionalized nanoparticle building blocks possessing well-defined gold cores.[13–17] Using 1.4 nm phosphine-stabilized particles as common precursors, thiol-stabilized nanoparticles can be prepared that contain nearly any omega functional group in the ligand shell. The metal core can be tuned from 1.4 to 10 nm in diameter through reaction of the phosphine precursor with various ligand molecules.[13,15] The properties of these materials (e.g., the small size, narrow dispersity, and enhanced stability of the 1.4-nm alkanethiol-stabilized nanoparticles) have been exploited in studies of nanoparticle-based nanoelectronic materials.[18–20] The phosphine-stabilized precursor is also finding use as a seed material for controlling the growth of other nanoscale materials. For example, it has been shown that the phosphine-stabilized gold particles can be used as seeds for growth of nearly monodisperse Bi, Sn, and In nanoparticles.[21]

An improved synthesis of this type of phosphine-stabilized gold nanoparticle was necessary to make them convenient for use as synthetic intermediates in materials applications. We developed a safer, more convenient, and more versatile synthesis[22] of phosphine-stabilized nanoparticles analogous to those originally reported by Schmid.[12] The synthesis described here eliminates the use of diborane and can be carried out quickly with minimal concern for inert conditions. The nanoparticle product is comparable to that prepared by the Schmid procedure, possessing a core diameter of 1.5 ± 0.4 nm and the same reactivity. In addition, the new synthesis can be adapted to permit the use of a variety of phosphines as passivating ligands and provide control over particle core size.[22]

Reagents

Hydrogen tetrachloroaurate(III) hydrate (99.9% Au) is available from Strem Chemicals, and all other reagents are available from Aldrich Chemical Co. Sodium borohydride is purchased as 98% or higher grade. Sodium hydroxide, a known decomposition product of sodium borohydride, increases the average particle size; therefore sodium borohydride that has had minimal exposure to air should be used. Deuterated chloroform is obtained from Cambridge Isotope Labs. All other organic solvents are purchased from Fischer Scientific. All chloroform, including the deuterated form used for NMR, must be filtered through basic alumina prior to use because acidic chloroform may cause decomposition of the product. Reagents and other organic solvents are used without further purification.

■ **Caution.** *Hydrogen tetrachloroaurate(III) hydrate is corrosive; plastic or Teflon implements should be used to handle this material. Gas evolution can be violent on addition of sodium borohydride, and the contents of the flask may* *overflow. Equipping the flask with a large funnel with a 24/40 ground-glass joint prevents the contents from spilling, and allows the reductant to be added rapidly.*

Procedure

All operations are carried out in a fume hood under ambient conditions. Hydrogen tetrachloroaurate(III) hydrate (1.00 g, 2.54 mmol) is added to 60 mL of deionized water in a 500-mL round-bottomed flask equipped with a magnetic stirbar. This results in a golden yellow solution. Once the hydrogen tetrachloroaurate(III) hydrate is completely dissolved, 60 mL of toluene is added. On addition of tetraoctylammonium bromide (1.40 g, 2.56 mmol), the aqueous layer becomes colorless and the organic phase turns red. The mixture is stirred vigorously for 5 min before triphenylphosphine (2.30 g, 8.76 mmol) is added to the reaction mixture. On addition, the organic phase turns cloudy white. The mixture is stirred vigorously for an additional 10 min. Sodium borohydride (2.00 g, 52.9 mmol) freshly dissolved in 10 mL of deionized water is added rapidly with stirring. The organic layer turns from red to dark reddish-brown on addition of the sodium borohydride. The reaction mixture is stirred for 3 h after addition of the borohydride.

The aqueous and organic layers are separated using a 500-mL separatory funnel. The organic layer is washed 3 times with 100 mL of deionized water. An emulsion commonly forms during this step, so saturated sodium chloride may be added to the mixture or used instead of water to aid in the separation of the layers. The organic layer is separated, filtered to remove precipitates, and evaporated to dryness under flowing nitrogen. The crude product is transferred to a 500-mL Erlenmeyer flask with 35 mL of chloroform. To this solution, 300 mL of pentane is added to precipitate the product.*

The suspension is filtered through a 150-mL medium-porosity fritted Büchner funnel to collect the solid, crude nanoparticles. The filtrate should be colorless. The product is triturated with the following solvent combinations to remove the tetraoctylammonium bromide:

- 2 × (100 mL hexanes followed by 100 mL 2 : 3 MeOH : H_2O)
- 2 × (100 mL hexanes followed by 100 mL 1 : 1 MeOH : H_2O)
- 100 mL hexanes

If a large amount of tetraoctylammonium bromide remains, the product is washed with 1 : 1 MeOH : H_2O followed by hexanes until only a small amount

*The checkers suggest adding the pentane slowly to avoid ligand dissociation and destabilization of the nanoparticles.

is detectable by ^{1}H NMR in $CDCl_3$. The characteristic resonance for this impurity appears at $\delta \sim 3.4$ ppm as a multiplet.

Next the product is washed on the same frit with the following solvent combinations to remove $(PPh_3)AuCl$:

- 2 × 150 mL 3 : 1 pentane : chloroform
- 2 × 150 mL 2 : 1 pentane : chloroform
- 2 × 150 mL 1 : 1 pentane : chloroform

During each pentane : chloroform wash, the product is agitated and allowed to soak in the wash solution for about 5 min.

The purified product can be rinsed through the frit with dichloromethane or carefully scraped off as a powder from the surface of the frit. The purity of the gold nanoparticles is confirmed by ^{1}H NMR spectroscopy in $CDCl_3$. When a relaxation delay of 5 s is used, a broad resonance at $\delta \sim 6$–8 ppm with a small multiplet at $\delta \sim 7.5$ ppm (from residual $(PPh_3)AuCl$) is observed. If the intensity of the resonance at δ 7.5 ppm due to $(PPh_3)AuCl$ is much greater than the broad resonance due to the bound triphenylphosphine ligand, then additional 1 : 1 pentane : chloroform washes are performed until the resonance is sufficiently reduced.

Purification yields 200–300 mg* of triphenylphosphine-stabilized gold nanoparticles, which should be stored cold (−20°C) in the solid state, or immediately converted to thiol- or amine-stabilized nanoparticles through subsequent reaction with the appropriate ligand. The particles decompose in solution; thus manipulation of the product in solution should be minimized.

Physical Properties and Reactivity

Samples of triphenylphosphine-stabilized gold nanoparticles with $d_{core} \sim 1.5$ nm are dark brown powders that are soluble in nonpolar organic solvents. These solutions appear reddish brown. The visible spectrum gives a broad, low-intensity, plasmon resonance between 400 and 600 nm. The ^{1}H NMR spectrum exhibits a broad resonance at δ 6–8. Transmission electron microscopy in conjunction with particle size analysis using Scion Image software† yields $d_{core} = 1.5 \pm 0.4$ nm. X-ray photoelectron spectroscopy yields atomic ratios of 19.1 : 4 : 1 Au : P : Cl. Thermogravimetric analysis of the product indicates that 75.5% of the sample is gold. The estimated composition of the average particle is Au_{101} $(PPh_3)_{21}$ Cl_5.[22]

*The checkers obtained a yield of 190 mg.

†This software package is available for download at *http://www.scioncorp.com.*

The 1.5-nm nanoparticles readily react with thiol or amine-terminated ligands under mild conditions to yield thiol- or amine-stabilized nanoparticles.[13,15,17] Triphenylphosphine-stabilized particles thermally decompose with the production of $(PPh_3)AuCl$ and metallic gold.

Acknowledgments

The authors thank Ms. Lauren Huffman for her assistance in testing and optimizing this preparation. This work was supported by the National Science Foundation, the Camille and Henry Dreyfus Foundation (JEH is a Camille Dreyfus Teacher Scholar), and the Department of Education GAANN program.

References

1. G. Schmid, *Chem. Rev.* **92**, 1709 (1992).
2. D. V. Averin and K. K. Likharev, *J. Low Temp. Phys.* **62**, 345 (1986).
3. G. Schoen, and U. Simon, *Colloid Polym. Sci.* **273**, 101 (1995).
4. R. P. Andres, T. Bein, M. Dorogi, S. Feng, J. I. Henderson, C. P. Kubiak, W. Mahoney. R. G. Osifchin, and R. Reifenberger, *Science* **272**, 1323 (1996).
5. D. L. Feldheim, K. C. Grabar, M. J. Natan, and T. E. Mallouk, *J. Am. Chem. Soc.* **118**, 7640 (1996).
6. L. Clarke, M. N. Wybourne, L. O. Brown, J. E. Hutchison, M. Yan, S. X. Cai, and J. F. W. Keana,. *Semicond. Sci. Technol.* **13**, A111 (1998).
7. G. Schmid, R. Pfeil, Boese, F. Brandermann, S. Meyer, G. H. M. Calis, and J. W. A. Van der Velden, *Chem. Ber.* **114**, 3634 (1981).
8. W. Vogel, B. Rosner, and B. J. Tesche, *Phys. Chem.* **97**, 11611 (1993).
9. C. J. McNeal, J. M. Hughes, L. H. Pignolet, L. T. J. Nelson, T. G. Gardner, J. P. Fackler, Jr., R. E. P. Winpenny, L. H. Irgens, G. Vigh, and R. D. Macfarlane, *Inorg. Chem.* **32**, 5582 (1993).
10. D. H. Rapoport, W. Vogel, H. Coelfen, and R. Schloegl, *J. Phys. Chem. B* **101**, 4175 (1997).
11. E. Gutierrez, R. D. Powell, F. R. Furuya, J. F. Hainfeld, T. G. Schaaff, M. N. Shafigullin, P. W. Stephens, and R. L. Whetten, *Eur. Phys. J. D* **9**, 647 (1999).
12. G. Schmid, *Inorg. Synth.* **27**, 214 (1990).
13. L. O. Brown and J. E. Hutchison, *J. Am. Chem. Soc.* **119**, 12384 (1997).
14. G. Schmid, R. Pugin, J. O. Malm, and J. O. Bovin, *Eur. J. Inorg. Chem.* 813 (1998).
15. L. O. Brown and J. E. Hutchison, *J. Am. Chem. Soc.* **121**, 882 (1999).
16. G. Schmid, R. Pugin, W. Meyer-Zaika, and U. Simon, *Eur. J. Inorg. Chem.* 2051 (1999).
17. M. G. Warner, S. M. Reed, and J. E. Hutchison, *Chem. Mater.* **12**, 3316 (2000).
18. L. Clarke, M. N. Wybourne, M. Yan, S. X. Cai, L. O. Brown, J. Hutchison, and J. F. W. Keana, *J. Vac. Sci. Technol. B* **15**, 2925 (1997).
19. M. N. Wybourne, J. E. Hutchison, L. Clarke, L. O. Brown, and J. L. Mooster, *Microelectron. Eng.* **47**, 55 (1999).
20. C. A. Berven, L. Clark, J. L. Mooster, M. N. Wybourne, and J. E. Hutchison, *Adv. Mater.* (Weinheim, Ger.) **13**, 109 (2001).
21. H. Yu, P. C. Gibbons, K. F. Kelton, and W. E. Buhro, *J. Am. Chem. Soc.* **123**, 9198 (2001).
22. W. W. Weare, S. M. Reed, M. G. Warner, and J. E. Hutchison, *J. Am. Chem. Soc.* **122**, 12890 (2000).

CONTRIBUTOR INDEX

Volume 34

Inorganic Syntheses, Volume 34, edited by John R. Shapley
ISBN 0-471-64750-0 2004 John Wiley & Sons, Inc.

SUBJECT INDEX

Volume 34

Prepared by Magdalena Pala, *UW-River Falls*

Names used in this Subject Index are based upon IUPAC *Nomenclature of Inorganic Chemistry*, Recommendations 1990, Blackwell Scientific Publications, London.

Inverted forms of the chemical names (parent index headings) are used for most entries in the alphabetically ordered index.

Inorganic Syntheses, Volume 34, edited by John R. Shapley
ISBN 0-471-64750-0

FORMULA INDEX

Volume 34

Prepared by MAGDALENA PALA, *UW-River Falls*

The formulas for compounds described in volume 34 are entered in alphabetical order. They represent the total composition of the compounds, e.g., $BF_{24}KC_{38}H_{21}$ for potassium tetra-3,5-bis(trifluoromethyl)phenylborate. The elements in the formulas are arranged in alphabetical order, with carbon and hydrogen listed last. All formulas are permuted on the symbols other than carbon and hydrogen representing organic groups in coordination compounds. Thus potassium tetra-3,5-bis(trifluoromethyl)phenylborate can be found under B, F, and K in this index.

Water of hydration and other solvents found in crystal lattice are not added into formulas of the compounds listed, e.g., $C_{36}H_{46}N_2O_4 \cdot (C_2H_5)_2O$.

Inorganic Syntheses, Volume 34, edited by John R. Shapley
ISBN 0-471-64750-0